LATIN* STUDENTS IN ENGINEERING

LATIN* STUDENTS IN ENGINEERING

An Intentional Focus on a Growing Population

EDITED BY

LARA PEREZ-FELKNER, SARAH L.
RODRIGUEZ, AND CIERA FLUKER

RUTGERS UNIVERSITY PRESS
New Brunswick, Camden, and Newark, New Jersey
London and Oxford

Rutgers University Press is a department of Rutgers, The State University of New Jersey, one of the leading public research universities in the nation. By publishing worldwide, it furthers the University's mission of dedication to excellence in teaching, scholarship, research, and clinical care.

Library of Congress Cataloging-in-Publication Data

Names: Pérez-Felkner, Lara, editor. | Rodriguez, Sarah L., editor. | Fluker, Ciera, editor.
Title: Latin* students in engineering : an international focus on a growing population / edited by Lara Perez-Felkner, Sarah L. Rodriguez, and Ciera Fluker.
Other titles: Latinx students in engineering
Description: New Brunswick : Rutgers University, [2024] | Includes bibliographical references and index.
Identifiers: LCCN 2023047872 | ISBN 9781978838680 (hardcover) | ISBN 9781978838673 (paperback) | ISBN 9781978838697 (epub) | ISBN 9781978838703 (pdf)
Subjects: LCSH: Engineering—Study and teaching (Higher)—United States. | Engineering students—United States. | Hispanic American college students. | BISAC: EDUCATION / Schools / Levels / Higher | TECHNOLOGY & ENGINEERING / Social Aspects
Classification: LCC T73 .L35 2024 | DDC 620.0071/2073—dc23/eng/20240415
LC record available at https://lccn.loc.gov/2023047872

A British Cataloging-in-Publication record for this book is available from the British Library.

References to internet websites (URLs) were accurate at the time of writing. Neither the author nor Rutgers University Press is responsible for URLs that may have expired or changed since the manuscript was prepared.

♾ The paper used in this publication meets the requirements of the American National Standard for Information Sciences—Permanence of Paper for Printed Library Materials, ANSI Z39.48-1992.

rutgersuniversitypress.org

Para el presente y el futuro. We dedicate this book to those who have uplifted us and continue to do so, those we hope will benefit from this work, and the community of colleagues, authors, and students we aim to contribute to with this book.

CONTENTS

FOREWORD

This is a long-awaited and much-needed edited collection. As a former program officer and co-lead for the National Science Foundation (NSF)'s Hispanic-Serving Institutions (HSIs) Program, I can attest firsthand that there is a persistent and important need to cultivate an *intentional focus on a growing population* of Latin* populations in engineering education.

This book is a response to the rapidly changing landscape of higher education given the increasing number of institutions emerging with federal designations as "Hispanic-Serving Institutions," coupled with massive investments from federal agencies to supplement resources for low-income students at traditionally under-resourced colleges and universities. In this volume, editors Lara Perez-Felkner, Sarah L. Rodriguez, and Ciera Fluker have produced a valuable collection of key concepts and cutting-edge scholarship, synthesizing recent findings on strategies for supporting Latin* students and offering thoughtful recommendations for future research in engineering education in particular. Shining a light on the particularities of engineering education is critical, given our concerted efforts as educators and leaders to build capacity for current and future students to design breakthrough engineering innovations. The corpus of scholarly insights assembled here amplifies the importance of issues affecting the diverse and rising numbers of Latin* students whose talents will support our country's economic and leadership aims.

The reason this book is necessary at this moment is because the number of Latin* students, and institutions that serve them, is growing. Unlike Historically Black Colleges and Universities (HBCUs) and Tribal colleges and universities (TCUs), the number of federally designated HSIs is dynamic. Each year the Department of Education adds and subtracts the names of HSIs to its list. The fluidity persists because some institutions attain this federal status anew; other institutions lose HSI designation when their Latin* student numbers fall below the number required by the federal definition. According to *Excelencia* in Education (2023), a DC-based think tank that advocates for Latino student success by analyzing and informing educational policies, "Latino student enrollment at HSIs has tripled in the last 28 years. In 2021–22, HSIs enrolled over 1.4 million Latino undergraduates, compared to 490,000 in 1994–95." As of 2022, there are 571 HSIs; these comprise about one-fifth of all institutions of higher education (IHE) and these HSIs enroll 62 percent of Latino undergraduates (ibid). In addition, 471 "emerging HSIs" (eHSIs), characterized by having between 15 and 24.9 percent undergraduate FTE Latin* enrollment, can be found across the country. Many Latin* students at eHSIs are children of immigrants drawn to work in regional industries in states such as Wyoming, Utah, and Iowa. Children born to immigrant poultry workers in rural Alabama and Missouri, for example, are of college age and entering campuses at eHSIs. The future of these students' social mobility can be shaped by colleges and universities dedicated to their success.

When I was serving at the NSF, many HSI proposals were rejected because they embedded a deficit-orientation in their projects. The focus was on how *students* needed to change, rather than on how the *institutions* could be reformed to better serve them. Increasingly, public and private initiatives are in motion to create a science and engineering enterprise that is attractive and accessible (National Science Board [NSB], 2020). Federal support to these institutions is available. Per the guidelines in 502 of the Higher Education Act of 1965 (20 U.S.C. 1101a), grants awarded to Hispanic-serving institutions of higher education aim to assist them "to plan, develop, undertake, and carry out programs to improve and expand the institutions' capacity to serve Hispanic students and other low-income students." Investments from NSF to build capacity for faculty and students and support innovative science, technology, engineering, and mathematics (STEM) education research are significant. For the HSI Program in particular (which seeks to enhance undergraduate education quality, promote success of STEM majors, and support innovative research in fields including engineering education at HSIs), the NSF budget request to Congress for 2024 was $60.5 million. Many of those principal investigators seeking to resubmit their proposals to the HSI Program would do well to incorporate findings from this book.

We know institutions make a difference (Ohland et al., 2011) in recruiting, retaining, and graduating engineering majors. Additionally, faculty, industry, and community partners continue to build coalitions to fortify partnerships in sustaining and serving undergraduate students who want to pursue advanced degrees. In the context of engineering education, however, without concerted effort and action, an engineering culture characterized by a lack of diversity and problems with teaching and learning will continue to reproduce systemic problems. Social scientists and engineering educators have documented structural patterns of exclusion in students' higher education experiences, including ostracism, overt and subtle racism and sexism, and backlash based on perceptions of affirmative action (e.g., Camacho & Lord 2013; Mejia et al., 2022; Mejia & Martin, 2023). Under such conditions, some students stumble through, stop out, drop out, migrate internally within engineering majors, or migrate out of engineering (Lord et al., 2019).

Supporting students during these periods of liminality is achievable by improving transparency and accountability among faculty and institutional leadership (Alvarez-Gutiérrez et al., 2020). Revolutionary change in engineering education is not only about adding constructive practices, creating networks to sustain student personal and professional development, and systematically growing macro and micro student support strategies. Additionally, moving the needle requires sustained efforts aimed at collectively confronting negative practices and gatekeeping patterns, and obliterating toxic behaviors in engineering departments.

This book moves us to consider the tremendous challenges that remain within the landscape of higher education, with the intent of closing equity gaps in degree completion rates. I am grateful to these authors for encouraging us to invest in and think deeply about their compelling findings. We need the full power of our current and future STEM students, including those at the intersections of various dimensions of identity—their talents will lend creativity to future engineering breakthroughs and discoveries.

Michelle M. Camacho
Dean, University of Utah

REFERENCES

Alvarez-Gutiérrez, L., Fukushima, A. I., & Gaytán, M. S. (2020). "Essential Latinx Educators: Teaching in A Time of Pandemic." *Latinx Talk*. Retrieved July 9, 2023, from https://latinxtalk.org/2020/07/06/essential-latinx-educators-teaching-in-a-time-of-pandemic/

Camacho, M., & Lord, S. (2013). *The borderlands of education: Latinas in engineering*. Lexington Books.

Excelencia in Education (2023). *28 years of HSIs*. https://storymaps.arcgis.com/stories/fc9d2e2e2274405c866ce3bd59dd17c4

Lord, S. M., Ohland, M. W., Layton, R. A., & Camacho, M. M. (2019). Beyond pipeline and pathways: Ecosystem metrics. *Journal of Engineering Education, 108*(1), 32–56.

Mejia, J. A., Alarcón, I. V., Mejia, J., & Revelo, R. (2022). Legitimized tongues: Breaking the traditions of silence in mainstream engineering education and research. *Journal of Women and Minorities in Science and Engineering, 28*(2), 53–77.

Mejia, J. A., & Martin, J. P. (2023). Critical perspectives on diversity, equity, and inclusion research in engineering education. *International Handbook of Engineering Education Research* (pp. 218–238). Routledge. https://doi.org/10.4324/9781003287483-13

National Science Board (NSB). (2020). *Vision 2030*. Retrieved June 7, 2023, from https://www.nsf.gov/nsb/publications/2020/nsb202015.pdf

National Science Foundation (NSF). (2023) *Fiscal year 2024 budget request to congress*. Retrieved July 7, 2023, from https://new.nsf.gov/about/budget/fy2024

Ohland, M. W., Brawner, C. E., Camacho, M. M., Layton, R. A., Long, R. A., Lord, S. M., & Wasburn, M. H. (2011). Race, gender, and measures of success in engineering education. *Journal of Engineering Education, 100*(2), 225–252. https://doi.org/10.1002/j.2168-9830.2011.tb00012.x

LATIN* STUDENTS IN ENGINEERING

1 · UPDATING THE NARRATIVE ON LATIN* ENGINEERING STUDENTS AND THE INSTITUTIONS THAT COULD BETTER SERVE THEM

LARA PEREZ-FELKNER, SARAH L. RODRIGUEZ, AND CIERA FLUKER

The growing population of engineering students who identify as Latina/e/o/x and/or Hispanic (here, inclusively termed "Latin*") have diverse identities and backgrounds, and they pursue engineering at a range of institutional types across and beyond the United States. This population is growing and already comprises the second largest racial/ethnic group in the United States, at 19 percent of all Americans (Funk & Lopez, 2022) and a fifth of postsecondary students enrolled across the country (Mora, 2022). Far from a rarity, Latin* engineering students have increased as a proportion of today's baccalaureate graduates, as high as 14.0 percent by 2020—up from 5.9 percent twenty years earlier (NCSES, 2023). Further, Latin* students also represent a rising share of engineering master's and doctoral degree enrollees and recipients (Anderson et al., 2018), but Latin* engineering students remain underrepresented. This is a problem because (1) there is a rising need to train US students in engineering skills to meet the demands of our increasingly technological workforce and (2) structurally excluding Latin* students from engineering hinders their economic and educational opportunities.

AUDIENCE, AIMS, AND SIGNIFICANCE OF THE WORK

We intend to reach audiences from engineering education, higher education, education research, and STEM education policy. Leveraging our collective multidisciplinarity, we invite engineering departments, colleges, agencies, and funders, as well as scholars and practitioners engaged in professional societies—ranging from the Association for the Study of Engineering Education (ASEE), American Educational Research Association (AERA), Association for the Study of Higher Education (ASHE), American Association of Hispanics in Higher Education (AAHHE), Society of Hispanic Professional Engineers (SHPE), and other interdisciplinary communities with interest in Latin* educational equity—to meaningfully broaden participation in STEM fields and engineering education.

There are limited spaces for research in this area, which we seek to bridge and advance, while also amplifying the voices and engagement of the scholars represented in this book. Our edited book is organized around addressing the problems identified previously: (a) the tension between the need for a more inclusive and larger engineering workforce and the to-date insufficient efforts to enhance Latin* students' access to and experiences in engineering; and (b) how departments, institutions, and broader systems can change and adapt to rectify the structural exclusion of Latin* students in engineering, facilitating greater student- and institutional-level success in recruiting, retaining, and training Latin* engineers. Because of our focus on engineering as a discipline and its comparatively rare presence in P–12 settings—especially those schools and programs attended by Latin* students—we focus on postsecondary settings while offering suggestions and implications relevant across P–20 and beyond.

This book fills an important need within this field by providing an evidence-based approach to enhancing the educational experiences of Latin* students in engineering through actionable applications of research. While much scholarship and attention has been given to broader STEM issues over time, too few resources explicitly demonstrate how practitioners can translate what we know about the engineering context to concrete strategies for enhancing practice, policies, and collaborations. Virtually none are centered on the needs of Latin* students in engineering. Scholars and practitioners can use this book to take an integrated approach to enhancing educational experiences for Latin* students in engineering disciplines, in line with the deep work emerging in professional scientific organizations and funding agencies, such as the National Science Foundation (NSF), SHPE, and other impactful spaces for education and engineering. To address these needs, we, as a collective of educational stakeholders, need to update the narrative on Latin* engineering students and the institution that could better serve them.

BOOK ORGANIZATION

The book is organized around two focal issues. In part 1 of the book, we present a series of chapters which theoretically and empirically examine the experiences of Latin* students in engineering education. Here, we focus on key theoretical frameworks[1] employed and their application across a series of emerging and established methodologies. Part 1 attends primarily to the student experience and improving our understanding of how to effectively support Latin* engineering students, before and during college. Part 2 extends beyond the student level and focuses on institutional and social structures, as well as the intersectionality of identities—including dis/ability, sexuality, race, and gender. The social structures investigated include, but are not limited to, institutional characteristics and institutional, departmental, and programmatic resources and policies to support students. Lastly, our final chapter synthesizes the book's overall findings, highlights emergent work, and offers considerations for future research, policy, and practice.

Our book uniquely focuses on Latin* engineering students, and through our intentional decision to publish this book as an edited volume, we bring together a community of innovative scholar-authors with the community of readers. Indeed, this book's chapters take an intentionally nuanced and intersectional approach to Latin* engineering experiences and potential avenues for structural change. We aim to facilitate positive change in Latin* students' pathways through engineering education to the present and future workforce.

Part I: Examining the Experiences of Latin* Students in Engineering

The first half of this book focuses on the training of Latin* engineering students. The past decade has witnessed an emerging cadre of scholars advancing theoretical developments on Latin* engineering student success, grounded in innovative studies across disciplines and methodologies (Wilson-Lopez et al., 2016). Notably, Latin* students in engineering navigate a hidden curriculum of knowledges and ways that are not explicitly taught in their coursework—and that, if left unlearned, can hinder their opportunities to succeed and advance in these fields (Villanueva et al., 2020). The understanding, development, and support of an engineering identity can be transformational for Latin* students (Kendall et al., 2019; Rodriguez et al., 2022). Notably, Latin* engineering students already face other compounding challenges, including stereotype threats based on their intersectional gender, class, race/ethnicity, and other identities, experienced in their engineering and related educational contexts (Banda, 2020; Revelo & Baber, 2018).

Part II: Structural Issues and Intersectionality for Latin* Students in Engineering

The second issue the book addresses is the structural exclusion of Latin* students and intersectional identities that may further constrain or enhance their opportunity to succeed and thrive in engineering. It is important to consider the structure of opportunity to participate and succeed in engineering (Simmons & Lord, 2019), given the influence of postsecondary institutions and the actors within them on Latin* students' access to equitable and engaging learning spaces in engineering. Structural barriers can inhibit access to curricular and extracurricular opportunities, such as dual credit programs in schools, which can enhance high school and community college students' access to on-ramps into postsecondary engineering programs—and even a greater sense of belonging once enrolled (Ozuna Allen et al., 2020). Our book focuses on and centers Latin* students but, in each chapter, we address implications that might enhance their access to and thriving in engineering environments.

Intended Impact: Implications from the Book and Its Chapters

Environments focused on inclusion at any institution may support the success of Latin* students in engineering and other fields, including but not limited to Hispanic-Serving Institutions and/or community colleges. Such spaces may enable Latin* students to leverage their community cultural wealth and funds of knowledge, assets they bring with them into educational environments; these can bolster their capacity to navigate engineering spaces which might otherwise be potentially hostile learning environments (Revelo & Baber, 2018; Verdín & Godwin, 2018; Wilson-Lopez et al., 2016). In fact, such skills may not only be assets for Latin* students themselves, but also for engineering innovation (Abreu et al., 2021). Accordingly, enhancing equitable access to engineering classrooms, engineering makerspaces, and mentored research opportunities in the field may help Latin* students' success and retention (see Fluker et al., 2022; Villanueva et al., 2021). As such, prior research and the work within this book come together to provide implications both for scholars and well as practitioners concerned with making individual-level and structural impacts to the experiences of Latin* engineering students.

REPRESENTATION OF THE FIELD

This book was written to and by our familia in this work. So often we are asked to compete against each other in academic spaces for resources or recognition, but we resist this notion. Instead, through this book, we sought to bring together a collective of professionals who care deeply about this issue and inspire those in our community to continuously push toward change. We assembled a community of scholar-authors immersed in the field in distinct ways and across professional communities to increase the validity and impact of this work. The editors and many of the authors have experience in multiple, varied roles across disciplinary boundaries and university, industry, and policy settings. Throughout the book, we intentionally draw on authors with diverse perspectives, ranging in career stage (e.g., intentional collaborations between senior and junior scholars), discipline (e.g., engineering education, higher education, other social science fields, and engineering practitioners), and institutional type and selectivity (e.g., research universities, Hispanic-Serving Institutions).

As editors of this volume, we intentionally recruited and convened a range of authors and perspectives to not only write about inclusive practices and offer evidence to inform potential interventions and new directions but also to practice inclusion in the development of an author community. Indeed, thirty-four distinct authors are represented in this book, from an array of states and regions as well as positional identities, including but not limited to ethnic, racial, and national origin identities. In addition to contributor biographies, each chapter includes a positionality statement, in which the authors explicitly and reflexively situate how they approach the work, given their identities and roles; this way, they model research transparency and offer further insights into the process of developing and analyzing evidence (see Secules et al., 2021).

IDENTIFYING AND NAMING TERMS

The book employs "Latin*" to represent the continued variety of linguistic terms and identities used by contemporary undergraduate students and scholars in the United States. Drawing on Latin* as a root, "Latinx" and another recently emerging terms "Latine" (sometimes "Latiné") signify Latin American origin while not being gender-specific (María Del Río-González, 2021; Merriam-Webster, n.d.; Salinas, 2019, 2020). An older term, "Hispanic"—developed for government purposes—is still commonly used today; it signifies an ethnic origin in Spain or the Spanish-speaking former colonies of Spain in Latin America. Latino (male) and Latina (female) are also used to signify Latin American origin and are associated with specific binary categories for biological sex and/or gender. While, in Spanish grammar, Latino is inclusive of all identities, it still employs "male" as the default category.

We choose Latin* (pronounced *lə-ˈtē-(ˌ)n* in Spanish) as the primary term to refer to Latin American origin students' race/ethnicity while recognizing that language is fluid and will continue to shift after the publication of this book. The asterisk (*) functions as a wildcard, to signify the range and multiplicity of identities employed by a group with shared identity characteristics but fluid use of umbrella terms to signal their shared community (see Steinmetz, 2018). Gender-inclusive terms used for other populations (see e.g., Nicolazzo, 2016 on trans* students) have used an asterisk in this manner; and Latin* is becoming a more widely used gender-inclusive term for Latina/e/o/x students (Salinas, 2020; Cuellar & Salinas, 2022). Indeed, there is emerging consensus among engineering education scholars that these terms remain fluid and that engineering educators and disciplines should engage gender-inclusive terms to represent the Latina/e/o/x and Hispanic community of engineers (Villanueva Alarcón et al., 2022; see also this book and its chapters' associated references).

We have a multiplicity of preferred terms among the authors represented in the book; for this reason, we selected a shared term for the book while granting authors the autonomy to choose the term(s) they prefer for each chapter. Even if terms continue to change in the coming decades, as they likely will, we hope this discussion will be a useful resource for scholars, faculty, administrators, and practitioners to understand why and how these terms vary.

SUMMARY

This book has truly been a labor of love, toward our care of the Latin* engineering community and towards each other, and we hope that you find value in the pages ahead. Rather than hypothesize about how the research could be of help to practitioners, this book as a whole and each of its chapters provide concrete, evidence-informed strategies for how to enhance educational experiences for Latin* students in engineering. It is our dream that in the coming years Latin* students will encounter more equitable engineering environments and that we all continue to press forward toward a more socially just world.

NOTE

1. Mindful that readers may be more interested in application over theory, the book's chapters purposefully explain and apply theoretical frameworks key for engineering education, which can be leveraged by practitioners as well as researchers (see, e.g., Magana, 2022).

REFERENCES

Abreu, A. N., Guardia, L., Perez, V. V. B., Hasbun, I. M., & Strong, A. C. (2021, October 13–16). *Towards a culturally responsive design experience: How students' community capital contributes to their design approach* [Paper presentation]. IEEE Frontiers in Education Conference, Lincoln, Nebraska.

Anderson, E. L., Williams, K., Ponjuan, L., & Frierson, H. T. (2018). *The 2018 status report on engineering education: A snapshot of diversity in degrees conferred in engineering.* Association of Public & Land-Grant Universities. https://www.aplu.org/library/the-2018-status-report-on-engineering-education-a-snapshot-of-diversity-in-degrees-conferred-in-engineering/

Banda, R. M. (2020). From the inside looking out: Latinas' intersectionality and their engineering departments. *International Journal of Qualitative Studies in Education, 33*(8), 824–839.

Cuellar, M. G., & Salinas, C. (2022). Latin* college students' diversity and intersectionality: Future directions for research and practice. In M. Bonous-Hammarth (Ed.), *Neighborhoods, communities, and urban marginality* (pp. 55–83). Springer Nature. https://doi.org/10.1007/978-981-16-8000-7_4

Fluker, C., Perez-Felkner, L., & McCoy, K. (2022). *Students' perceptions of their engineering identity development and REU summer internship program experiences: An equity-centered analysis.* American Society for Engineering Education Annual Conference [Peer proceedings]. Minneapolis, MN. https://peer.asee.org/40969

Funk, C., & Lopez, M. H. (2022). *Hispanic Americans' trust in and engagement with science.* Pew Research Center. https://www.pewresearch.org/science/2022/06/14/a-brief-statistical-portrait-of-u-s-hispanics/

Kendall, M. R., Denton, M., Choe, N. H., Procter, L. M., & Borrego, M. (2019). Factors influencing engineering identity development of Latinx students. *IEEE Transactions on Education, 62*(3), 173–180.

Magana, A. J. (2022). The role of frameworks in engineering education research. *Journal of Engineering Education, 111*(1), 9–13. https://doi.org/10.1002/jee.20443

María Del Río-González, A. (2021). To Latinx or not to Latinx: A question of gender inclusivity versus gender neutrality. *American Journal of Public Health, 111*(6), 1018–1021. https://doi.org/10.2105/AJPH.2021.306238

Merriam-Webster. (n.d.). Latine. In *Merriam-Webster.com dictionary.* Retrieved April 13, 2023, from https://www.merriam-webster.com/dictionary/Latine

Mora, L. (2022). *Hispanic enrollment reaches new high at four-year colleges in the U.S., but affordability remains an obstacle.* Pew Research Center. https://www.pewresearch.org/fact-tank

/2022/10/07/hispanic-enrollment-reaches-new-high-at-four-year-colleges-in-the-u-s-but
-affordability-remains-an-obstacle/

National Center for Science and Engineering Statistics. (2023). *Women, minorities, and persons with disabilities in science and engineering.* https://ncses.nsf.gov/wmpd

Nicolazzo, Z. (2016). *Trans* in college: Transgender students' strategies for navigating campus life and the institutional politics of inclusion.* Stylus Publishing.

Ozuna Allen, T., Thompson, M. L., & Collins, S. (2020). How do Latinx dual credit earners describe their sense of belonging in engineering programs? *Journal of College Student Retention: Research, Theory & Practice, 24*(1), 3–19. https://doi.org/10.1177/1521025119898153

Ozuna Allen, T., Thompson, M. L., & Martinez-Cosio, M. (2019). Message, hope, and reality. *High School Journal, 103*(1), 38–52. https://doi.org/10.1353/hsj.2020.0002

Revelo, R. A., & Baber, L. D. (2018). Engineering resistors: Engineering Latina/o students and emerging resistant capital. *Journal of Hispanic Higher Education, 17*(3), 249–269. https://doi.org/10.1177/1538192717719132

Rodriguez, S. L., Doran, E. E., Sissel, M., & Estes, N. (2022). Becoming la ingeniera: Examining the engineering identity development of undergraduate Latina students. *Journal of Latinos and Education, 21*(2), 181–200. https://doi.org/10.1080/15348431.2019.1648269

Salinas, C., & Lozano, A. (2019). Mapping and recontextualizing the evolution of the term Latinx: An environmental scanning in higher education. *Journal of Latinos and Education, 18*(4), 302–315. https://doi.org/10.1080/15348431.2017.1390464

Salinas, C. (2020). The complexity of the "x" in Latinx: How Latinx/a/o students relate to, identify with, and understand the term Latinx. *Journal of Hispanic Higher Education, 19*(2), 149–168. https://doi.org/10.1177/1538192719900382

Secules, S., McCall, C., Mejia, J. A., Beebe, C., Masters, A. S., L. Sánchez-Peña, M., & Svyantek, M. (2021). Positionality practices and dimensions of impact on equity research: A collaborative inquiry and call to the community. *Journal of Engineering Education, 110*(1), 19–43. https://doi.org/10.1002/jee.20377

Simmons, D. R., & Lord, S. M. (2019). Removing invisible barriers and changing mindsets to improve and diversify pathways in engineering. *Advances in Engineering Education. 7*(2), 1–22.

Steinmetz, K. (2018, April 3). The Oxford English dictionary added "trans*." Here's what the label means. *Time.* https://time.com/5211799/what-does-trans-asterisk-star-mean-dictionary/

Verdín, D., & Godwin, A. (2018). Exploring Latina first-generation college students' multiple identities, self-efficacy, and institutional integration to inform achievement in engineering. *Journal of Women and Minorities in Science and Engineering, 24*(3), 261–290. https://doi.org/10.1615/JWomenMinorScienEng.2018018667

Villanueva, I., Di Stefano, M., Gelles, L., Youmans, K., & Hunt, A. (2020). Development and assessment of a vignette survey instrument to identify responses due to hidden curriculum among engineering students and faculty. *International Journal of Engineering Education, 36*(5), 1549–1569.

Villanueva, I., Downey, R. J., Nadelson, L., Choi, Y. H., Bouwma-Gearhart, J., & Tanoue, C. (2021). Understanding equity of access in engineering education making spaces. *Social Sciences, 10*(10), 384. https://doi.org/10.3390/socsci10100384

Villanueva Alarcón, I., Mejia, J. A., Mejia, J., & Revelo, R. (2022). Latiné, Latinx, Latina, Latino, or Hispanic: Problematizing terms often used in engineering education. *Journal of Engineering Education, 111*(4), 735–739. https://doi.org/10.1002/jee.20486

Wilson-Lopez, A., Mejia, J. A., Hasbún, I. M., & Kasun, G. S. (2016). Latina/o adolescents' funds of knowledge related to engineering. *Journal of Engineering Education, 105*(2), 278–311. https://doi.org/10.1002/jee.20117

PART 1 EXAMINING THE EXPERIENCES OF LATIN* STUDENTS IN ENGINEERING

2 · THE IMPORTANCE OF ENGINEERING IDENTITY FOR LATINX STUDENTS

SARAH L. RODRIGUEZ, MARIA L. ESPINO, MORGAN NICHOLS, AND BRIAN D. LE

Today, there are more Latinx students interested in and pursuing engineering careers than ever before (Esquinca & Mucino, 2021; Johnson, 2019; Wilson-Lopez & Acosta-Feliz, 2021). However, engineering as a field remains racially and ethnically homogenous and continues to be overwhelmingly dominated by white men (Rodriguez et al., 2019; Turochy et al., 2021). While the share of engineering bachelor's degrees awarded to Latinx students has increased over the past decade, Latinx students made up only around 12 percent of all engineering graduates in 2021 (National Center for Science and Engineering Statistics [NCES], 2021). Latina women earned a mere 2.85 percent of all engineering bachelor's degrees, a percentage that has only slightly increased over the past decade (NCES, 2021). Furthermore, individuals from Latinx backgrounds currently represent only 9 percent of all full-time employed engineers (NCES, 2021).

These inequities are the result of historically oppressive racist and other structures embedded in engineering. Latinx students continue to be racially and ethnical marginalized within engineering environments. Because these oppressive structures are inescapable, to persist toward degree completion in engineering, Latinx students must operate within systems of oppression—with their success often dependent on their navigation skills. Rather than force students to operate within these oppressive structures or rely on students to improve their own environments, educational stakeholders from across the United States are (and have been) seeking to improve pathways and learning experiences for Latinx students pursuing engineering careers. One approach to doing this work within higher education settings is to focus on how students begin to embrace their engineering identity and how they think about the various pathways and learning experiences as processes of developing that identity (Denton, et al., 2022; Johnson, 2019; Villa et al., 2020. The next section provides a broad overview of how scholars have defined engineering identity and its major constructs, and then reviews the wider scope of literature on the subject.

WHAT IS ENGINEERING IDENTITY?

Although the term's exact origins are unclear, the word "identity" can be traced to the French word *identité*: "to denote a sameness or oneness" (Harper, n.d.). Ultimately from the Latin word *identitas*, it may derive from the root *idem*, or "same," and *identidem*, meaning something that is done repeatedly (Simpson et al., 1989). "Identity," when used in reference to a person, often denotes who someone is and the characteristics, feelings, and beliefs they hold; it may also refer to feeling similar to others and to being able to understand others who have these similarities (Simpson et al., 1989). With regard to the learning context, Lave (1998) proposes that students are active participants within their experiences, actively moving toward—or perhaps away from—their identities: "Learning is, in this purview, more basically, a process of coming to be, of forging identities in activity in the world. In short, learners are never only that, but are becoming certain sorts of subjects with certain ways of participating in the world" (p. 3). In this way, we can think of engineering identity development as a process of becoming, with Latinx students learning and refining how they participate or exit the engineering world. It is how one understands and positions oneself within engineering contexts (e.g., classrooms, peers, faculty, and student organizations), and it encourages Latinx students to see themselves in particular roles, such as that of an engineer. Within this paradigm, students are active participants in their identity development process, and the interactions that they have—both with those in their immediate engineering community as well as the larger community—can contribute to, or push them away from, a career in engineering.

Engineering identity (and STEM identity, more broadly) has been conceptualized and studied in numerous ways (Godwin et al., 2013a, 2013b; Godwin, 2016; Patrick & Borrego, 2016). Some scholars have conceptualized engineering identity as the way in which a student "defines the self through a role or performance in engineering" (Jones, 2014, p. 1342), while others have drawn attention to the way a student might describe how they connect to a future engineering career role (Matusovich et al., 2011). Scholars have conceptualized engineering identity as a stage theory in which development takes place over time (Meyers et al., 2012), as well as a socially conscious perspective in which the engineering environment is a "figured world" (Holland, 2001) in which engineering identity is socially and culturally produced (Godwin et al., 2013a, 2013b; Godwin, 2016; Tonso, 2006).

Four main elements (see table 2.1) appear that may encompass the development of an engineering identity: interest, competence, performance, and recognition. Some scholars (e.g., Godwin et al., 2013a, 2013b; Hazari et al., 2010) acknowledge the importance of students' perceptions of all four of these elements,

TABLE 2.1   Engineering Identity: Major Constructs

Construct	Meaning
Interest	Desire to know more and engage in engineering; can propel students to overcome barriers and challenges
Competence	Skills and knowledge, including content area information and specialized tools
Performance	Ability to apply the knowledge acquired as engineering students
Recognition	Ability to recognize and have others recognize oneself as an engineer (especially engineering peers, faculty)
Socially constructed	
Process of becoming	
Possibilities of association or alienation	

while others (e.g., Carlone & Johnson, 2007) emphasize the importance of some, while leaving out or combining others.

1. *Interest* is the desire of Latinx engineering students to want to know more about concepts and the field of engineering. Engineering students demonstrate interest by taking opportunities to be engaged with engineering content and overcome challenges that may arise (e.g., difficult concepts and rigorous assignments) within their engineering experiences.
2. *Competence* centers on the skills and various forms of knowledge that a Latinx engineering student possesses. Competence provides a space of growth from which to learn more, build confidence, and apply the information they know. Competence can be derived from both formal and informal engineering settings.
3. *Performance* refers to a Latinx student's ability to express behaviors or actions that are associated with the field of engineering. Similar to competence, which focuses on the skills and forms of knowledge needed, performance refers to the more social act of expressing those elements through behaviors and actions. As students who are often marginalized by engineering contexts, it is essential that Latinx engineering students are provided with opportunities to take on an engineering identity, "perform" tasks, and apply their acquired knowledge (Godwin et al., 2013a, 2013b; Godwin, 2016; Kendall et al., 2019). Closely related, these engineering identity "performances" are important rites of recognition for Latinx students.
4. Lastly, *recognition* refers to the way in which Latinx students come to see themselves as the kind of individuals who engage in engineering work. Often, such self-recognition is the result of meaningful others (e.g., family, faculty, staff, or industry members) recognizing the skills, knowledge, and capabilities

of the students. Students may find it hard to see themselves as engineers due to norms projected by their engineering faculty, peers, and organizations.

WHY IS ENGINEERING IDENTITY IMPORTANT?

Engineering identity can be described as a concept by which a person experiences similarities and a sense of connectedness with the role of an engineer—not merely once but repeatedly, over time (Godwin et al., 2013a, 2013b; Godwin, 2016; Rodriguez et al., 2021). This person feels a sense of sameness, oneness, or connectedness with the engineering environment and its actors. Furthermore, engineering identity can help educational stakeholders determine how students become associated with or alienated from engineering communities and their norms and culture, which can be key to ensuring equity, access, and success for historically marginalized populations.

Over time, the scholarship on engineering identity has grown; however, it remains a largely unexplored area. Research has shown that developing a strong sense of engineering identity is important to professional formation and taking on the role of an engineer (e.g., Foor et al., 2007; Pierrakos et al., 2009). Engineering identity encompasses the ability to feel, be interested in, and possess the skills and knowledge to actively engage in engineering practices (Godwin, 2016). A strong sense of engineering identity allows students to refine and develop the toolset they need for the engineering industry (Godwin, 2016; Godwin et al., 2013a, 2013b). Students can also recognize themselves—and be recognized by others—as engineers (Rodriguez et al., 2019).

This process of becoming a person who identifies as an engineer connects to the socialization and engagement of engineering opportunities through internships in industry, student organizations, and the classroom. Prior research reveals that empowering collegiate experiences are crucial to identifying with the engineering major (Jones et al., 2014)—particularly those in first-year engineering courses; these experiences include making decisions and being given choices, and having instructors who explain the usefulness of and generate interest in the material. In addition, exposure to tinkering, design, and analysis may be linked to students having stronger engineering identities (Choe et al., 2019; Verdín, 2024 [ch. 5 in this volume]). Participation in undergraduate research also positively influences engineering identity development (Fluker et al., 2022). Moreover, families—especially fathers—are key in shaping engineering identity development (Fluker et al., 2022). Students who lack positive engineering identity-building experiences may face challenges within the engineering environment and be vulnerable to transitioning out of the major altogether (Tonso, 2014).

While scholars have extended what we know about engineering identity, little is known about how Latinx students in engineering build and maintain their engineering identities. To meet the needs of the current Latinx students in engineering, as well as encourage the next generation of potential students, it is important to understand the nuanced experiences of these students. The next section addresses prior and current literature on engineering identity for Latinx engineering students on college campuses, including a focus on which environments cultivate (or discourage) engineering identity and who might be major contributors (or detractors) to their development.

WHAT DO WE KNOW ABOUT ENGINEERING IDENTITY FOR LATINX STUDENTS?

Scholars have articulated the importance of engineering identity development, particularly for Latinx students. Engineering identity is an important predictor of success, especially for students who are traditionally racially and ethnically marginalized within an engineering context (Kendall et al., 2019; Rodriguez et al., 2021). As literature and research continues to grow on this topic, Latinx engineering identity development provides a variety of lenses, experiences, and outcomes for scholars to examine these unique experiences (Smith et al., 2022; Villa et al., 2020; Wilson-Lopez & Acosta-Feliz, 2021). Scholars have also suggested that intentionally developing Latinx students' engineering identity is key to building a strong engineer. Latinx students need the space to engage in engineering learning experiences, both independently and collaboratively (Flores et al., 2021; Mejia, 2022). These performances and subsequent moments of recognition, especially in front of engineering faculty and peers, are essential to seeing oneself as an engineer, particularly in light of the various intersections of identities within the Latinx student population.

Prior research on engineering identity experiences for Latinx students tells us that while many students develop engineering interests and competence at a young age, for some, those interests are not discovered or nurtured until college (see figure 2.1; Denton et al., 2022; Flores et al., 2021; Mein et al., 2020; Turochy et al., 2021).

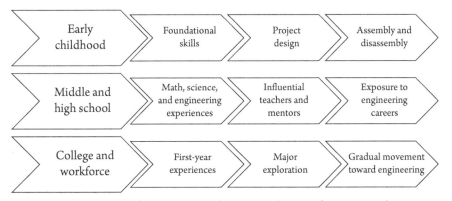

FIGURE 2.1. Opportunities for Engineering Identity Development for Latinx Students

TABLE 2.2   Potential Influences on Engineering Identity

Student characteristics	Student precollege and college experiences	Institutional types and pathways
• Race	• Belonging	• Predominantly White Institution (PWI), Hispanic-Serving Institution (HSI)
• Stereotypes	• Mattering	
• Socioeconomic status	• Family	• Community college, four-year university)
• Gender	• Peers	
• Sexuality	• Faculty	• . . . and many more
• . . . and many more	• Identity-based organizations	
	• Resistance and community connections	
	• . . . and many more	

Some Latinx students discovered their interest in engineering in early childhood by being good at building things with blocks, designing projects, and assembling and disassembling items. Others recall developing their engineering interests and competence later in childhood by taking high school engineering courses, participating in math and science projects, having influential teachers and mentors, or being exposed to individuals in engineering careers (Kendall et al., 2019; Mein et al., 2020). Latinx students develop their engineering identities in a variety of ways and from a range of experiences and sources (see table 2.2; Esquinca et al., 2021; Flores et al., 2021; Mejia, 2022; Turochy et al., 2021).

Student Characteristics

Latinx engineering students hold a variety of identities and backgrounds (table 2.2) that may inform their engineering identity experiences, including but not limited to race, socioeconomic status, gender, and sexuality. These intersections may complicate engineering identity development due to existing oppressions and marginalization (Esquinca & Mucino, 2021; Mein et al., 2020; Revelo et al., 2017; Turochy et al., 2021). A variety of stereotypes may exist that are detrimental to the development of Latinx engineering students. For example, in terms of race, anti-Black stereotypes may complicate engineering identity development for Afro-Latinx engineering students (Turochy et al., 2021). While engineering education literature has acknowledged the harms of anti-Blackness in engineering (Holly & Quigley, 2022), it is often less explicit in terms of how anti-Blackness might shape engineering identity experiences within the Latinx community.

Socioeconomic status (e.g., family income, education, and occupation) can also have a significant impact on the experiences of engineering students (e.g., Major, 2022). The cost of pursuing an engineering degree (including tuition, textbooks, and specialized equipment or software) can be a significant financial burden for many students, especially those from low-income Latinx families. Financial constraints can limit students' ability to participate in extracurricular activities, attend conferences or workshops, and pursue internships or research opportunities—all of which can influence engineering identity development. Additionally, students from low-income backgrounds may need to work part- or full-time to support themselves or their families while in school, potentially impacting their engineering identity journey.

First-generation college status may be a particularly important aspect of a Latinx engineering student's educational experience and engineering identity sensemaking (e.g., Coronella, 2018, 2020; Kendall et al., 2019). First-generation Latinx engineering students may bring a different perspective and set of experiences to their work in engineering, which can be valuable in problem-solving and innovation. Yet, being the first person in their immediate family to pursue higher education can come with a unique set of challenges and opportunities. For many first-generation students, pursuing a career in engineering can be especially challenging due to the rigorous and technical nature of the field, as well as the fact that educational stakeholders may hold deficit views of Latinx engineering students (Coronella, 2018). First-generation Latinx engineering students, particularly those who are Latina, often struggle to feel a sense of belonging (Verdín & Godwin, 2018). Latina engineering students often experience a lack of both compositional diversity and faculty who care about their students' success and promote competition (Banda, 2020). First-generation Latina engineering students often utilize familial and community aspirations and goals to help them persist in their degree (Coronella, 2020).

Engineering identity experiences can also be complicated by a range of issues affecting members of the lesbian, gay, bisexual, transgender, queer (LGBTQ+) community (Day et al., 2022; Haverkamp et al., 2021). While research regarding Latinx engineering students who identify as LGBTQ+ is limited, this work is becoming more prominent—both in terms of general engineering populations and looking at the identity experiences of Latinx engineering students specifically. Of the work that has been published, we know that gay Latino men may face challenges around navigating issues of masculinity and whether they can bring their authentic selves to engineering spaces (Rodríguez-Simmonds, 2022). For all Latinx engineering students, overcoming these challenges and succeeding in the field can be especially meaningful and empowering, as it can pave the way for future generations of their families and communities.

Student Precollege and College Experiences

Latinx engineering students possess a range of precollege and college experiences (table 2.2), including but not limited to belonging, mattering, family, identity-based organizations, peers, faculty, and resistance and community connections. Latinx engineering students with a stronger engineering identity and sense of belonging tend to be more associated with engineering careers (Andrews et al., 2023). Latinx students who do not have strong engineering identity risk feeling and being isolated, and may be ultimately pushed out of engineering (Tonso, 2006; Turochy et al., 2021). To enhance their experiences and gain a stronger sense of engineering identity, engineering students often create their own spaces for belonging and mattering (Perez-Felkner et al., 2020). Students may lean on peers (particularly other Latinx engineering students), faculty, and identity-based organizations for support along the way (Perez-Felkner et al., 2020; Verdín & Godwin, 2018; Banda & Flowers, 2017; Rodriguez et al., 2019).

Latina engineering students may seek out additional support, as they tend to experience further marginalization due to sexism within engineering spaces (Rodriguez et al., 2019; Turochy et al., 2021). As such, Latina engineering students tend to develop professional identities through interactions with family members (especially their fathers and brothers) and engagement with identity-based organizations (e.g., women in engineering and Latinx students in engineering; Rodriguez et al., 2019; Villa et al., 2020. In considering the experiences of Latinx engineering students, researchers have called for a wider use of asset-based and culturally affirming frameworks, including funds of knowledge and Anzaldúa's framework of *conocimiento* (Mejia et al., 2022; Garza & Rodriguez, 2022; Wilson-Lopez et al., 2016). These funds of knowledge often include familial and community ways of knowing that students apply to engineering interactions and thinking (Wilson-Lopez, 2016). In terms of identity-based organizations, of special note are the Society of Hispanic Professional Engineers (SHPE)[1] and the Society of Women Engineers (SWE),[2] both of which are recognized as spaces that promote belonging and engineering identity (Banda & Flowers, 2017; Rodriguez et al., 2019). Furthermore, across the nation, engineering is continuing to see the proliferation of SHPEtinas initiatives, such as the annual conference and campus chapters that hold the promise to enhance how Latina students view themselves as engineers.

Relatedly, Latinx engineering students may develop their engineering identities through their interactions with the communities around them, in classroom settings, and beyond (e.g., personal, group, and internship), yet many of those exchanges are fraught. For example, we know that engineering is often a collaborative field, necessitating the ability to work effectively with others; engineers must be able to communicate clearly and respectfully and work collaboratively with colleagues from a range of disciplines and backgrounds. However, emerging research demonstrates that the more Latinx engineering students like collaborative work, the less likely they are to move toward engineering careers, which may indicate that Latinx students may not equate engineering with collaborative work or they might have experienced marginalizing (e.g., racist, sexist, or classist) team settings (Andrews et al., 2023). Research has shown that Latinx students find a range of avenues for resisting exclusionary practices and environments in engineering, including outreach to their communities and engaging in collective resistance (Revelo & Baber, 2017).

Institutional Types and Pathways

Developing an engineering identity as a Latinx student is a multipart, recursive process based on the interactions between individuals, and it is highly dependent on context. For example, Latinx engineering students at predominantly white institutions (PWIs) may have very different engineering identity experiences than those at Hispanic-serving institutions (HSIs; Kendall et al., 2019). Meanwhile, at PWIs, Latinx engineering students may experience challenges related to building academic networks, particularly depending on whether they integrate into predominantly white networks, segregate within predominantly Latinx communities, or socially adapt in order to connect with a variety of racial networks (Johnson, 2019). This could, in turn, influence their engineering identity development. HSIs may also provide a more supportive environment for engineering identity development; the literature demonstrates that Latinx engineering students at HSIs who have a stronger engineering identity were positively correlated with engineering-focused career plans (Andrews et al., 2023). Scholarship has revealed that HSI students tend to perceive themselves as having stronger engineering identities in comparison to PWI students (Choe et al., 2019). However, Latinx engineering students at HSIs may still struggle to form a strong engineering identity due to systemic oppressions, such as gendered racism in the case of Latina students (e.g., Garriott et al., 2019).

Beyond institutional type, other contextual factors may also influence engineering identity for Latinx students: for example, their enrollment in community colleges and four-year institutions. Prior research has mostly centered on engineering students' experiences in four-year university settings. However, there is an emerging scholarship that encompasses community college students' engineering experiences (Allen & Zhang, 2016; Ogilvie & Knight, 2020). Community college pathways can provide meaningful learning and impactful mentoring experiences with peers and faculty, which are essential to transferring to a four-year institution (e.g., Allen & Zhang, 2016). For Latinx students, in particular, community colleges may represent an affordable and flexible option that best fits their professional goals and non-academic commitments (Ogilvie & Knight, 2020). All these experiences contribute to shaping an engineering identity for Latinx students and may be considerably different from the identity experiences of their four-year institution counterparts.

IMPLICATIONS FOR ENGINEERING IDENTITY RESEARCH

While the knowledge base for engineering identity for Latinx students has certainly grown over the past decade, there are still a few key areas that scholars need to address (see table 2.3).

TABLE 2.3   Implications for Research on Latinx Students' Engineering Identity

Research area	Possible explorations

Understanding individual elements of engineering identity	• Interest
	  ∘ Early childhood engineering identity formation; later pathways for engineering identity maintenance and development
	• Recognition
	  ∘ Effective practices
	  ∘ Latinx communities and families
	• Competence
	  ∘ Challenging areas
	  ∘ Capacity of assignments and team projects; culturally sustaining pedagogies
	• Performance
	  ∘ Making meaning of racialized experiences
	  ∘ Performing in engineering contexts

Investigating further the influence of institutional contexts and types	• Subdisciplinary investigations of identity development • PWIs, HSIs, and (HBCUs) • Community colleges and four-year institutions • Online and hybrid learning modalities
Engaging in longitudinal studies alongside Latinx engineering students	• Engineering identity over time • Engagement research with Latinx students or current engineers

Understanding Individual Elements of Engineering Identity

First, researchers need to understand the individual elements of engineering identity to understand how we might best address each of these elements in a practical manner for Latinx engineering students. For example, studies might seek to understand how early and late childhood interests in engineering form and what specific connections bridge to Latinx student engineering pathways and identity development. Relatedly, scholars might seek to investigate interests over time and how they are created, managed, refined, or lost as a result of undergraduate educational experiences—and on into graduate education and the workforce. Longitudinal research focused on engineering identity could help scholars understand how to better encourage engineering interests at all levels and also how to manage transitions between experiences in a more nuanced way.

While prior research has taken up identity recognition as an important aspect of the engineering identity development process, future studies might also seek to understand which forms of engineering identity recognition are most effective for Latinx students. In particular, research demonstrates that Latinx communities and families (especially fathers) are important for this development, but future research might delve deeper into this to more fully understand how these relationships can be leveraged. Similarly, researchers know that faculty, staff, and peers are important for recognition; however, more could be uncovered about which specific practices are effective for Latinx engineering students, as well as how those strategies might be institutionalized for structural change.

In terms of competence and performance, future studies might focus on how Latinx engineering students build and perform the technical knowledge needed to be successful: for example, exploring which areas of technical knowledge are particularly challenging or how assignments and team projects have the capacity to affirm engineering identities. Relatedly, scholars also might seek to understand how culturally sustaining pedagogies influence how Latinx engineering students develop their identities. And doing so might shed more light on how students make meaning of their racialized experiences or experience salience of their Latinx versus other identities—as well as perform those identities for others within engineering contexts.

Investigating Further the Influence of Institutional Contexts and Types

Future researchers might examine the influence of institutional context on Latinx engineering identity development. Investigating the differences in engineering identity experiences for Latinx students within individual engineering disciplines is important. As disciplinary-based research has demonstrated in the past, each engineering discipline retains its own set of norms and socialization processes. For example, the identity experiences of students within electrical and computer engineering disciplines, which have fewer Latinx engineering students in general, might be very different than the experiences of students in mechanical engineering, which tends to have greater numbers of Latinx students. From this, it stands to reason that students in each of the disciplines may also have different experiences with engineering identity development. Understanding this process might allow scholars to better understand the experiences of undeclared engineering majors or students seeking to transfer disciplines within engineering.

Furthermore, the wide range of institutional contexts and types throughout the United States means that Latinx engineering students may have an equally diverse set of engineering experiences. Scholars might seek to understand the experiences of Latinx students in terms of differences in geographical region, urbanicity/rurality, size, and mission, among others. Similarly, scholars might also seek to understand differences by institutional type, including public versus private institutions or teaching versus research-intensive institutions. The possibilities in this area are endless for understanding the existing nuances for Latinx students' engineering identity development. For example, Latinx students at community colleges are likely to have different engineering identity development experiences than those at universities. Similarly, students at HSIs are likely to have contrasting experiences to those at PWIs or historically Black colleges and universities (HBCUs). Furthermore, experiences may differ at institutions with strong colleges or departments of engineering or historical roots in this area, as well as institutions with other areas of focus (such as those with a land-grant mission.) Scholars may also choose to consider the growing influence of online and hybrid learning modalities offered to Latinx students; while the COVID-19 pandemic may have forced institutions into these modalities, there will undoubtedly be sustained and growing conversations about which elements of engineering programs can be translated into online and hybrid learning modalities. Understanding how these contexts influence engineering identity could be a crucial component to those conversations, particularly if there are elements of the contexts that do or do not contribute to identity development.

Engaging in Longitudinal Studies Alongside Latinx Engineering Students

Finally, future scholarship might seek to engage in longitudinal studies with Latinx engineering students to understand how their engineering identities develop over time. While mentioned previously as an important implication for understanding engineering interests, longitudinal studies could be crucial for understanding all elements of engineering identity development. Rather than merely capturing one snapshot in time, or even a couple, engineering identity development might be best understood longitudinally. Connectedly, studies that position Latinx students or current engineers as coresearchers with more participatory action roles could be helpful in understanding the Latinx engineering identity development process and could be empowering to students, engineers, and communities. By allowing participants to engage as equals in the research process, participants may be more likely to be truthful and open in their responses and able to craft more meaningful research questions and approaches to working with their communities during and after the research is completed.

As researchers ourselves, we realize that nuanced efforts to understand engineering identity development will be the most fruitful in the future. Without a dedication toward understanding the components of engineering identity and situating that development within institutional contexts for sustained periods of time, we are only capturing a small range of experiences. To broaden our understanding of engineering identity development for Latinx students, it remains important to consider the vast diversity of not only the Latinx student population within engineering but the wide range of contexts with which these students engage on a day-to-day basis.

Recommendations for Practitioners, to Enhance Latinx Students' Engineering Identity

To enhance engineering identity for Latinx students, practitioners might look at a variety of strategies for building and sustaining engineering interests, such as what brought them to like engineering, how they developed competence and understanding of the field, as well as ways to encourage engineering performances and recognitions (see table 2.4).

TABLE 2.4   Promising Practices for Implementing Engineering Identity
Practices with Latinx Students

Practice area	Promising practices
Interest	• Career discovery • Engineering environment immersion • In-class, hands-on experiential learning • Connections with industry partners • Increased interest in college-level engineering • Academic pathways • Early and ongoing advising • Structured pathways
Competence	• Early experiences • Provide opportunity for foundations skills • Technical learning opportunities (continuous learning) • Scaffolding learning activities • Give students a chance to build on prior knowledge • Students gradually become more independent
Performance	• Academic supports • On-site or online tutoring • Supplemental instruction, group peer tutoring, study tables • Mentoring programs • Undergraduate research • Exposure to labs and engineering environments • Cocreation of knowledge with engineering professionals • Team-based learning
Recognition	• Institutional-level partnerships • Become part of the engineering experience from the beginning (K–12/community college/four-year) • Opportunities to be recognized • See engineering possibilities that align with their identities • Integrate into the classroom experience • Participate in internships/co-ops • See themselves, be recognized

Building and Sustaining Engineering Interests

To build and sustain engineering interests, practitioners might consider encouraging career discovery within Latinx communities and strengthening academic pathways. In particular, practitioners might look to engineering environment immersion activities; these include in-class, hands-on experiential learning as well as implementing early and ongoing immersive advising procedures and more strengthened, structured engineering pathways that will more effectively encourage Latinx students to see themselves as engineers. Further, engineering modules can help increase interests and provide a different lens for students to understand engineering on a broader spectrum and an opportunity to see themselves become one (Denton et al., 2022).

Similarly, practitioners might look for ways to implement early and ongoing engineering advising that is focused on promoting and refining engineering interests. This might mean helping students articulate what it is about engineering as a field or certain engineering disciplines that sparks their interests or drives their curiosity or desire to learn. As students progress through their programs, it might also mean encouraging them to continue refining their interest areas with additional internships or co-ops, or by engaging with identity-based organizations to find others who share similar interests and backgrounds. In addition, practitioners might also consider how engineering pathways are structured, including which markers help students define their interest areas and stay engaged with the content throughout their program.

Encouraging Engineering Competence

To encourage a foundation of engineering competence, practitioners might focus on shaping early engineering experiences and scaffolding learning activities. Practitioners might consider a variety of foundational and technical learning opportunities and gradually scaffold activities that allow Latinx students to not only to build their knowledge and skills but incrementally become more independent in their engineering work—thereby allowing them to develop their engineering identities more fully. Creating ongoing technical learning opportunities and a culture of continuous learning might encourage Latinx students to further develop their engineering identities. Lastly, practitioners, particularly faculty members, might consider examining technical knowledge and skills for opportunities for culturally sustaining activities. Doing so might mean considering which elements of the curriculum are steeped in white, masculine, or other oppressive norms or asking what forms of knowledge, skills, and experiences are valued over others. This examination invites the infusion of culturally sustaining activities throughout the curriculum to allow Latinx students to see their cultures reflected in institutional structures and utilize ways of knowing from their homes and communities.

Enhancing Opportunities for Engineering Performance

To enhance engineering identity development, practitioners might consider strategies that support engineering identity performance and opportunities for recognition for Latinx students. Practitioners might consider how academic supports, undergraduate research, and team-based learning could be utilized to support Latinx engineering identity development. Similarly, institutional-level partnerships and opportunities for recognition can encourage Latinx students in engineering to see themselves as engineers over time and gain opportunities to be recognized as the kind of individuals who engage with engineering.

Students might have the opportunity to perform their engineering identities using both within- and outside-of-class settings by participating in well-facilitated group activities, serving as an engineering course tutor or supplemental instructor, or engaging in engineering projects with peers, faculty, or industry partners. Moreover, developing mentoring programs—whether with students further along in the program or professionals who are in the students' respective engineering field—can yield benefits in performance. In addition, practitioners might look toward expanding undergraduate research activities, helping students get exposure to specific labs and engineering environments that encourage the cocreation of knowledge with engineering professionals. Practitioners can also provide team-based learning activities for students to develop their performance, competence, and interests at an early age, such as an opportunity through the K–12 system.

Creating Recognition Experiences

Opportunities like these might allow for instances of engineering identity to be recognized, as the institution creates spaces and partnerships that enable Latinx students to see themselves in the role of an engineer. Practitioners might consider structuring institutional-level partnerships so that Latinx students become a part of the engineering experience from the very beginning of their educational journeys. Doing so might mean partnering with K–12, community-college, or four-year university partners to make transitions seamless and so that Latinx engineering students can envision themselves in engineering academic pathways and careers. Similarly, practitioners might consider integrating opportunities for students to be recognized throughout the curriculum as well as the overall engineering journey. Ideally, practitioners might find ways to align projects with Latinx student interests, values, and career trajectories, encouraging students to bring their whole selves to engineering. Practitioners could consider how recognition opportunities may be infused into the classroom experience, as well as how they might work with industry partners to ensure that Latinx students are positioned in meaningful internships and co-ops—particularly those that have connections to Latinx communities.

CONCLUSION

We write this chapter as four scholars representing a variety of backgrounds and experiences. As scholars from historically marginalized groups, we are highly invested in efforts to uplift these communities and deconstruct the exclusionary nature of engineering spaces. As a collective, we espouse an unapologetic commitment to diversity, equity, and inclusion, which is reflected in our writing and our hopes that such a chapter can be useful to both scholars and practitioners who are similarly committed to the enhancement of experiences for Latinx engineering students. Together, we have a history of working in both K–12 and higher education settings to address issues related to identity and transition, as well as collaborating on large-scale funded research and implementation projects. As such, we recognize the importance of identity development, particularly for students who are historically and systematically marginalized by educational institutions. We have a definitive research-to-practice orientation that compels us not only to write about engineering identity development but also to offer solutions for implementing best practices across the college experience.

Thanks to a range of efforts from a variety of educational stakeholders, a great deal of media attention, scholarly work, and funding has been dedicated to broadening participation for Latinx students in engineering (Anderson et al., 2018; Flores et al., 2021; Turochy et al., 2021). With broadening participation for Latinx engineering students, these stakeholders continue to support and push for the increased presence of diverse engineers (Denton et al., 2022; Godwin, 2016; Mejia, 2022).

This chapter highlights the need to structure Latinx student educational experiences around engineering identity development. A strong engineering identity means that Latinx students have an opportunity to build their interests and competence as well as to perform and be recognized as the type of individuals who engage in engineering work. Latinx students come from a myriad of backgrounds and possess a variety of identities influencing how they show up in engineering spaces and experience a range of privileges and oppressions. As a non-monolithic group, scholars can delve more deeply into their understanding of engineering identity development, especially as it relates to translating research to practice. Practitioners share in this opportunity as they implement research-based approaches to serve Latinx students in engineering, to help them develop their engineering identities and work toward the elimination of structural barriers to the educational process.

NOTES

1. More information on SHPE and SHPEtinas can be found at: https://shpe.org/.
2. More information on SWE can be found at: https://swe.org/.

REFERENCES

Allen, T. O., & Zhang, Y. (Leaf). (2016). Dedicated to their degrees: Adult transfer students in engineering baccalaureate programs. *Community College Review, 44*(1), 70–86. https://doi.org/10.1177/0091552115617018

Anderson, E. L., Williams, K. L., Ponjuan, L., & Frierson, H. (2018). *The 2018 status report on engineering education: A snapshot of diversity in degrees conferred in engineering.* Association of Public & Land-grant Universities. https://www.aplu.org/our-work/5-archived-projects/access-and-diversity/2018-status-report-on-engineering-education/

Andrews, M. E., Kendall, M. R., Rodriguez, S. L., & Borrego, M. (2023). Career plans of Latinx mechanical engineering undergraduates studying at Hispanic serving institutions. *Journal of Women and Minorities in Science and Engineering, 29*(1), 45–85. https://doi.org/10.1615/JWomenMinorScienEng.2022040409

Banda, R. M. (2020). From the inside looking out: Latinas' intersectionality and their engineering departments. *International Journal of Qualitative Studies in Education, 33*(8), 824–839. https://doi.org/10.1080/09518398.2020.1735565

Banda, R. M., & Flowers III, A. M. (2017). Birds of a feather do not always flock together: A critical analysis of Latina engineers and their involvement in student organizations. *Journal of Hispanic Higher Education, 16*(4), 359–374. https://doi.org/10.1177/1538192716662 9

Carlone, H. B., & Johnson, A. (2007). Understanding the science experiences of successful women of color: Science identity as an analytic lens. *Journal of Research in Science Teaching, 44*(8), 1187–1218. https://doi.org/10.1002/tea.20237

Choe, N. H., Martins, L. L., Borrego, M., & Kendall, M. R. (2019). Professional aspects of engineering: Improving prediction of undergraduates' engineering identity. *Journal of Professional Issues in Engineering Education and Practice, 145*(3), 04019006. https://doi.org/10.1061/(ASCE)EI.1943-5541.0000413

Coronella, T. (2018). Transforming academic advising: Implementing validating advising experiences with first-generation Latina engineering students. *SoJo Journal: Educational Foundations and Social Justice Education, 4*(2), 57–71.

Coronella, T. T. (2020). First-generation Latina engineering students' aspirational counterstories. In E. Gonzalez, F. Fernandez, & M. Wilson (Eds.), *An asset-based approach to advancing Latina students in STEM* (pp. 115–130). Routledge.

Day, J. K., Goldberg, A. E., Toomey, R. B., & Beemyn, G. (2022). Associations between trans-inclusive resources and feelings of inclusion in campus LGBTQ+ groups: Differences for trans students of color. *Psychology of Sexual Orientation and Gender Diversity.* Advance online publication. https://dx.doi.org/10.1037/sgd0000616

Denton, M., Chambers, B., Saleh, N., & Kirisits, M. (2022, June 26–29). *Extended engagement in an engineering outreach program at a predominately Latinx high school (RTP, Diversity)* [Conference presentation]. American Society for Engineering Education Annual Conference & Exposition, Minneapolis, MN, United States. https://strategy.asee.org/41182

Esquinca, A., Mein, E., & Mucino, H. (2021, January 24–28). *Latinx students' sense of belonging in engineering/computer science at an HSI* [Conference presentation]. Collaborative Network for Engineering and Computing Diversity Virtual Conference. https://strategy.asee.org/36105

Flores, L. Y., Navarro, R. L., Lee, B. H., Hu, X., Diaz, D., & Martinez, L. (2021). Social cognitive predictors of Latinx and white engineering students' academic satisfaction and persistence intentions: Exploring interactions among social identities and institutional context. *Journal of Vocational Behavior, 127,* 103580. https://doi.org/10.1016/j.jvb.2021.103580

Fluker, C., Perez-Felkner, L., & McCoy, K. (2022, June 26–29). *Students' perceptions of an REU summer internship program and their engineering identity development: An equity-centered analysis.* American Society for Engineering Education Annual Conference [Peer proceedings]. Minneapolis, MN, United States. https://peer.asee.org/40969

Foor, C. E., Walden, S. E., & Trytten, D. A. (2007). "I wish that I belonged more in this whole engineering group": Achieving institutional diversity. *Journal of Engineering Education, 96*(2), 103–115. https://doi.org/10.1002/j.2168-9830.2007.tb00921.x

Garriott, P. O., Navarro, R. L., Flores, L. Y., Lee, H. S., Carrero Pinedo, A., Slivensky, D., Muñoz, M., Atilano, R., Lin, C.-L., Gonzalez, R., Luna, L., & Lee, B. H. (2019). Surviving and thriving: Voices of Latina/o engineering students at a Hispanic serving institution. *Journal of Counseling Psychology, 66*(4), 437–448. https://doi.org/10.1037/cou0000351

Garza, N., & Rodriguez, S. (2022, June 26–29). *Conocimientos and the borderlands of identity from Mexican American women in Engineering and Computer Science* [Conference paper]. American Society for Engineering Education Annual Conference & Exposition, Minneapolis, MN, United States. https://peer.asee.org/conocimientos-and-the-borderlands-of-identity -from-mexican-american-women-in-engineering-and-computer-science-work-in-progress.pdf

Gee, J. P. (2000). Identity as an analytic lens for research in education. *Review of Research in Education, 25*(1), 99–125. https://doi.org/10.3102/0091732X025001099

Godwin, A. (2016, June 26–29). *The development of a measure of engineering identity* [Conference presentation]. American Society for Engineering Education Annual Conference & Exposition, New Orleans, LA, United States. https://par.nsf.gov/biblio/10042227-development -measure-engineering-identity

Godwin, A., Potvin, G., & Hazari, Z. (2013a, June 23–26). *The development of critical engineering agency, identity, and the impact on engineering career choices* [Conference presentation]. American Society for Engineering Education Annual Conference, Atlanta, GA, United States. https://peer.asee.org/22569

Godwin, A., Potvin, G., Hazari, Z., & Lock, R. (2013b, October 23–26). *Understanding engineering identity through structural equation modeling* [Conference presentation]. Institute of Electrical and Electronics Engineers Frontiers in Education Conference, Oklahoma City, OK, United States. https://ieeexplore.ieee.org/document/6684787

Harper, D. (n.d.). Etymology of identity. In *Online Etymology Dictionary*. Retrieved June 6, 2022, from https://www.etymonline.com/word/identity

Haverkamp, A., Bothwell, M., Montfort, D., & Driskill, Q. L. (2021). Calling for a paradigm shift in the study of gender in engineering education. *Studies in Engineering Education*, $1(2)$, 55–70. https://doi.org/10.21061/see.34

Hazari, Z., Sonnert, G., Sadler, P. M., & Shanahan, M.-C. (2010). Connecting high school physics experiences, outcome expectations, physics identity, and physics career choice: A gender study. *Journal of Research in Science Teaching*, $47(8)$, 978–1003. https://doi.org/10.1002/tea .20363

Holland, D. C. (2001). *Identity and agency in cultural worlds*. Harvard University Press.

Holly Jr, J., & Quigley, L. T. (2022). Reckoning with the harm of anti-Blackness in engineering education: A reparatory justice research approach. *Journal of Women and Minorities in Science and Engineering*, $28(2)$, 95–110. https://doi.org/10.1615/JWomenMinorScienEng.2022036667

Johnson, A. M. (2019). "I can turn it on when I need to": Pre-college integration, culture, and peer academic engagement among Black and Latino/a engineering students. *Sociology of Education*, $92(1)$. https://doi.org/10.1177/0038040718817064

Jones, B. D., Osborne, J. W., Paretti, M. C., & Matusovich, H. M. (2014). Relationships among students' perceptions of a first-year engineering design course and their engineering identification, motivational beliefs, course effort, and academic outcomes. *International Journal of Engineering Education*, $30(6)$, 1340–1356.

Kendall, M. R., Denton, M., Choe, N. H., Procter, L. M., & Borrego, M. (2019). Factors influencing engineering identity development of Latinx students. *IEEE Transactions on Education*, $62(3)$, 173–180. https://doi.org/10.1109/TE.2019.2909857

Lave, J. (1998). The culture of acquisition and the practice of understanding. In D. Kirshner & J. A. Whitson (Eds), *Situated cognition* (pp. 17–36). Erlbaum.

López, E. J., Basile, V., Landa-Posas, M., Ortega, K., & Ramirez, A. (2019). Latinx students' sense of *familismo* in undergraduate science and engineering. *Review of Higher Education*, $43(1)$, 85–111. https://doi.org/10.1353/rhe.2019.0091

Major, J. C. (2022). *More than income: Socioeconomic inequality, trauma, and the pathways of low-income undergraduate engineering students* [Unpublished doctoral dissertation]. Purdue University.

Matusovich, H. M., Barry, B. E., Meyers, K., & Louis, R. (2011, June 26–29). *A multi-institution comparison of identity development as an engineer* [Conference presentation]. American Society for Engineering Education, Vancouver, BC, Canada. https://doi.org/10.18260/1-2--17351

Mein, E., Esquinca, A., Monarrez, A., & Saldaña, C. (2020). Building a pathway to engineering: The influence of family and teachers among Mexican-origin undergraduate engineering students. *Journal of Hispanic Higher Education*, $19(1)$, 37–51. https://doi.org/10.1177/15381927 18772082

Mejia, J. (2022, June 26–29). *Conocimiento as a framework: Promoting a culturally affirming identity development for Latinx engineers* [Conference presentation]. American Society for Engineering Education Annual Conference & Exposition, Minneapolis, MN, United States. https://sftp.asee.org/conocimiento-as-a-framework-promoting-a-culturally-affirming-identity-development-for-latinx-engineers.pdf

Mejia, J. A., Alarcón, I. V., Mejia, J., & Revelo, R. A. (2022). Legitimized tongues: Breaking the traditions of silence in mainstream engineering education and research. *Journal of Women and Minorities in Science and Engineering, 28*(2). https://doi.org/10.1615/JWomenMinorScienEng.2022036603

Meyers, K. L., Ohland, M. W., Pawley, A. L., Silliman, S. E., & Smith, K. A. (2012). Factors relating to engineering identity. *Global Journal of Engineering Education, 14*(1), 119–131.

National Center for Science and Engineering Statistics. (2021). *Women, minorities, and persons with disabilities in science and engineering: 2021* (Special Report NSF 21–321). National Science Foundation. https://ncses.nsf.gov/wmpd

Ogilvie, A. M., & Knight, D. B. (2020). Engineering transfer students' reasons for starting at another institution and variation across subpopulations. *Journal of Hispanic Higher Education, 19*(1), 69–83. https://doi.org/10.1177/1538192718772659

Patrick, A., & Borrego, M. (2016, June 26–29). *A review of the literature relevant to engineering identity* [Conference presentation]. American Society for Engineering Education Annual Conference & Exposition, New Orleans, LA, United States.

Perez-Felkner, L. C., Shore, C., Dickens, T., & Yang, M. D. (2020, June 22–26). *Engineering resilience through research mentorship: Manufacturing pathways to careers* [Conference presentation]. American Society for Engineering Education Virtual Annual Conference. https://doi.org/10.18260/1-2--34557

Pierrakos, O., Beam, T. K., Constantz, J., Johri, A., & Anderson, R. (2009, October 18–21). *On the development of a professional identity: Engineering persisters vs engineering switchers* [Conference presentation]. Frontiers in Education 39th Annual Conference, San Antonio, TX, United States. https://ieeexplore.ieee.org/document/5350571

Revelo, R. A., & Baber, L. D. (2017). Engineering resistors: Engineering Latina/o students and emerging resistant capital. *Journal of Hispanic Higher Education, 17*(3), 249–269. https://doi.org/10.1177/1538192717719132

Revelo, R. A., Mejia, J. A., & Villanueva, I. (2017, June 24–28). *Who are we? Beyond monolithic perspectives of Latinxs in engineering* [Conference presentation]. American Society for Engineering Education Annual Conference & Exposition, Columbus, OH, United States. https://peer.asee.org/who-are-we-beyond-monolithic-perspectives-of-latinxs-in-engineering.pdf

Rodríguez-Simmonds, H. E. (2022). Transcending the margins and boundaries as Latin-American engineer. In K. J. Cross, S. Farrell, & B. Hughes (Eds.), *Queering STEM culture in U.S. higher education* (pp. 36–56). Routledge.

Rodriguez, S. L., Doran, E. E., Friedensen, R. E., Martinez-Podolsky, E., & Hengesteg, P. S. (2019). Inclusion & marginalization: How perceptions of design thinking pedagogy influence computer, electrical, and software engineering identity. *International Journal of Education in Mathematics, Science and Technology, 8*(4), 304–317. https://doi.org/10.46328/ijemst.v8i4.952

Rodriguez, S. L., Doran, E. E., Sissel, M., & Estes, N. (2019). Becoming *la ingeniera*: Examining the engineering identity development of undergraduate Latina students. *Journal of Latinos and Education, 21*(2), 181–200. https://doi.org/10.1080/15348431.2019.1648269

Rodriguez, S. L., Espino, M. L., Le, B., & Cunningham, K. (2021). The influence of policy implementation in the Midwest: How an SSTEM program broadens participation and enhances engineering identity for community college students. *Education Policy Analysis Archives, 29*(29). https://doi.org/10.14507/epaa.29.5429

Simpson, J. A., Weiner, E. S. C., & Oxford University Press. (1989). *The Oxford English Dictionary*. Clarendon Press.

Smith, K. C., Poleacovschi, C., Feinstein, S., & Luster-Teasley, S. (2022). Ethnicity, race, and gender in engineering education: The nuanced experiences of male and female Latinx engineering undergraduates targeted by microaggressions. *Psychological Reports*. https://doi.org/10.1177/00332941221075766

Tonso, K. (2014). Engineering identity. In A. Johri & B. Olds (Eds.), *Cambridge handbook of engineering education research* (pp. 267–282). Cambridge University Press. https://doi.org/10.1017/CBO9781139013451

Tonso, K. L. (2006). Teams that work: Campus culture, engineer identity, and social interactions. *Journal of Engineering Education, 95*(1), 25–37. https://doi.org/10.1002/j.2168-9830.2006.tb00875.x

Turochy, E., Perez, M. A., Poleacovschi, C., & Doran, E. (2021, July 26–29). *Types of stereotype threats that Latinx students experience in undergraduate engineering education* [Conference presentation]. American Society for Engineering Education Virtual Annual Conference. https://peer.asee.org/types-of-stereotype-threats-that-latinx-students-experience-in-undergraduate-engineering-education-research.pdf

Verdín, D., & Godwin, A. (2018). Exploring Latina first-generation college students' multiple identities, self-efficacy, and institutional integration to inform achievement in engineering. *Journal of Women and Minorities in Science and Engineering, 24*(3), 261–290. https://doi.org/10.1615/JWomenMinorScienEng.2018018667

Verdín, D. (2024). Examining the funds of knowledge that support Latinx students engineering identity development and career certainty. In L. Perez-Felkner, S. Rodriguez, & C. Fluker (Eds.), *Latin* engineering students: An intentional focus on a growing population*. Rutgers University Press.

Villa, E. Q., Esquinca, A., Hampton, E., & Guerra, H. M. (2020). "Is engineering for me?": Examining Latinas' narratives of resilience and agency to confront enduring struggles and challenges in undergraduate engineering studies. *Peace and Conflict: Journal of Peace Psychology, 26*(4), 403–413. https://doi.org/10.1037/pac0000427

Wilson-Lopez, A., & Acosta-Feliz, J. (2021). Transnational Latinx youths' workplace funds of knowledge and implications for assets-based, equity-oriented engineering education. *Journal of Pre-College Engineering Education Research, 11*(1). https://doi.org/10.7771/2157-9288.1289

Wilson-Lopez, A., Mejia, J. A., Hasbún, I. M., & Kasun, G. S. (2016). Latina/o adolescents' funds of knowledge related to engineering. *Journal of Engineering Education, 105*(2), 278–311.

3 · "I DON'T THINK THIS IS THE PLACE FOR YOU"

The Belonging Cues That Latina/o/x Students Receive as They Navigate within and across Engineering Environments

TONISHA B. LANE, BLANCA RINCÓN, AND RENÉ HERNANDEZ

The Latina/o/x population is one of the fastest-growing racially minoritized groups in the United States. In 2019, the Census Bureau estimated that they represented 18.4 percent of the US population. Due in part to their increasing presence in the United States, they are also increasingly represented in the engineering workforce. Over the last four years, Latina/o/x professionals in the engineering workforce have increased from 7 to 9 percent (NCSES, 2017; Pew Research Center, 2021). Despite these gains, they still only represent a fraction of the overall engineering workforce (9 percent Latina/o/x compared to 71 percent white; Pew Research Center, 2021). More research is needed about the pathways to and through engineering for Latina/o/x college students.

Researchers continue to point to the culture of engineering fields as a rationale for why some Latina/o/x students do not persist in these disciplines (Banda & Flowers, 2018; Rincón & Rodriguez, 2020). Many have expressed concerns about the competitive nature of engineering (Banda, 2020), exclusionary practices (Camacho & Lord, 2011), and lack of representation of individuals of color in the content and among the faculty (Dika and Martin, 2018). As such, there are a number of challenges and barriers that Latina/o/x students must navigate in engineering, given their diverse and intersectional identities. Anti-immigration sentiments, disdain for accented language, and various "othering" practices create unique experiences for negotiating identity in these spaces, compounded with the challenge of navigating engineering curricula and culture (Smith et al., 2022).

When Latina/o/x students do not perceive a sense of belonging, they may be less likely to fully immerse themselves in the engineering field. This may result in a lost opportunity to foster relationships with faculty and peers, join study groups, contribute to student organizations, and complete their degrees. Alternatively, when Latina/o/x students are validated, cared for, and made to feel like they matter, they are more likely to thrive in engineering (Abrica et al., 2022). Thus, engineering education programs need to know more about environmental factors, such as the implicit and explicit messages Latina/o/x students receive about the perceived alignment of their identities in engineering contexts. Studying Latina/o/x students' microsystems and mesosystems may be key to understanding engineering contexts in which they may or may not perceive a sense of belonging.

Microsystems—such as engineering courses, co-ops and internships, peer groups, and clubs and organizations—are critical environments where students assess their belonging in the discipline (Rodriguez et al., 2022). As such, positive or negative experiences that are aligned with these microsystems could impact Latina/o/x engineering students' persistence, retention, and degree completion (Montelongo, 2019). Additionally, mesosystems entail the interactions between two or more microsystems (Bronfenbrenner, 2005). While students may receive positive messaging from one microsystem, they may encounter negativity about their fitness in engineering in another. These competing messages could cause students to question their belongingness in engineering. Alternatively, suppose a student is affirmed for their decision to pursue engineering in multiple microsystems (e.g., peer group and faculty). In that case, the effect of this mesosystem may strengthen their resolve to remain in the major. Consequently, the meaning Latina/o/x students ascribe to these experiences could affect their connection to and sense of value in engineering (Rodriguez et al., 2022).

Therefore, the research question that guided this study was as follows: How do Latina/o/x students in engineering experience and make meaning of belonging cues within academic and nonacademic environments? This question is important because belonging has been tied to strong academic performance (Strayhorn, 2018), health and well-being (Walton & Brady, 2017), and student retention (Strayhorn, 2018). In this chapter, we refer to these messages as "belonging cues." While there is a robust literature base on the sense of belonging among Students of Color in STEM, we still know little about how Latina/o/x students, in particular, experience belonging, especially in the context of engineering—and, more importantly, how they receive cues about their belonging (or lack thereof) in these contexts.

LITERATURE REVIEW

Our literature review draws from empirical studies investigating the experiences of Latina/o/x students in engineering. We begin by discussing their experiences with racial and gender discrimination, noting differences between men and women. Next, we use the literature to explain interactions with faculty and peers that can be supportive or create barriers for Latina/o/x engineering students. We also incorporate literature on STEM enrichment programs (SEPs), because the engineering students in the current study were members of SEPs on their respective campuses. From this literature, we observe that SEPs create environments that are conducive for the successful navigation of engineering programs. We conclude this section with a discussion of the importance of belonging in engineering contexts.

Experiences with Racial and Gender Discrimination

A recent study found that Latina/o/x students in engineering experience microaggressions due to their perceived foreignness, assumed criminality, and presumed lack of intelligence (Smith et al., 2022). Using a psychological, phenomenological approach, in which semi-structured interviews were the primary data source, Smith et al. (2022) uncovered that Latinos were more likely to encounter hostile microaggressions than Latinas. The researchers also note that, while the number of microaggressions for Latinas was greater than for Latino men, the types of microaggressions experienced by Latino men were more egregious. Despite these findings, gender discrimination is most often cited as a site of contention for Latinas navigating engineering education (Banda & Flowers, 2018), and some research suggests that their gender identities may be more salient than their racial identities.

In Banda and Flowers's (2018) critical qualitative study of Latinas in engineering, they uncovered that gendered experiences were more pronounced than any other identities the participants held. Specifically, they discussed the pervasiveness of sexism in engineering (e.g., classrooms, student organizations, labs, and internships). Instead of being viewed as intellectual counterparts, the Latina participants perceived that they were viewed as objects of affection or empty vessels that have nothing to offer. Because of the range in critical consciousness surrounding systemic and structural oppression among the participants, they responded to sexism encountered in engineering departments by becoming more assertive, working harder, "playing tough," or engaging in acceptance or complacency (Banda & Flowers, 2018).

Verdín (2021) also notes that Latinas who may be at the intersection of multiple marginalized identities have to exert substantial effort to construct a sense of belonging and persist in engineering. Verdín's (2021) study of a first-generation, nontraditional (i.e., older than twenty-five years) Latina student shows the unique ways Latinas must negotiate belonging in environments that were historically not designed for them. Hegemonic and white, masculine normative engineering spaces communicated to the participant that she was an outsider to engineering. Consequently, the participant created her own all-women study groups, as well as enrolled in online courses to avoid the toxicity of in-person classroom environments. Moreover, the extent to which racial and gender discrimination disrupts belongingness for Latina/o/x students and how this may be amplified within and across their microsystems (such as peers and faculty groups) warrants further study. As such, in the next sections, we focus on research concerning faculty and peer interactions.

Student–Faculty Interactions

Studies show that faculty transmit belonging cues when they demonstrate a willingness to foster relationships with students (Abrica et al., 2022; Dika & Martin, 2018); however, burgeoning literature highlights that Latina/o/x students may struggle to develop such relationships due to faculty disinterest (Foxx, 2014; Banda, 2020). In a qualitative study with eight Black and Latino men, participants reported both positive and negative interactions with engineering faculty (Foxx, 2014). However, most participants found engineering faculty to be indifferent. While some faculty were open and caring, others were distant and unapproachable. Consequently, participants built connections with faculty outside of their respective engineering departments to recoup what they were not receiving in their fields. In another qualitative study with eleven Latinas in engineering, Banda (2020) underscores that engineering departments need to improve how they create opportunities for Latinas to have meaningful interactions with faculty. Of note, participants expressed concerns with the paucity of faculty who demonstrated care for students' needs and aspirations.

While the aforementioned studies highlight the impact of unsatisfactory faculty–student interactions, other research points to positive attributes of those relationships. Dika and Martin's (2018) quantitative study of 288 Latina/o/x engineering students reinforces that interacting with engineering faculty is critical to student persistence. In fact, college support was more influential than precollege support in predicting persistence in engineering for Latino men. College support tended to focus on engineering-related resources, such as accessing internships. For students who are not allowed these opportunities, feeling a sense of belonging in engineering may be more complicated (Banda, 2020; McGee, 2021).

Peer Interactions

Studies show that Latina/o/x students in engineering benefit from connecting with other Latina/o/x students. Revelo and Baber (2018) discovered that Latina/o/x students resisted racialized stereotypes by serving as role models for other Latina/o/x students. In another study, researchers showed that having a community of Latino men in engineering from similar socioeconomic backgrounds proved to be beneficial for these students (Abrica et al., 2022). These peers were supportive when the time came to form study groups and countered the competitive and racially isolating engineering environment. In a qualitative study with eleven Latina senior engineering students, participants reported that participating in student organizations helped them adjust to college (Banda & Flowers, 2017). Reasons for joining these organizations included meeting academic and social needs, establishing a sense of belonging, and connecting with students with similar racial backgrounds. Despite the many benefits of peer interactions, encounters with some white peers reflect a different reality.

Research shows that Latina/o/x engineering students experience a number of microaggressions from their white peers (McGee, 2021; Smith et al., 2022). This may include being ignored, made to feel like an outsider, or treated like they are incapable of performing engineering-related tasks (Rincón & Rodriguez, 2020; Smith et al., 2022). In response to these behaviors and attitudes, Smith et al. (2022) found that Latina/o/x engineering students sought out minoritized peers, family, and organizations to discuss these challenges. However, some students reported remaining silent when they experienced microaggressions, while others confronted their peers.

The Importance of STEM Enrichment Programs

For more than five decades, SEPs have been instrumental in supporting students of color to navigate STEM contexts—both socially and academically—as well as enhancing their belongingness within these environments (Abrica et al., 2022; George et al., 2019; Lane et al., 2020). SEPs provide wraparound services geared toward facilitating access to and success within the STEM disciplines. Program administrators leverage their knowledge of the historical and systemic challenges encountered by communities of color within STEM to allocate resources and cultivate communities, with the goal of supporting their academic, personal, and professional success (Rincón & George-Jackson, 2016). To this end, numerous studies point to SEPs as beneficial for bolstering the existing intellectual capabilities and harvesting the potential of students of color in STEM (Lane & Id-Deen, 2023; Maton et al., 2016). Specifically, Latina/o/x students credit SEPs with helping them establish community in their local institutional contexts, extending their social capital, and making them feel important in STEM (Garcia et al., 2021; Lane et al., 2020). Without the refuge of these programs, Latina/o/x students lament that departmental climate can be hostile and toxic (Abrica et al., 2022; Rincón & George-Jackson, 2016). As such, previous literature supports that SEPs, as a microsystem, may play an integral role in Latina/o/x engineering students' success and support their capacity to establish relationships with peers and institutional agents.

Why Belonging Matters

In the context of the college environment, Strayhorn (2018) defines "sense of belonging" as "students' perceived social support on campus, a feeling or sensation of connectedness, and the experience of mattering or feeling cared about, accepted, respected, valued by, and important to the campus community or others on campus such as faculty, staff, and peers" (p. 4). Because of its significance in the lives of students, a burgeoning body of work underscores the criticality of belonging for Latina/o/x students in engineering. Having a sense of belonging has been linked to seamless transitions into college, positive academic and social adjustment, persistence, and degree attainment. For example, in a qualitative study exploring

Latino men navigating an engineering retention program, Abrica et al. (2022) found that faculty involved with the program were instrumental in Latino students experiencing a sense of belonging and establishing a community within engineering.

In contrast, Latina/o/x students may sense their lack of value or acceptance when their identities are not represented in engineering (Verdín & Godwin, 2018). For example, in Verdín and Godwin's study, Latina students were aware of the erasure of women (e.g., among the faculty, book authors, and lab instructors, as well as recruiters' expectations that engineering is/will be male). Resultantly, many Latina students found spaces outside of engineering that were more welcoming to their identities (Verdín & Godwin, 2018). Because of the potential negative consequences of incongruence with one's ethnic identities amid various groups and environments, a more in-depth understanding of the cues that Latina/o/x engineering students receive about belongingness, or the lack thereof, is warranted. Further, exploring how these messages are conveyed explicitly or implicitly in students' micro- and mesosystems may aid in minimizing harmful environmental factors and developing more inviting engineering climates.

CONCEPTUAL FRAMEWORK

We integrated concepts from Bronfenbrenner's (2005) ecological systems theory and the notion of belonging cues to analyze the emergent data in this study. The ecological systems theory enables researchers to examine individual development within a given context. Central to the theory is the process–person–context–time (PPCT) model. This model illuminates interactions among individuals, settings that directly involve them, and contexts that indirectly involve them but are relevant to their development. Because we were interested in how students experience and make meaning of belonging cues in engineering, we focused our analysis on the micro- and mesosystems for two reasons: (1) these spaces and relationships are closest to the individual within the ecological systems model; and (2) the interpersonal interactions established in these environments are crucial to understanding belongingness in engineering.

Four systems comprise the PPCT model: microsystems, mesosystems, exosystems, and macrosystems. Because this study solely focused on the micro- and mesosystems, we will describe those next. Microsystems contain the individual. In the context of engineering, these systems may entail families, student organizations, laboratories, classrooms, co-ops, and internships. Each individual will be affected by these contexts differently, depending on their developmentally instigative characteristics. Mesosystems are the interaction of two or more microsystems. The interactions of mesosystems may inhibit or facilitate development or belongingness in certain contexts, which may have consequences for their engineering identity development.

Within various engineering environments, we were also interested in how belonging cues signal to students their value, their importance, and the level of care they should expect (or the lack thereof). The notion of belonging cues is derived from constructs found in social identity threat, sense of belonging, and identity-safety cues. Murphy et al. (2007) describe social identity threat as "a threat that occurs when people recognize they may be devalued because of one of their social identities" (p. 879). They argue that individuals may be vulnerable to these threats—contributing to diminished performance, low self-esteem, and attrition—"even when targets [of the threat] are interested, confident, and proven achievers in the relevant domain" (p. 879).

In contrast, identity-safety cues communicate to minoritized groups that their identity will be valued and respected in a given context (Pietri et al., 2019). Similarly, belonging cues signal to students that, in environments where generating a sense of belonging is valued (e.g., inclusive syllabus language, representation in lectures), they will receive social support, connection, and acceptance (Strayhorn, 2018). However, they differ from identity-safety cues in that belonging cues are multidirectional. Environments or behaviors from actors in the environment may signal negative or positive belonging to minoritized individuals (Abrica et al, 2022). Furthermore, the combination of the PPCT framework and the concept of belonging cues enabled us to explore the messages students encounter in engineering environments that support or hinder their connection to the field.

METHODOLOGY

To explore the ecological systems of Latina/o/x engineering students, this study drew on a subset of data gathered from a research project funded by the National Science Foundation. Using phenomenology, this sub-study centered on the lived experiences and meaning-making processes of Latina/o/x undergraduate students pursuing majors in engineering. In particular, we examined the belonging cues that students received as they navigated within and between various engineering contexts.

Data Sources and Participants

Study participants were undergraduate students enrolled at one of four northeastern institutions. These institutions represent a variety of postsecondary contexts, including large, public land-grant universities; a private urban university; and a STEM-focused private urban university. However, all are predominantly white institutions (PWIs). While participants attended a variety of institutional contexts, they all participated in the Louis Stokes Alliance for Minority Participation (LSAMP) program—a federally funded program that aims to increase the number of underrepresented racially minoritized students matriculating into, and successfully completing, high-quality undergraduate degrees in STEM.

Student interviews were the primary source of data collected. In the 2017 to 2018 academic year, LSAMP students received an email asking them to participate in a one-on-one interview that took place in person or via phone. Two years later,

TABLE 3.1   Latina/o/x Engineering Participant Profiles (N = 10)

Pseudonym	Institution	Gender	Race/ethnicity
Angel	B	Man	Latina/o/x
David	B	Man	Black/Latina/o/x
Emilio	D	Man	Latina/o/x
Jazmine	C	Woman	Latina/o/x
Jordan	C	Woman	Latina/o/x/White
Josie	B	Woman	Latina/o/x/White
Lorenzo	B	Man	Latina/o/x
Luisana	C	Woman	Latina/o/x
Marisela	A	Woman	Black/Latina/o/x
Ricardo	A	Man	Latina/o/x

students were invited to participate in a follow-up interview in the spring semester of their fourth year in college. The interviews were conducted using a semi-structured protocol that included questions about college students' experiences, involvement, and interpersonal relationships. All interviews were audio recorded with the consent of participating students and ranged between forty to eighty-one minutes in length. This study draws on interview data from ten engineering majors who self-identified as Latina/o/x. Of these students, four identified with two racial and/or ethnic identifiers, and an equal number of students identified as women and men (see table 3.1).

Data Analysis

To begin data analysis, members of the research team transcribed each audio recording and reviewed each transcript for accuracy. Then, we de-identified the transcripts by replacing students' names and other identifying information with pseudonyms. Next, we uploaded the transcripts to Dedoose, an online software program used for analyzing qualitative data. To begin the coding process and to ensure a shared understanding of the codes, all members of the research team worked collaboratively to develop a codebook. We developed a set of descriptive codes reflecting the microsystems of college students (e.g., engineering faculty, registered student organizations [RSOs], and co-ops) as well as deductive codes reflecting the types of belonging cues students encountered (verbal vs. nonverbal; positive vs. negative). As a multimember team, continuous peer debriefing and the use of analytic memos capturing "emergent patterns, categories and subcategories, themes, and concepts" (Saldaña, 2016, p. 44) served as critical components in our analytical process.

Positionality

This research study was conducted by three people of color: one Black woman, one Latina woman, and one Salvadoran man. All are first-generation college students, and two pursued undergraduate STEM degrees. Two authors are higher-education faculty at large public universities, and one is a doctoral student. All three are researchers who broadly study the experiences of marginalized groups in STEM, with their principal areas of focus being identity development, SEPs, and the assets that college students of color use to navigate and persist in STEM. Further, one researcher brings professional experience working with a minority engineering program at a PWI that primarily served Black and Latina/o/x students.

As researchers, we hold that our shared experiences as people of color with academic and professional experiences in STEM are strengths that were leveraged throughout this study. For example, the second author's shared identities with our participants allowed her to build rapport and connect with the students during data collection, while the first author's identities allowed her to serve as an added measure of accountability throughout the analysis process. To this end, we drew on experiential and practical knowledge, given the shared identities of some authors and relevant professional experiences.

FINDINGS

In this study, we explored how Latina/o/x students in engineering navigate complex and dynamic environmental contexts that relay messages, or cues, of belonging. Our findings suggest that, as students navigate within and between microsystems, they receive negative and positive messages that influence their sense of fit and inform the extent to which they perceive engineering as a place for them. Importantly, while Latina/o/x students largely received messages that they did not belong from non-minoritized peers and faculty, these messages were often counteracted by Latina/o/x engineering peers and faculty.

Microsystems

Students' descriptions of their engineering microsystems included both negative and positive experiences in RSOs, engineering classrooms, co-ops, and SEPs (see figure 3.1). Importantly, Latino men were more likely to describe instances of racial discrimination within these engineering environments, whereas Latinas often encountered both racial and gender discrimination. Students' negative experiences occurred almost exclusively when interacting with white peers in formal learning settings (e.g., co-ops, classroom environments). Conversely, positive experiences were primarily described as occurring in informal learning environments, such as RSOs and SEPs.

Engineering peers and faculty. Latina/o/x students in engineering, and especially Latinas, described receiving messages about belonging from microsystems that included faculty and peers in engineering. Several students relayed that they had been the target of racist and sexist comments from classmates. Luisana recounted an incident in which her white male classmate made an offhand comment about

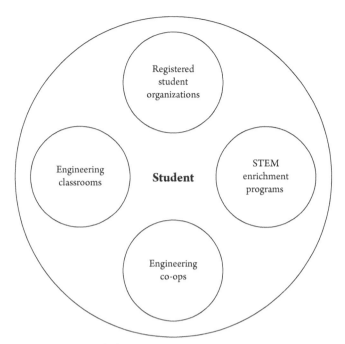

FIGURE 3.1. Latina/o/x Microsystems

her intellectual capacity: "In my [pre-engineering] chemistry class, we have to do labs and my partner is a white guy. He wouldn't let me do any of the lab parts. He was like "You just write" because he didn't want me to mess it up." Because some peers viewed Luisana as an "affirmative action" admit—as a woman and racial minority—they perceived her as being intellectually inferior, which became the basis by which they excluded her from actively participating in class group projects. Consequently, Luisana was forced to become an observer in the learning environment.

Belonging cues were also transmitted nonverbally. Latinas described how the mere absence of other women in engineering perpetuated messages of not belonging. In particular, students noted the absence of others who looked like them in the classroom, and shared being hyperaware of their tokenization by engaging in "counting exercises." Marisela explained, "I decided to count how many girls were in my class—there's 6 in a class of 60. It's just like, it's kind of a lot sometimes." Similarly, Luisana recounted, "I'm one of three girls in my aerospace class, which is crazy." While Marisela's account emphasizes the psychological toll of tokenization in engineering classrooms, Luisana's account demonstrates the cumulative effects of structural and interpersonal forms of sexism and racism within a microsystem.

Most students did not report positive or negative experiences in regard to their interactions with engineering faculty on their campuses. Luisana—who had aspirations to join the faculty ranks in engineering—described a situation in which a positive meeting with a faculty member took an unexpected turn, once she shared her GRE scores:

> I did not perform as well in the GRE as I wanted to. Like I barely got average, and [the state school] had cutoffs. [...] And I wanted to go interview at [the state school]. Because I just drove there, the professor asked me to come. And it was going great until he asked me about my scores. And I told him, and his face completely dropped. And he was like, "You're not going to get in here. You're not going to get in anywhere." That ruined my spirit. I was hesitant about applying at that point.

This faculty's conclusion that Luisana would not be admitted "anywhere" began to sow seeds of doubt as to whether she should apply to graduate school.

Engineering co-ops. Latina/o/x students in engineering recounted having had mixed experiences (both positive and negative) when engaging in cooperative learning experiences. Lorenzo shared how, at his first co-op experience, "every Thursday the engineering team, especially the younger guys, went out for sushi." These interactions with coworkers outside of the work environment created a sense of community that translated into a better work environment for Lorenzo. Conversely, Lorenzo described his second co-op experience as a "nightmare" rife with "racial and political tension" and constant clashing with coworkers. Lorenzo shared the indignity of having his intelligence and abilities questioned in an environment that was meant for learning to practice engineering skills and concepts in a real-world environment: "I remember, one of our first weeks there, we got into an argument because he [a co-worker] was just saying, 'Oh, how come you don't know how to do any of this?' I'm like, that's the reason I'm on co-op—to learn." These interactions eventually came to a head when one of his coworkers blatantly communicated, "I don't think this is a place for you. I think you should consider working somewhere else." Ultimately, Lorenzo stayed in the co-op, but this experience became a shadow that began to cast doubt on whether he would pursue a career in engineering upon graduation. Whereas Lorenzo found a sense of belonging with coworkers who he described as "younger guys" (i.e., gender and age) in his first co-op experience, his racial identity became the criterion for exclusion in his second co-op experience.

Registered student organizations. Students' identities (e.g., gender, race, age, and engineers) made them gravitate toward certain campus microsystems and not others. Every student in the study discussed the importance of peer relationships established within RSOs, especially culturally based student organizations. Angel described how he came to join the Society for Hispanic Professional Engineers (SHPE), a national organization committed to advancing the success of Latina/o/xs in engineering, after learning about the RSOs at the summer bridge program. He explained, "SHPE, I joined because they came to summer bridge. I remember a few clubs came to summer bridge. That's one cause I'm half Puerto Rican so I was like, 'Oh, that'll be pretty cool to join this,' so I just showed up."

Similarly, Josie shared the importance of women-only spaces in engineering: "I do think [Women in Engineering] is a very important club to have on campus obviously because women are very underrepresented in engineering." And when these spaces weren't available, students often created counter-spaces for themselves and others. Luisana recounted, "I started the [Women in Engineering] group. Because I felt there was a need to [. . .] create a sense of community for these women. Because it's tough. Like I've had experiences where people just disregard me. Like they think I can't do the work." Men in the study often discussed the importance of joining ethnicity-based organizations, whereas women almost exclusively described the importance of women in engineering groups.

Belonging cues were also transmitted when Latinas were simply ignored and relegated to "secretarial" work in RSOs. Despite her initial excitement to get involved in cocurricular activities on campus, Josie was left wondering what she could have done differently to insert herself into the "boys" club.

> The main club that I was sort of disenchanted with was [Bullet Rail]. Not because of their mission, which I still think is incredible. But there wasn't really a spot for me there. It was a lot of fifth years. I don't know if it relates or not, but just a lot of guys who were friends, who were older students, who were incredibly intelligent and worked on this cool stuff [. . .]. I went for a semester, and I helped a little bit. I would post on their Facebook for them, but not really anything else.

Taken together, peer interactions influenced students' sense of fit and whether they perceived there was a place—physical or psychological—for them in these microsystems.

STEM enrichment programs. SEPs served as vital microsystems in which students developed a sense of community with peers who shared their backgrounds as women, Latina/o/xs, and students in engineering. Josie, a member of LSAMP, related, "I've met all my current roommates, and future roommates, and past roommates through the program. Just people that were very like-minded people that are very studious. Especially women, minorities as well." Lorenzo also discussed the importance of LSAMP programs in bringing together other Students of Color on campus and validated their belonging in engineering. He explained, "Without LSAMP I wouldn't have met individuals who are the same skin color and have gone through the same experiences such as myself or have the same background as me and have the same dreams as me and are intellectually capable and excited about the same things that I'm also excited about. I feel like I wouldn't have met any of them, or at most, most of them if it wasn't for [LSAMP]." Notably,

SEPs facilitated the meeting of same-race peers who would become close friends, roommates, study mates, and extended kin. These peer groups cultivated through SEPs often provided important socio-emotional support that affirmed their identities as engineers. As Marisela notes, "They would ask me about my career goals. They were there to support me at all my research presentations, every poster presentation. My roommate cheered me on as I wrote my first paper. I think they were also super important in making sure I had not only the emotional support that I needed, but they were there in a way that I felt like my family couldn't be because they understood more so what was going on with the process."

The importance of SEPs extended beyond students' campus community and to other institutions where students participated in undergraduate summer research programs. In addition to building community with peers that shared common backgrounds, these programs often connected students to other Latina/o/x faculty via research experiences; these faculty affirmed their abilities and talents in engineering. Luisana shared,

So, I got really close to the MIT [Massachusetts Institute of Technology] professor. Partly because he was Mexican, and we talked a lot in Spanish. [. . .] Yeah, I don't know why, but I always connect with people, or it's easier to connect with people if—I guess if you have a similar background. And for me, it was that being Mexican and Spanish. I feel more comfortable when I can be myself like that. And he complimented me at the end of the program, and he asked me personally if he could write my letters [of recommendation].

Latina/o/x students often participated in multiple SEPs, both on their campuses and at other institutions. These SEPs mutually reinforced their belonging in engineering by connecting students to other peers and faculty who could serve as role models and could recognize their abilities as engineers. More importantly, SEPs also served as a tie that bound students to culturally based RSOs and that led to the development of new same-race and same-gender microsystems (e.g., roommates).

Mesosystem (When Microsystems Collide)

Students received conflicting and reinforcing messages about the extent to which they fit in engineering across various microsystems. Figure 3.2 depicts examples of how microsystems collide within the mesosystem. Latina/o/x engineering students mostly received positive belonging cues from other Latina/o/x peers and faculty in engineering that counteracted the negative messages from other white peers and faculty. And while some microsystems reinforced positive belonging cues, these microsystems were often composed of the same individuals, given the tokenization of Latina/o/x students at these highly selective institutions.

Luisana described several negative encounters with engineering peers and faculty. Classmates and faculty alike questioned her academic abilities and created

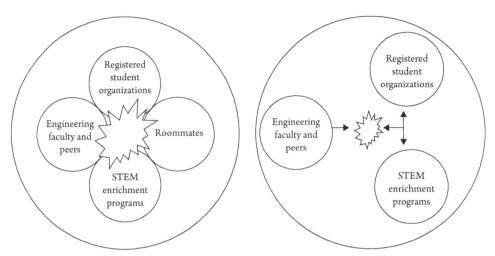

FIGURE 3.2. Latina/o/x Engineering Mesosystems

barriers to her participation in engineering by excluding her from group assignments and suggesting that she was not graduate school material. Counteracting
these messages, however, was the advice she received from a mentor from a
national Latina/o/x STEM organization, who encouraged her to apply anyway,
reassuring her that schools would see past her low test scores. And, as we saw previously, this messaging was affirmed by a Mexican MIT professor who recognized
her academic talents and asked to write her letters of recommendation for her
admission to doctoral engineering programs. Luisana would later decide to apply
to graduate school and received acceptance into one of the top aerospace engineering programs in the United States.

In another example of counteracting messages across microsystems, Latina
students who described instances of tokenization in the classroom and via RSOs
shared that they participated in or created new microsystems (i.e., RSOs) that
centered on women's inclusion and mattering in engineering. Josie and Luisana
both described the negative verbal and nonverbal cues that they received about
women—and women of color—not belonging in engineering from white men
peers, both in the classroom and within RSOs. Both women sought refuge within
RSOs aimed at supporting women's participation in engineering.

Microsystems also coalesced in ways that provided positive and reinforcing
messages of belonging for Latina/o/x students in engineering. However, for most
students in the study, including Josie and Lorenzo, there was a great amount of
overlap among individuals who comprised these multiple microsystems. That is,
students often learned about RSOs through SEPs, friends, mentors, study group
members, and/or roommates—and vice versa. And while this created strong
bonds with peers and allowed students to move freely within and between these
microsystems, it also created challenges for some students, as it reduced their

sphere of support (and subsequent belonging cues) on which they could count. Lorenzo described this phenomenon as he struggled to find a study group during a challenging academic semester: "So, my roommates that I thought I was going to do all my academic semesters with, we never talked about our co-op patterns and as a result both of them are on a different pattern than me. So, they're now both on co-op and those are my closest friends and now I'm here by myself." Lorenzo's roommates were his study mates and closest friends. And when they failed to plan to have their co-ops during the same semester, Lorenzo found himself in one of the most difficult engineering courses without anyone to study with—pointing to the precarity of Latina/o/x microsystems, as well as the lack of diversity in engineering programs.

DISCUSSION

The purpose of this study was to understand how Latina/o/x engineering students make meaning of belonging cues within academic and nonacademic environments. Using phenomenology as a methodological approach, we uncovered that Latina/o/x students receive positive and negative messages between and across their mesosystems, and these messages influenced how they perceived their fitness for engineering. We also found that negative messages were disproportionately relayed by non-minoritized faculty and peers. However, because of positive messages from Latina/o/x engineering peers and faculty, they were able to resist harmful messages and use their agency and networks to chart self-defined paths to persist in these fields. In the sections that follow, we situate these findings in a larger context of what is at risk if faculty, staff, and administrators continue to neglect the climate of engineering and fail to prioritize a sense of belonging for Latina/o/x students.

In the study, Lorenzo resisted the co-op peer's assertion that he did not belong in that environment due to his limited knowledge about a particular task demonstrating how Latina/o/x students must activate measures to coexist in engineering environments. As such, Lorenzo keenly pointed out that the very reason he was engaged in that co-op experience was to build knowledge and skills. While this phenomenon of proving others wrong about one's intellectual prowess is not new (Moore III et al., 2003), the level of stereotype management that Students of Color must enact can be taxing and add undue stress amid a cognitively rigorous academic environment (McGee & Martin, 2011). It also points to why students may not want to move beyond microsystems that have proven to affirm their engineering identities, even when other opportunities may exist.

For example, within the mesosystem, we found that participants received belonging cues from the same people in different settings (e.g., Latina/o/x peers within and outside classroom settings). Though this created some consistency across their microsystems, it also posed a problem when these members of their microsystems were temporarily or permanently removed. As a result, Lorenzo discussed being impacted when his roommates, who were his study partners and closest friends, left to complete their co-ops during the same semester. Consequently, Lorenzo had to find new study partners among students who either implicitly or explicitly made him feel like he did not belong in engineering. A similar concern arises when considering faculty or staff turnover. Though our participants did not discuss faculty or staff turnover as an issue in this study, other studies have pointed out that students are affected when this happens (Garcia et al., 2021). Consequently, students must find new groups or members for their microsystems that can support their success in engineering. They may also need to try to start the process all over again, determining who transmits positive belonging cues.

This study also underscores the continued relevance of structural diversity in engineering contexts. As previous scholarship shows (Pietri et al., 2019), belonging cues are communicated through representation of peers, professionals, and curriculum that reflect individual's identities. In the absence of such examples and role models, students may lose their sense of place. This may also contribute to students experiencing imposter phenomenon. While early research advanced that imposter phenomenon was largely a psychological phenomenon grounded in internalized self-doubt, more recent scholarship suggests a need to examine the systemic and structural factors that prompt the onset of the phenomenon (McGee, 2021). For example, many negative belonging cues, grounded in the historical legacy of inclusion and exclusion in engineering, are symptomatic of a more substantial problem in engineering that will not simply go without a strategy to address climate issues and "policies governing social interactions" (Strayhorn, 2018, p. 3).

A lack of structural diversity was also illuminated with students having to start new groups that would speak to their intersectional identities. Luisana's creation of a student group that centered Latinas demonstrates the ongoing struggle to find place and space at PWIs (Verdín, 2021). These actions also highlight the amount of energy racially minoritized students exert to create spaces where they belong, when this time may be better spent performing engineering-related tasks and building competencies in their fields. Engaging in the latter is critical to cultivating engineering identities and being recognized as someone who embodies the identity of an engineer, which are also important for staying motivated and persisting in these fields (Rodriguez et al., 2019).

Finally, our conceptual framework was helpful for investigating how belonging cues manifest in the micro- and mesosystems of Latina/o/x students in engineering. Future research should use this framework to explore how belonging cues manifest in other systems and the ways in which they communicate belongingness—or the lack thereof. For example, the exosystem illuminates policies and practices that are external to students' microsystems, but they are still a part of their ecosystem. One area to consider may be to investigate how differential tuition policies in engineering inadvertently push students out of engineering, since the tuition increases as they move into upper-division courses (George-Jackson et al., 2012); or how anti-immigration laws—which appear in the macrosystem of Bronfenbrenner's model—impact access to financial aid, and what this signals to students about belonging in engineering. Further application of the integration of ecological systems theory and belonging cues may aid in learning more about the various signals and messages that Latina/o/x students receive as they navigate engineering education. Further study may also help with generating empirical evidence to better unpack and nuance belonging cues in engineering environments.

IMPLICATIONS

Findings from this study suggest that Latina/o/x engineering students expend large amounts of emotional energy and time to process and make meaning of negative interpersonal interactions with peers, faculty, and coworkers. Further, students are often burdened with having to locate or create counter-spaces to build their own (and others') sense of belonging at institutions that are not equipped to meet their needs. This points to the racialized and gendered equity labor that marginalized students expend when institutions of higher education lack the infrastructure and staff to help students navigate and address hostile climates in engineering (Lerma et al., 2020). Moreover, findings from this study point to the precarity of microsystems that provide important positive belonging cues. Because of the lack of Latina/o/x faculty in engineering, Latina/o/x students relied heavily on same-race or -gender peers for support. These peers often overlapped across multiple microsystems—and their absence was felt across these various microsystems after their departure. The need to hire faculty and staff of color, and who are women, cannot be overstated (Abrica et al., 2022; Montelongo, 2019).

Additionally, findings suggest that Latinas are hyperaware of their tokenization in engineering. While the students in this study actively sought out and created spaces within engineering, it is likely that other students may seek out peer groups and majors outside of engineering to find a better fit. To reduce isolation and tokenization, colleges and universities should consider admitting Latina students in clusters, or families, and cultivating these relationships early on in their college careers. Likewise, higher-educational institutions should institutionalize and administer equitable resources for identity-based, professional organizations such as SHPE, to provide support and community to Latina/o/x students, as discussed by the participants in this study. Such organizations can be spaces where students may experience congruence between their ethnic and professional identities.

The findings from this study also point to negative interactions that students are experiencing outside of the college environment through university-sponsored programs such as co-ops. Because women and students of color are more likely to experience workplace discrimination, especially within engineering (Strayhorn & Johnson, 2016; Smith & Gayles, 2018), co-op and internship placements should be vetted, bias training should be provided for supervisors and mentors, and discussions of workplace discrimination (e.g., reporting) should be provided ahead of the co-op experience.

CONCLUSION

A sense of belonging has been linked to a positive, seamless transition into college, academic and social adjustment, persistence, and degree attainment—and is especially important for Latina/o/x students. Yet, researchers consistently find that minoritized students are more likely to encounter unwelcoming environments on college campuses. The Latina/o/x engineering students in our study described receiving negative belonging cues that positioned them as imposters in engineering. Importantly, these students also received positive and affirming belonging cues, often from ingroup peers and faculty that counteracted these messages. However, we argue that if higher education institutions want to increase representation of Latina/o/x students in engineering, engineering education faculty and administrators must consider the environmental factors that communicate whether these student populations belong and are valued in these spaces and eliminate negative messages.

Authors' Note

Tonisha B. Lane ORCID: 0000-0002-1063-2744

We have no conflicts of interest to disclose. This work was funded by the National Science Foundation (Award: 1619629). Any opinions, findings, and conclusions expressed in this article are those of the authors and do not necessarily reflect the views of the National Science Foundation.

REFERENCES

Abrica, E. J., Lane, T. B., Zobac, S., & Collins, E. (2022). Sense of belonging and community building within a STEM intervention program: A focus on Latino male undergraduates' experiences. *Journal of Hispanic Higher Education*, 21(2), 228–242. https://doi.org/10.1177/1538192720974884

Banda, R. M. (2020). From the inside looking out: Latinas intersectionality and their engineering departments. *International Journal of Qualitative Studies in Education*, 33(8), 824–839. https://doi.org/10.1080/09518398.2020.1735565

Banda, R. M., & Flowers III, A. M. (2017). Birds of a feather do not always flock together: A critical analysis of Latina engineers and their involvement in student organizations. *Journal of Hispanic Higher Education*, 16(4), 359–374.

Banda, R. M., & Flowers, A. M. (2018). Critical qualitative research as a means to advocate for Latinas in STEM. *International Journal of Qualitative Studies in Education*, 31(8) 769–783. https://doi.org/10.1080/09518398.2018.1479046

Bronfenbrenner, U. (2005). Making human beings human: Bioecological model of human development. In R. M. Lerner & W. Damon (Eds.), *Handbook of child psychology: Theoretical models of human development* (pp. 793–828). Hoboken, NJ: John Wiley & Sons.

Camacho, M. M. & Lord, S. M. (2011). Quebrando fronteras: Trends among Latino and Latina undergraduate engineers. *Journal of Hispanic Higher Education, 10*(2), 134–146. https://doi.org/10.1177/1538192711402354

Dika, S. L., & Martin, J. P. (2018). Bridge to persistence: Interactions with educators as social capital for Latina/o engineering majors. *Journal of Hispanic Higher Education, 17*(3), 202–215. https://doi.org/10.1177/1538192717720264

Foxx, K. A. (2014). A cultural wealth perspective on success and persistence in engineering for Black and Latino students: Exploring interactions with faculty and other institutional agents [Unpublished doctoral dissertation]. University of North Carolina at Charlotte.

Garcia, A., Lane, T. B., & Rincón, B. (2021). Cultivating graduate STEM pathways: How STEM enrichment programs broker opportunity for students of color. Special Issue on New developments in pathways towards diversity and inclusion in STEM: A United States perspective. *Frontiers in Education. 6.* https://doi.org/10.3389/feduc.2021.667976

Garcia, A., Rincón, B., & Hinojosa, J. K. (2020). "There was something missing": How Latinas construct compartmentalized identities in STEM. In E.M. Gonzalez, F. Fernandez, & M. Wilson (Eds.), *An asset-based approach to advancing Latina students in STEM: Increasing resilience, participation, and success* (pp. 181–192). Routledge.

George, C.E., Castro, E. L., & Rincón, B. (2019). Investigating the origins of STEM intervention programs: An isomorphic analysis. *Studies in Higher Education, 44*(9), 1645–1661. https://doi.org/10.1080/03075079.2018.1458224

George-Jackson, C. E., Rincon, B., & Martinez, M. G. (2012). Low-income engineering students: Considering financial aid and differential tuition. *Journal of Student Financial Aid, 42*(2), 4–24. https://doi.org/10.55504/0884-9153.1009

Lane, T. B., & Id-Deen, L. (2023). Nurturing the capital within: A qualitative investigation of Black women and girls in STEM summer programs. *Urban Education, 58*(6), 1298–1326. https://doi.org/10.1177/0042085920926225

Lane, T. B., Morgan, K., Lopez, M. M. (2020). "A bridge between high school and college": A case study of a STEM intervention program enhancing college readiness among underserved students. *Journal of College Student Retention: Research, Theory, & Practice, 22*(1), 155–179. https://doi.org/10.1177/1521025117729824

Lerma, V., Hamilton, L. T., & Nielsen, K. (2020). Racialized equity labor, university appropriation and student resistance. *Social Problems, 67,* 286–303. https://doi.org/10.1093/socpro/spz011

Maton, K. I., Beason, T. S., Godsay, S., Sto. Domingo, M. R., Bailey, T. C., Sun, S., & Hrabowski III, F. A. (2016). Outcomes and processes in the Meyerhoff Scholars Program: STEM PhD completion, sense of community, perceived program benefit, science identity, and research self-efficacy. *CBE—Life Sciences Education, 15*(3), ar48. https://doi.org/10.1187/cbe.16-01-0062

McGee, E. O. (2021). *Black, brown, bruised: How racialized STEM education stifles innovation.* Harvard Education Press.

McGee, E. O., & Martin, D. B. (2011). "You would not believe what I have to go through to prove my intellectual value!" Stereotype management among academically successful Black mathematics and engineering students. *American Educational Research Journal, 48*(6), 1347–1389. https://doi.org/10.3102/0002831211423972

Mein, E., Guerra, H. M., & Herrera-Rocha, L. (2020). Latina undergraduates in engineering/computer science on the US–Mexico border: Identity, social capital, and persistence. In *An Asset-Based Approach to Advancing Latina Students in STEM* (pp. 131–146). Routledge.

Moore III, J. L., Madison-Colmore, O., & Smith, D. M. (2003). The prove-them-wrong syndrome: Voices from unheard African-American males in engineering disciplines. *The Journal of Men's Studies, 12*(1), 61–73. https://doi.org/10.3149/jms.1201.61

Montelongo, R. (2019). Cultural factors that impact Latina/o college student success at predominantly-white institutions: Past and current knowledge for educators. *Diversity, Social Justice, and the Educational Leader,* 3(1), 11–25.

Murphy, M. C., Steele, C. M., & Gross, J. J. (2007). Signaling threat: How situational cues affect women in math, science, and engineering settings. *Psychological science,* 18(10), 879–885.

National Center for Science and Engineering Statistics (NCSES). 2017. Women, Minorities, and Persons with Disabilities in Science and Engineering: 2017. Special Report NSF 17–310. National Science Foundation, Arlington, VA. https://www.nsf.gov/statistics/2017/nsf17310 /digest/occupation/hispanics.cfm

Pew Research Center. (April, 2021). *STEM jobs see uneven progress in increasing gender, racial, and ethnic diversity.* https://www.pewresearch.org/science/2021/04/01/stem-jobs-see -uneven-progress-in-increasing-gender-racial-and-ethnic-diversity/

Pietri, E. S., Drawbaugh, M. L., Lewis, A. N., & Johnson, I. R. (2019). Who encourages Latina women to feel a sense of identity-safety in STEM environments? *Journal of Experimental Social Psychology,* 84, 103827. https://doi.org/10.1016/j.jesp.2019.103827

Revelo, R. A., & Baber, L. D. (2018). Engineering resistors: Engineering Latina/o students and emerging resistant capital. *Journal of Hispanic Higher Education,* 17(3), 249–269. https://doi .org/10.1177/1538192717719132

Rincón, B., Fernández, E., & Hinojosa, J. K. (2020). "I want to follow in her footsteps": Activating, nurturing, and extending community cultural wealth for Students of Color entering STEM Pathways. *Teachers College Record,* 122(9), 1–36. https://doi.org/10.1177/0161468120 12200903

Rincón, B., & George-Jackson, C. E. (2016). Examining departmental climate for women in engineering: The role of STEM interventions. *Journal of College Student Development,* 57(6), 742–747. https://doi.org/10.1353/csd.2016.0072

Rincón, B., & Lane, T. B. (2017). Latin@s in science, technology, engineering, and mathematics (STEM) at the intersections. *Journal of Equity and Excellence in Education,* 50(2), 182–195. https://doi.org/10.1080/10665684.2017.1301838

Rincón, B., & Rodriguez, S. (2020). Latinx students charting their own STEM pathways: How community cultural wealth informs their STEM identities. *Journal of Hispanic Higher Education,* 20(2), 149–163. https://doi.org/10.1177/1538192720968276

Rodriguez, S., Cunningham, K., Jordan, A. (2019). STEM identity development for Latinas: The role of self- and outside recognition. *Journal of Hispanic Higher Education,* 18(3), 254–272. https://doi.org/10.1177/1538192717739958

Rodriguez, S. L., Doran, E. E., Sissel, M., & Estes, N. (2022). Becoming *la ingeniera:* Examining the engineering identity development of undergraduate Latina students. *Journal of Latinos and Education,* 21(2), 181–200. https://doi.org/10.1080/15348431.2019.1648269

Saldaña, J. (2016). *The coding manual for qualitative researchers* (3rd ed.). Sage.

Smith, K. C., Poleacovschi, C., Feinstein, S., & Luster-Teasley, S. (2022). Ethnicity, race, and gender in engineering education: The nuanced experiences of male and female Latina/o/x engineering undergraduates targeted by microaggressions. *Psychology Reports,* 126(5), 2345-2382. https://doi.org/10.1177/00332941221075766

Smith, K. N., Gayles, J. G. (2018). "Girl power": Gendered academic and workplace experiences of college women in engineering. *Social Sciences,* 7(1), 11. https://doi.org/10.3390/socsci 7010011

Strayhorn, T. L. (2018). *College students' sense of belonging: A key to educational success for all students.* Routledge.

Strayhorn, T. L., & Johnson, R. M. (2016, June 26–29). *What underrepresented minority engineering majors learn from co-ops and internships* [Conference paper]. American Society for Engineering Education Annual Conference & Exposition, New Orleans. LA.

Verdín, D. (2021). The power of interest: Minoritized women's interest in engineering fosters persistence beliefs beyond belongingness and engineering identity. *International Journal of STEM Education, 8*(1), 1–19. https://doi.org/10.1186/s40594-021-00292-1

Verdín, D., & Godwin, A. (2018). Exploring Latina first-generation college students' multiple identities, self-efficacy, and institutional integration to inform achievement in engineering. *Journal of Women and Minorities in Science and Engineering, 24*(3), 261–290. http://doi.org/10.1615/JWomenMinorScienEng.2018018667

Walton, G. M., & Brady, S. T. (2017). The many questions of belonging. In A. J. Elliot, C. S. Dweck, & D. S. Yeager (Eds.), *Handbook of competence and motivation: Theory and application* (pp. 272–293). The Guilford Press.

4 · BEYOND ETHNIC IDENTITY

The Intersectional Experience of Latina/o/x Students in Engineering with Stereotype Threats

ERIN DORAN, ELIZABETH TUROCHY, MICHAEL A. PEREZ, CRISTINA POLEACOVSCHI, AND TIMOTHY YUEN

There is a well-documented need to increase the number of Latina/o/x engineers in the United States (Martin-Vega, 2006; Mendoza Diaz, 2014; Rodriguez et al., 2022). The National Science Foundation ([NSF], 2021) has reported that, in 2019, only about 8 percent of engineering occupations were held by a Latina/o/x person in the United States. This is in stark contrast to the nearly 68 percent of engineers who identify as white (NSF, 2021). Given the growth of the Latina/o/x demographic in the United States (Passel et al., 2022), numerous studies and reports have pointed out that the future demand for engineers and workers for other fields in science, technology, and mathematics (together known as STEM fields) is unlikely to be met without dramatically increasing Latina/o/x participation in these areas (Banda, 2020; Crisp & Nora, 2012; Revelo & Baber, 2018).

Yet, despite increased attention on these education and employment disparities, there remain barriers that keep Latina/o/x students from pursuing engineering degrees. Camacho and Lord (2013) found that, while Latina/o/x students matriculate through engineering programs at roughly the same rate as their white counterparts, recruiting Latina/o/x students into these programs is the greatest challenge. The authors made recommendations for recruiting Latina/o/x students into these fields, including better partnerships with K–12 pre-engineering programs, high school math courses that incorporate culturally relevant curricula, and programs that specifically target Latinas.

Once Latina/o/x students make the decision to enroll in undergraduate engineering programs, their experiences may not be positive or affirming. Carlone and Johnson (2007) point out that engineering programs are often seen as white and masculine cultures that may negatively impact students' sense of belonging or ability to thrive, especially if they identify as women or racially/ethnically minoritized. Students may also encounter stereotype threats in engineering and other STEM fields (Beasley & Fischer, 2012; Revelo & Baber, 2018). A stereotype threat is a fear of inadvertently confirming a negative stereotype about oneself or an identity group of which one is a part (racial/ethnic group, gender, etc.; Steele, 2010). Latina/o/x students in engineering may experience various forms of stereotype threat in the undergraduate programs that influence their academic performance and persistence. These stereotype threats, which can originate from instructors as well as other students, may include references not just to race/ethnicity and gender but to Latina/o/x students' language (i.e., use of Spanish and/or accented speech), their countries of origin, or a conflation of diverse Latina/o/x identities.

The purpose of this chapter is to describe the stereotype threats beyond race/ethnicity and gender experienced by Latina/o/x undergraduate students in engineering programs at three universities. The institutions include a land-grant, Predominantly White Institution (PWI) in the Midwest, a land-grant PWI in the Southeast, and a Hispanic-Serving Institution (HSI) in the South. The data for this present study include responses to surveys (with Likert-scale and open-ended questions) and qualitative interviews across the three institutions. This study, which was funded by the NSF, was guided by the following research question: How do Latina/o/x undergraduate students in engineering programs experience intersectional, identity-based stereotype threats beyond gender and race/ethnicity?

While previous findings from this larger research project found that Latinas were more likely to experience stereotype threat about their gender (Turochy et al., 2023), the present study focused on a more nuanced understanding of stereotype threats as they relate to students' skin color, countries of origins, language, and citizenship status. This adds a deeper understanding of the types of microaggressions and stereotype threat that Latinx students in engineering may or may not face. Several researchers have pointed out that studies on the challenges students face in engineering have lumped together students from racially minoritized backgrounds (Camacho & Lord, 2013; Revelo & Baber, 2018); in doing so, this research fails to offer a nuanced understanding of the experiences of students from different racial/ethnic backgrounds. In order to more fully explore and discuss the identities of Latina/o/x undergraduate students in engineering programs, we utilized a multilevel model of intersectionality, as discussed by Núñez (2014) and applied specifically to Latina/o/x students. We found that Latina/o/x students at three institutions experienced various stereotype threats that intersect with multiple dimensions of their Latina/o/x identities.

LITERATURE REVIEW

Our framework—Núñez's (2014) multilevel model of intersectionality—is explained in more detail in the next section. However, first, a brief discussion of the intergroup diversity of Latinas/os/xs in the United States is needed, to better explain why more nuanced considerations of this group are greatly needed.

Diversity within the Latina/o/x Community

As of 2021, there were approximately sixty million Latinas/os/xs in the United States; this ethnic category, as defined by the US Census Bureau, is a "person of Cuban, Mexican, Puerto Rican, South or Central American, or other Spanish culture or origin regardless of race" (U.S. Census Bureau QuickFacts, 2021). Mexicans comprise the largest origin group, at nearly two-thirds of all Latinas/os/xs in the United States, and the Pew Research Center reports that Venezuelans, Dominicans, and Guatemalans have experienced the largest population growth between 2010 and 2019 (Noe-Bustamante, 2019). As the federal definition suggests, Latina/o/x identity may transcend racial categories and itself not be a racial group (at least according to federal documents), but Gallegos and Ferdman (2011) discuss the ways that Latina/o/x represents its own racialized experience outside of the traditional Black–white binary. Building on other scholars' work on race and ethnicity, Gallegos and Ferdman (2011) call for Latina/o/x identity to be understood as ethnoracial, "especially given the persistence and apparent intractability of race consciousness and racialized language in the current U.S. society" (p. 54). Gallegos and Ferdman also draw on intersectionality to make the case that Latina/o/x identity may be contextually influenced and adaptive—meaning that Latina/o/x cannot be explained as or reduced to one thing.

Disaggregated education outcomes. One example of how diversity among Latinas/os/xs manifests is within educational outcomes. Torres (2004) clearly reminds us that much of the research that uses the terms "Latinos" or "Hispanics" really means "Mexican American." Torres describes how immigration, educational achievement, and economic attainment are interrelated experiences for non-Mexican Latinos/as/xs, resulting in differing pathways to and through postsecondary education. In another study, Núñez and Crisp (2012) compared the pathways to postsecondary education for Mexican American and Puerto Rican students. They found that Puerto Rican students were more likely to begin their higher education journeys at four-year institutions. In both groups, students' decisions to enroll in higher education were influenced by a variety of factors, including age, first-generation student status, citizenship status, and various types of capital (e.g., familial, social, and financial). This body of research to disaggregate Latina/o/x subgroups further serves as an important reminder that students' identities and experiences influence their educational journeys, both regarding how they arrive at an institution and their experiences within it.

In engineering fields, it is important to note that Latina/o/x enrollment statistics and outcomes may vary widely. Camacho and Lord (2013) argued that empirical research tends to conflate racially minoritized students into a single group and also tends to present students' experiences across science and engineering as a singular pathway. However, it is worth noting that Latina/o/x presence and representation can vary widely among engineering subdisciplines. In the 2015 to 2016 academic year, more than three thousand bachelor's degrees were conferred to Latina/o/x students in mechanical engineering, compared to fewer than two hundred in environmental engineering (Anderson et al., 2018). Therefore, Latina/o/x students' abilities to interact with other Latina/o/x students in their classes may depend heavily on what respective engineering major students pursue.

Stereotypes Experienced by Latinas/os/xs

As previously mentioned, the complexity of Latina/o/x identity has kept it from being fully understood as a race, an ethnicity, or both. Ramirez and Peterson (2020) argue that the lack of understanding about the racialization of Latinas/os/xs in the United States has contributed to persistent and pervasive stereotypes and stigmas about them. As a result, Latinas/os/xs are frequently stereotyped as newly arrived immigrants (Gallegos & Ferdman, 2007), criminals and drug lords (Ramirez & Peterson, 2020), and uneducated and/or uninterested in education (Valencia, 2002; Villenas & Deyhle, 1999).

In engineering contexts, various identity markers (like language) may be a source of microaggressions. In one study, a Latina participant reported that her non-Latina/o/x peers' comments about her English negatively impacted her motivation and self-efficacy (True-Funk et al., 2021). In effect, she felt self-conscious about her ability to communicate with her instructor or to seek help through tutoring or other support services. A growing body of research is devoted to the role that skin color plays in discrimination, both inside and outside the Latina/o/x community (Haywood, 2017; Quiros & Araujo Dawson, 2013). Haywood (2017) describes this succinctly: "While Latinos experience discrimination, the degree of discrimination, how often it occurs, and the outcomes vary considerably based on skin color stratification or colorism" (p. 761). Haywood's work underscores the racism that Afro-Latina/o/x college students experience within and outside the Latina/o/x community, even in their own families, which is influenced by a complicated mix of both anti-Black racism and white supremacy. Understanding the range of identities that Latina/o/x students may bring with them to engineering contexts necessitates a theoretical lens that could be reflective of these multiple dimensions of identity and how they intersect.

THEORETICAL FRAMEWORK

Intersectionality draws from the field of critical legal studies and critical race feminism, starting with the work of Crenshaw, who describes the multiple dimensions of oppression Black women face in the United States along both racial and gendered lines. Collins (2000) later writes, "As opposed to examining gender, sexuality, race, class, and nation as separate systems of oppression, the construct of intersectionality references how these systems mutually construct one another"

(p. 47). Rincón and Lane (2017) applied intersectionality as their framework to look at how the intersections of race/ethnicity, gender, and socioeconomic status influenced the experiences of Latin@ students in STEM programs at six universities across the Midwest.

In this study, we employed a version of intersectionality conceptualized by Núñez (2014). Specifically, Núñez used research on Latino/a/x im/migrant high school students in California to develop her multilevel model and called for research to analyze three different levels: social categories or relations; multiple arenas of influence; and historicity. Regarding the first level, Núñez describes the complex nature of Latina/o/x identities and that this racial category is often intertwined with others, including language, citizenship, im/migration status, countries of origin, race, and so forth. She argues, "Because social categories are not neatly bounded due to individuals' multiple identities, within- and between-group comparisons are appropriate when exploring influences on societal inequalities" (p. 87). With the understanding that the Latina/o/x population is not monolithic, we attempt to be multidimensional in our consideration of identities and know that individual identities cannot ever be separated from the rest, in any given person. Regarding the second dimension, multiple arenas of influence, Núñez calls for a broader understanding of the societal structures that may contribute to inequality. She names examples like anti-im/migrant policies, the dismantling of bilingual education programs, and anti-affirmative action policies—particularly in education—that create barriers to Latina/o/x access to various education sectors. Beyond formal policy and laws, Núñez describes how popular culture and media can also influence public perceptions of groups, which may have material consequences on the lives of members of these groups. Finally, regarding the third dimension (historicity), Núñez calls on readers to understand the interlocking structures of power that evolve over time.

Concerning the first dimension of Núñez's (2014) multilevel model of intersectionality (social categories), we found great variation in how the students who participated in this study described and made sense of their Latina/o/x and related identities. Their experience with stereotype threats came from those around them—namely, students in their classes and the professors teaching in their programs—and these stereotypes were often mediated by popular culture representations; local, state, and federal laws; and public figures. This finding speaks to the second dimension (multiple arenas of influence). Finally, we know that stereotypes often become codified into stereotypes once they have been baked into the collective psyche over time, adding to their sense of historicity (the third dimension).

We found Núñez's (2014) concept of intersectionality useful for this study and to answer our research question, as it considers the complexity of Latina/o/x identities and how these are shaped within individuals, in groups, and at the societal level. What is interesting about this study is that students at Roble University, an HSI in a Latina/o/x-majority city, experienced many of the same stereotype threats as students at other PWIs in other geographical areas.

METHODS

This study utilized a general qualitative approach to study (a) the sources of stereotype threat experienced by Latina/o/x undergraduate students in engineering programs at three different universities and (b) how they coped with these threats. There are two main sources of data: First, a survey instrument was distributed among engineering students at the universities. While this survey did include Likert-scale questions, the present study focused on the open-ended questions that respondents were asked, such as "Please list a stereotype you have experienced based on your ethnicity." Second, we drew from survey respondents to conduct more in-depth qualitative interviews. The multilevel model for intersectionality was used to analyze this data.

Institutional Contexts of This Study

It is noteworthy that the data for this study were collected at three separate and seemingly different universities: Roble University is a large HSI located in the south-central United States. It has a student population of over thirty thousand, over half of whom identify as Latina/o/x. Morrill University is a land-grant university located in the southeastern United States. In the fall 2021 semester, Morrill enrolled just under thirty thousand students, with approximately one thousand Latina/o/x undergraduate students (Morrill University Fact Book, 2021). Hilton University is the smallest of the three institutions (approximately twenty-six thousand students, with about 1,800 Latina/o/x undergraduates in 2021) and is a land-grant university in the Midwest.

Surveys

A survey was designed and distributed via Qualtrics for this study to explore students' experiences with stereotype threat. This survey was distributed through the researchers' networks, primarily by asking departmental advisors to distribute the call for survey respondents to their students. The survey received 261 responses, and of those, 185 were complete and useable for this specific sample. There were fifty-two responses from Morrill University, eighty-three from Hilton University, and fifty from Roble University. The electronic survey used both Likert-scale questions as well as open-ended questions where students could type in their answers. Respondents were able to skip questions or choose "I prefer not to answer" for each question.

The survey consisted of several parts. First, we collected demographic information about the respondents (e.g., gender, racial/ethnic identity, and major). The second part of the survey asked students about their confidence in themselves as engineering students and in their choice to major in engineering. For example, students were asked to rate their response to the statement "I have a strong sense of belonging to the engineering student community." The third part asked about how students felt they were perceived by others based on their identities: specifically, their ethnicity and gender, including their physical appearance. Through open-ended questions, this section also asked students to describe a specific moment where they were stereotyped because of their ethnicity or gender, as well as how they coped with stereotypes. Finally, they were asked about general engineering experiences (e.g., participation in peer tutoring or mentoring and engagement with engineering student groups and/or honor societies), which were also open-ended questions.

Student Interviews

Students who completed the survey were invited to participate in a qualitative interview with the research team. There were twenty-seven students from across the institutions who completed an interview. The interviews were audio recorded, and Otter.ai was used for the transcription. The transcripts were edited by a member of the research team who matched them with the audio file to ensure maximum accuracy. The interview protocol included questions such as "Have you had any negative experiences in your STEM courses based on your race?" and "Have you ever felt the need to defend your reputation as an engineering major?" As most of the interviews were conducted in the summer and fall of 2020, in the midst of the COVID-19 pandemic, interviews were completed virtually via Zoom.

Data Analysis

As previously noted, our research team had already focused on findings related to racial/ethnicity and gender (see Poleacovschi et al., 2022; Turochy et al., 2021). For example, we found that female students were more likely to report stereotypes and moments of discrimination based on their gender rather than their race and/or ethnicity. Our previous work on this data set tended to focus more on the analysis of the surveys, particularly the scaled answers that were quantitatively analyzed. In sifting through the data, we decided to more intently hone in on the open-ended questions of the survey responses as well as the interview data.

The first round of analysis took an inductive approach and used open coding (Saldaña, 2016) to see what stereotypes the Latina/o/x students in the sample reported, aside from racial/ethnic and gender-based stereotypes. Examples of some of these codes are "language," "colorism," and "country of origin." The second round of axial coding (Saldaña, 2016) served to reorganize some of these themes into larger categories, as well to add nuance. For instance, the code "colorism" had sub-codes for "white passing," "Black," and "white privilege."

Researchers' Positionality

Erin Doran is a multi-ethnic woman (Mexican American and white), who is an educational researcher focused on creating equitable spaces for teaching and learning. Elizabeth Turochy is a Caucasian female, born and raised in the southern United States, and is currently a graduate student in engineering. Michael Perez is a first-generation US citizen who was born and raised in a primarily Hispanic community in South Florida. He is a licensed professional engineer and a civil engineering professor. Cristina Poleacovschi is a white female assistant professor in civil engineering, whose research focuses on social justice related to infrastructure and engineering education. Finally, Timothy Yuen approaches his research through the lens of having been a first-generation college student and the son of immigrants. All of us were motivated to understand more about the experiences of Latina/o/x students in engineering and how this understanding could be used to advocate for more equitable opportunities for them—and ultimately all students—to participate, learn, and succeed in engineering programs.

FINDINGS

Our team's previous findings from the data show that engagement with stereotypes often happened along race/ethnicity and gendered lines, with gendered stereotypes occurring more frequently (Poleacovschi et al., 2022; Turochy et al., 2021). It should also be acknowledged that not all students reported experiencing stereotypes from their peers or others with whom they engaged in their programs. Among the twenty-seven interviews, eight students did not report any stereotype threats, one female student talked about the racism she experienced through a racist joke directed at her outside the classroom, and five female students described their experiences of gender-based stereotypes only. Two students who did not report any stereotype threats were first-time freshmen who were taking all their courses online, and one of them acknowledged that this may have made a difference.

In presenting these findings, we made the decision to present the survey and interview data findings separately. This was done since two separate (albeit related) sets of questions were asked, and the findings yielded somewhat different themes.

Survey Findings

The two most common stereotypes uncovered in the open-ended questions of the survey were related to academic performance and language. Related to the first stereotype, eight different students across all institutions used the word "lazy" to describe non-Latinas/os/xs' perceptions of Latinas/os/xs and thus how their non-Latina/o/x peers viewed them. One male student from Roble University wrote that others perceive him as "lazy because he is Mexican." A female from Roble also shared, "I am a Hispanic, therefore, I am not smart enough to be an engineer." Other similar words and sentences captured in the written responses included "unintelligent" (female, Morrill University), "not smart because I am Mexican" (male, Morrill University), and "[I'm] struggling because [I'm] a minority" (female, Hilton University).

There were also responses in the survey that reflected a perception that Latina/o/x students were only allowed in their university or the engineering program because of affirmative action-like programs. Three students (one from Morrill University and two from Hilton) specifically wrote this in their response. A female student from Morrill wrote, "I only was accepted to meet a diversity quota." Similarly, a female student at Hilton submitted, "[t]hat I have a full ride because I'm Hispanic, [and] that I'll get a job easily because I'm a minority." Finally, a nonbinary student from Hilton shared, "[p]eople [believe] I have made it this far in my education due to being a minority in this major." While racial quotas may have been the practice of some selective universities' admissions offices, it is important to note that the use of quotas were ruled to be unconstitutional by the US Supreme Court in 2003 in its decision on the *Grutter vs. Bollinger* case (Poon & Segoshi, 2018). Five students specifically used the word "dumb" to describe others' stereotypes of Latinas/os/xs in relation to their intelligence and/or academic performance. Other comments included "I can't read/speak properly" (male, Roble University), "probably below average student" (male, Hilton University), and "[t]hat I do not understand the language and therefore, I do not understand the topics I'm studying" (male, Morrill University).

On the topic of language, the findings indicate some nuance about Latina/o/x students—their use of Spanish, in particular. A number of students reported that they were assumed to know Spanish because of their Latina/o/x identity: it is unclear whether that meant that they did not know Spanish and were disparaged for it (perhaps by other Latinas/os/xs) or that they did and considered this a stereotype. A female Hilton student specifically stated that she had experienced discrimination due to her accent. A male student from Roble described his own accent as "thick" and said that, in response, he usually felt compelled to "try and speak louder and slower." Another male student from Roble replied with a longer example of discrimination in which language stereotypes were invoked: "An older gentleman approached me and a friend, asked us a question but we weren't sure about where the classroom he was asking us about was and [we] got called stupid and [he] told us to learn English." Finally, a nonbinary student from Hilton expressed that they typically only experienced a stereotype after others heard them speaking fluent Spanish, especially since they are white passing.

Among other stereotypes that received different responses included assumptions of Mexican background/heritage, stereotypes of criminality, and assumptions of poverty. A female student at Hilton shared that others "[assumed she's] from Mexico and speak[s] Spanish." She did not clarify which subgroup she identified with, but her answer implied that she did not identify as Mexican, and in reality, we know that Latinas/os/xs may speak a variety of languages, including Indigenous languages (e.g., Nahuatl), Portuguese, and French. Among the stereotypes of criminality include others' beliefs that associate Latina/o/x subgroups with drug and/or with being in the United States illegally. One male student from Morrill said he experienced "things [stereotypes] related to cocaine since [he's] from Colombia." Two students reported that they encountered assumptions that they or members of their families were "illegal" or "undocumented." In response, a male from Morrill shared, "I carry a passport card with my citizenship on it"—presumably to prove his citizenship status to anyone who questioned it. Finally, the findings point to assumptions that Latinas/os/xs live in poverty. A male student from Morrill stated, "I am Nicaraguan, and many people stereotype us as poor." Finally, one male from Roble disclosed that others assumed he "[lives] in a rough neighborhood" because of his identity.

Coping mechanisms described in the surveys. The two most reported coping mechanisms that students across all three institutions evenly shared were ignoring stereotypes (twenty-six responses) and/or using stereotypes to prove others wrong (eleven responses). A male student from Hilton described his coping as "[s]pecifically avoid[ing] participating in those stereotypes, calling out if someone is imposing that stereotype on [him]." A male junior from Roble said that his response to others' behavior was to "[t]ry to achieve titles like undergraduate researcher, straight A's, Phi Kappa Phi honor society, student leader" to prove he was not "lazy because [he] is Mexican." A female from Morrill said that she ignored others to focus on her own work and pointed out, "I also have a strong support system at home that encourages me and reminds me I am not the things people say." Eight students at Hilton and Roble reported engaging in various therapeutic strategies, including walking their dog, journaling, and seeing a mental health specialist like a counselor. Interestingly, only three students (one from Morrill and two from Hilton) mentioned building solidarity or community with other minoritized students in their classes.

Interview Findings

Though roughly a third of the interview participants did not report any stereotype threats, others described being faced with various stereotypes, as well as microaggressions and more overt racist acts (e.g., racist jokes directed at them). For some students, these negative interactions made students feel a great deal of responsibility to not only succeed but to help combat against stereotypes of Latinas/os/xs or women in engineering. A male student at Hilton revealed the pressure he felt as the only person of color in a room: "You feel like you have to, like, represent your entire [group.] I had to be very careful in my actions, and that's when I really try and put in a lot of work—and so people understand that."

The most prominent finding in the interviews related to colorism. Specifically, five students attributed their lack of discrimination or perception as a Latina/o/x stereotype directly to the fact that they were considered white passing. For example, when a male Roble student was asked about any stereotypes he experienced based on his race, he explained, "I feel that I'm quite pale, if you've noticed. And I do think that people treat me differently because of that. A lot of people don't generally know I am Hispanic. So I feel like a lot of stereotypes . . . people don't associate with me just because I don't look [like] the stereotypical Hispanic. I don't think a lot of stereotypes have been placed on me just because people don't know what I am." Similarly, a female student at Morrill pointed out, "My dad is Mexican, and my mom is white. Like I definitely look very white. So, unless you like, specifically

[ask] or get to know me well, most people will assume that I'm just white." As such, this student went on to say, "I feel like there are more stereotypes that come with being a woman in engineering than I've seen being Hispanic in engineering."

Another female student at Morrill described feeling like her non-Latina/o/x peers tended to erase her Latina identity. Describing her negative experiences based on her race as "subtle," she said,

> Because I'm considered white passing by my American peers, it's like they'll say things to me not directed at me because they don't recognize me as being Hispanic. I really feel like it's kind of, I feel like they ignore my ethnicity, like it's not something they bring up. And even though that isn't like a hateful act from them, it does affect me because it feels like they only highlight or people only highlight the negative stereotypes. And when I succeed, it's not highlighted. It's not really valued. I've been told, people have been like "You're white, you're basically white." Like to my face, it's like they just kind of disregard my identity.

Most of the interactions described in both the surveys and interviews were presumed to take place between students. On the flip side of colorism, a Morrill student from Colombia told of an incident with an older white instructor: "I would be like in office hours, and I remember one time you [the interviewer] were with me and this African American girl, we were asking the professor some questions. And he was like, 'You two are like too dark to understand,' something like that. And I just looked at her and then she just laughed. So, I guess he jokes like that. I mean, he's like an 80-year-old dude from Alabama—I wouldn't expect anything different from that."

Coping strategies. In the interviews, Latina/o/x participants tended to describe the same coping strategies as the student survey responses: ignoring and/or dispelling stereotypes. A male Morrill student imparted a sort of internal mantra: "I always try to tell myself, like we're all engineers. Like, it doesn't really matter what race we are, it's just our character. It doesn't matter that I'm Hispanic or Asian. I'm just the same as everybody else." However, a male student from Hilton reported that his coping strategy for dealing with stereotypes was effectively to act white or according to the white norms he perceived as "professional": "I've had to separate . . . Hispanic stuff . . . and just be professional or act professional. It's very hard, especially like right now, I'm doing all these job interviews and all this study, and I have to come off as—you know, not only is somebody unique, but somebody professional. And the way I see it, and that's not to be rude or anything, but it's very whitewashed, very, you know, white-male-dominated, the industry especially."

Summary of Findings

In the surveys, the most prominent stereotypes Latina/o/x students reported were around assumptions about their language proficiency, having an accent, and their academic performance. The latter includes assumptions from other students that they (the Latina/o/x students) were there because they benefited from affirmative action programs that privilege minoritized students over white students. Within the interviews, students spoke of colorism—in most cases, the way that being white passing shielded them from dealing with the stereotypes other Latinas/os/xs might face. To deal with these stereotypes, most respondents chose to either ignore them or to channel them in order to prove others wrong. In the next section, we discuss how this study's findings fit into the scholarly conversation on Latina/o/x students in engineering programs.

DISCUSSION

This study builds on previous scholarship that has used intersectionality as a framework for understanding minoritized students' experiences in engineering (e.g., Banda, 2020; True-Funk et al., 2021). Though our previous work added specifically to the work that focuses on the intersection of race/ethnicity and gender, as others have done (e.g., Banda & Flowers, 2017; Camacho & Lord, 2013; Martin et al., 2013; Poleacovschi et al., 2022; Rodriguez et al., 2022; Turochy et al., 2021), the purpose of the present study was to more deeply investigate other identities held by Latina/o/x students that may contribute to their experience with stereotype threat. Given the identities that we found have been understudied and/or unexplored in the engineering and higher education literature, such as language and color, this study reinforces the need for more research that does not treat Latina/o/x students as a monolithic group.

Interestingly, from the outset of our research activities, we assumed that we would find less engagement with stereotypes at Roble University, the HSI. As a university that is majority Latina/o/x, we assumed that there would be a critical mass of Latina/o/x students on campus that might mitigate some of the racism and stereotypes that students may face on other campuses (e.g., PWIs, which Morrill and Hilton Universities are; see Hagedorn et al. [2007] for a discussion on "critical mass"). This perhaps further supports the work of Comeaux and colleagues (2021), who argue that, even on racially diverse campuses like HSIs where diversity may be actively promoted, campus environments can still be hostile for minoritized students, and these campuses are not immune to racism. While our ability to compare Latina/o/x experiences across institutional types may be limited, several observations can be made. Students across all three institutions reported the perception that others saw them as "dumb" or "lazy" because of their Latina/o/x identity. However, only students at Morrill and Hilton (the PWIs) reported that they only made it to college because of their minoritized status and perhaps benefited from affirmative action-like programs.

What is notable about this study is that students did not necessarily have to hold certain identities or experiences to be perceived as having them. Simply by being Latina/o/x, these students may have been perceived by others to be uneducated, lazy, less than proficient in English, undocumented, and so forth. These negative stereotypes held by others have the potential to negatively influence interpersonal interactions within a field of study that heavily promotes group work and collaboration.

Limitations

No study is without limitations, and we acknowledge that there are some ways in which we could have improved this study. First, we did not ask specifically what Latinx subgroup(s) students were from. Some students disclosed their backgrounds in open-ended questions on their survey responses or in the interviews. We therefore know that we had participants who identified as Salvadoran and Colombian, Puerto Rican, half-Korean and half-Ecuadorian, and Honduran, as well as a number of participants who identified as half-white and half-Latina/o/x. We likely had more diversity than we realized. We also had one female who participated in an interview who was Brazilian and who self-identified as Latina. We chose to include her data in the study because of her self-identification as Latina, though we acknowledge that there are differing opinions on whether or not Brazilians are considered "Latinas/os/xs," if the latter term is dependent on use of the Spanish language (see Marrow [2003] for a broader discussion of this).

In addition to not fully exploring the Latina/o/x subgroups of this study, we acknowledge that there were many identity categories that were not explored in the questions we asked. For instance, there were three students who disclosed their gender identity as nonbinary, as well as one trans female, all of whom responded to the survey. One nonbinary student indicated that they are often misgendered by their classmates, but the survey did not allow us to explore that further. The limitation of any survey is the lack of ease in asking follow-up questions, especially if students do not volunteer their contact information.

Implications for Practice

In this study, the most common coping strategy Latina/o/x students reported for dealing with stereotypes was to ignore them. Depending on their frequency and severity, we argue that this approach may only go so far and may exacerbate students' overall stress levels, mental and emotional health, and desire to persist in their program. Support programs and affinity groups on campus may help support students and provide them with ways to manage their reactions to stereotypes and to directly confront others, if needed.

Given that the main source of stereotypes in this study appeared to be other students, this makes us think about the way diversity and working within groups with others who hold different identities may be effectively taught to all students in engineering. We note that the adherence to upholding diversity, equity, and inclusion is included in the Code of Ethics for the National Society of Professional Engineers (NSPE), which states that, "Engineers shall treat all persons with dignity, respect, fairness, and without discrimination" (Brooks, 2020, para. 3). While it is virtually impossible for programs and individual instructors to monitor every student interaction in classrooms and lab spaces, it is important to think about how programs may instill the importance of these values to students, as part of their induction into the engineering profession. Given that the NSPE's own Code of Ethics treats inclusion as a professional value, this might be reified in engineering programs through coursework (such as first-year experience programs that introduce students to the profession, engineering ethics courses, and so on). As Banda (2020) notes, the lack of diversity throughout engineering programs (e.g., among students, faculty, staff, and administrators) signals to students that diversity is not prioritized; this lack may negatively impact the minoritized students in these programs. To the extent that diversity and inclusion can be treated as a necessary part of engineering work, this may teach all students the need to respect everyone they encounter and create welcoming, inclusive spaces in small groups, labs, classrooms, and—ultimately—in the workplace.

Implications for Research

One lesson we also learned is that definitions of race and ethnicity may be blurred for Latinas/os/xs, a point that has been discussed by Gallegos and Ferdman (2007; 2011) and served as justification for their term "ethnoracial." A typical example of this within our data is that students answered "Hispanic" as their race and a country of origin (e.g., "Ecuadorian") as their ethnicity. In fact, one male student from Morrill stated, "I don't know really how to say like race versus ethnicity. I just think of myself as Hispanic and that is a catch-all term for me." Researchers, including ourselves, who do work on Latina/o/x identity should be aware of this complexity and how it might influence the ways survey and/or interview questions are asked to participants. Though outside the scope of this study, more research might be conducted on Latina/o/x identities and the ways Latina/o/x engineers make their way in industry as Latinas/os/xs. Understanding the stereotypes they may encounter in the field postgraduation might impact the ways that programs and mentors can prepare students for the realities of the profession.

We also note that students in this study spoke of being white passing, and they understood that this may have shielded them from certain stereotype threats. The scholarship on colorism has grown in recent years (see Haywood, 2017; Quiros & Araujo Dawson, 2013). We could not find previous research on colorism in engineering, especially for those with darker skin tones. More work should be undertaken to understand what, if any, impact colorism and white privilege has on the experiences of racially minoritized students in engineering.

CONCLUSION

This study took an intersectional approach to detailing the experiences of Latina/o/x undergraduate students as well as their coping strategies for dealing with stereotypes. Through understanding these stereotypes, we can learn how to address them and to hopefully better retain the Latina/o/x students who arrive at college with the intention of attaining an engineering degree. By improving the student experience in combating stereotypes or more intentionally helping Latina/o/x students deal with them, the hope is that programs can boost the graduation rates of these students and help build the Latina/o/x workforce in engineering fields.

Authors' Note

This chapter is based on work supported by the National Science Foundation under grant no. 1949691. Any opinions, findings, conclusions, or recommendations expressed in this material are those of the authors and do not necessarily reflect the views of the National Science Foundation.

REFERENCES

Banda, R. M. (2020). From the inside looking out: Latinas' intersectionality and their engineering departments. *International Journal of Qualitative Studies in Education, 33*(8), 824–839. https://doi.org/10.1080/09518398.2020.1735565

Banda, R. M., & Flowers III, A. M. (2017). Birds of a feather do not always flock together: A critical analysis of Latina engineers and their involvement in student organizations. *Journal of Hispanic Higher Education, 16*(4), 359–374. https://doi.org/10.1177/1538192716662966

Beasley, M. A., & Fischer, M. J. (2012). Why they leave: The impact of stereotype threat on the attrition of women and minorities from science, math, and engineering majors. *Social Psychology of Education, 15*, 427–448. https://doi.org/10.1007/s11218-012-9185-3

Brooks, J. (2020, July/August). *Why should I care about diversity in engineering?* National Society of Professional Engineers, *PE Magazine.* Retrieved from https://www.nspe.org/resources/pe-magazine/july-2020/why-should-i-care-about-diversity-engineering

Camacho, M. M., & Lord, S. M. (2013). Latinos and the exclusionary space of engineering education. *Latino Studies, 11*, 103–112. https://doi.org/10.1057/lst.2012.57

Carlone, H. B., & Johnson, A. (2007). Understanding the science experiences of successful women of color: Science identity as an analytic lens. *Journal of Research in Science Teaching, 44*(8), 1187–1218. https://doi.org/10.1002/tea.20237

Cobas, J. A., & Feagin, J. R. (2008). Language oppression and resistance: The case of middle-class Latinos in the United States. *Ethnic and Racial Studies, 31*(2), 390–410. https://doi.org/10.1080/01419870701491945

Collins, P. H. (2000). Gender, Black feminism, and Black political economy. *Annals of the American Academy of Political and Social Scientists, 568*, 41–53.

Comeaux, E., Grummert, S. E., & Cruz, N. A. (2021). Strategies of resistance among racially minoritized students at a Hispanic-serving institution: A critical race theory perspective. *The Journal of Higher Education, 92*(3), 465–498. https://doi.org/10.1080/00221546.2020.1851569

Crisp, G., & Nora, A. (2012). *Overview of Hispanics in science, mathematics, engineering and technology (STEM): K-16 representation, preparation, and participation.* Hispanic Association of Colleges and Universities. http://hdl.handle.net/10919/83073

Gallegos, P. V., & Ferdman, B. M. (2007). Identity orientations of Latinos in the United States: Implications for leaders and organizations. *The Business Journal of Hispanic Research, 1*(1), 26–41.

Gallegos, P. V., & Ferdman, B. M. (2011). Latina and Latino ethnoracial identity orientations: A dynamic and developmental perspectives. In C. Wijeyesinghe and B. W. Jackson (eds.), *New perspectives on racial identity development: Integrating emerging frameworks* (2nd ed., 51–80). New York University Press.

Hagedorn, L. S., Chi, W., Cepeda, R. M., & McLain, M. (2007). An investigation of critical mass: The role of Latino representation in the success of urban community college students. *Research in Higher Education, 48*, 73–91. https://doi.org/10.1007/s11162-006-9024-5

Haywood, J. M. (2017). "Latino spaces have always been the most violent": Afro-Latino collegians' perceptions of colorism and Latino intragroup marginalization. *International Journal of Qualitative Studies in Education, 30*(8), 759–793. https://doi.org/10.1080/09518398.2017.1350298

Lippi-Green, R. (2012). *English with an accent: Language, ideology, and discrimination in the United States* (2nd ed.). Routledge.

Lopez, M. H., Krogstad, J. M., & Flores, A. (2018, April 2). *Most Hispanic parents speak Spanish to their children, but this is less the case in later immigrant generations.* Pew Research Center. Retrieved from https://www.pewresearch.org/fact-tank/2018/04/02/most-hispanic-parents-speak-spanish-to-their-children-but-this-is-less-the-case-in-later-immigrant-generations/

Marrow, H. (2003). To be or not to be (Hispanic or Latino): Brazilian racial and ethnic identity in the United States. *Ethnicities, 3*(4), 427–464. https://doi.org/10.1177/1468796803003004001

Martin, J. P., Simmons, D. R., & Yu, S. L. (2013). The role of social capital in the experiences of Hispanic women engineering majors. *Journal of Engineering Education, 102*(2), 227–243. https://doi.org/10.1002/jee.20010

Mendoza Diaz, N. V. (2014, June 15–18). *Hispanics in engineering* [Conference paper]. 121st ASEE Annual Conference & Exposition, Indianapolis, IN. https://doi.org/10.18260/1-2--20559

Martin-Vega, L. A. (2006). Hispanics in engineering. *ASEE Prism, 15*(6), 46.

National Science Foundation. (2021). *Women, minorities, and persons with disabilities in science and engineering.* National Center for Science and Engineering Statistics. Retrieved from https://ncses.nsf.gov/pubs/nsf21321/report

Noe-Bustamante, L. (2019, September 16). *Key facts about U.S. Hispanics and their diverse heritage.* Pew Research Center. Retrieved from https://www.pewresearch.org/fact-tank/2019/09/16/key-facts-about-u-s-hispanics/

Núñez, A-M. (2014). Employing multilevel intersectionality in educational research: Latino identities, contexts, and college access. *Educational Researcher, 43*(2), 85–92. https://doi.org/10.3102/0013189X14522320

Núñez, A-M., & Crisp, G. (2012). Ethnic diversity and Latino/a college access: A comparison of Mexican American and Puerto Rican beginning college students. *Journal of Diversity in Higher Education, 5*(2), 78–95. https://doi.org/10.1037/a0026810

Quiros, L., & Araujo Dawson, B. (2013). The color paradigm: The impact of colorism on the racial identity and identification of Latinas. *Journal of Human Behavior in the Social Environment, 23*(3), 287–297. https://doi.org/10.1080/10911359.2012.740342

Passel, J. S., Lopez, M. H., & Cohn, D. (2022, February 3). *U.S. Hispanic population continued its geographic spread in the 2010s.* Pew Research Center. Retrieved from https://www.pewresearch.org/fact-tank/2022/02/03/u-s-hispanic-population-continued-its-geographic-spread-in-the-2010s/

Poon, O. A., & Segoshi, M. S. (2018). The racial mascot speaks: A critical race discourse analysis of Asian Americans and Fisher vs. University of Texas. *The Review of Higher Education, 42*(1), 235–267. https://doi.org/10.1353/rhe.2018.0029

Ramirez, M. D., & Peterson, D. A. M. (2020). *Ignored racism: White animus toward Latinos.* Cambridge University Press.

Revelo, R. A., & Baber, L. D. (2018). Engineering resistors: Engineering Latina/o students and emerging resistant capital. *Journal of Hispanic Higher Education, 17*(3), 249–269. https://doi.org/10.1177/1538192717719132

Rincón, B. E., & Lane, T. B. (2017). Latin@s in science, technology, engineering, and mathematics (STEM) at the intersections. *Equity & Excellence in Education, 50*(2), 182–195. https://doi.org/10.1080/10665684.2017.1301838

Rodriguez, S. L., Doran, E. E., Sissel, M., & Estes, N. (2022). Becoming *la ingeniera*: Examining the engineering identity development of undergraduate Latina students. *Journal of Latinos and Education, 21*(2), 181–200. https://doi.org/10.1080/15348431.2019.1648269

Saldaña, J. (2016). *The Coding Manual for Qualitative Researchers* (3rd ed.). SAGE.

Steele, C. M. (2010). *Whistling Vivaldi and other clues to how stereotypes affect us.* W.W. Norton.

Torres, V. (2004). The diversity among us: Puerto Ricans, Cuban Americans, Caribbean Americans, and Central and South Americans. *New Directions for Student Services, 105*, 5–16. https://doi.org/10.1002/ss.112

True-Funk, A., Poleacovschi, C., Jones-Johnson, G., Feinstein, S., Smith, K., & Luster-Teasley, S. (2021). Intersectional engineers: Diversity of gender and racial microaggressions and their effects in engineering education. *Journal of Management in Engineering, 37*(3). https://doi.org/10.1061/(ASCE)ME.1943-5479.0000889

Turochy, E., Perez, M. A., Poleacovschi, C., & Doran, E. (2021, July 26–29). *Types of stereotype threats that Latinx students experience in undergraduate engineering education* [Conference presentation]. American Society for Engineering Education Virtual Annual Conference. https://peer.asee.org/37944

Turochy E., Ballesteros, L., Nelson, T.N.T., Perez, M., Estes, K., Poleacovschi, C., Doran, E., & Yuen, T. (2023). Stereotypes and stereotype threats experienced by Latinx undergraduate engineering students. *International Journal of Engineering Education, 39*(5), 1181–1195.

U.S. Census Bureau. (2021). *QuickFacts: United States.* Washington, DC. Retrieved from https://www.census.gov/quickfacts/fact/table/US/PST045221

Valencia, R. R. (2002). "Mexican Americans don't value education!" On the basis of the myth, mythmaking, and debunking. *Journal of Latinos and Education, 1*(2), 81–103. https://doi.org/10.1207/S1532771XJLE0102_2

Villenas, S., & Deyhle, D. (1999). Critical race theory and ethnographies challenging the stereotypes: Latino families, schooling, resilience and resistance. *Curriculum Inquiry, 29*(4), 413–445. https://doi.org/10.1111/0362-6784.00140

5 · EXAMINING THE FUNDS OF KNOWLEDGE THAT SUPPORT LATINX STUDENTS' ENGINEERING IDENTITY DEVELOPMENT AND CAREER CERTAINTY

DINA VERDÍN

The choices people make and their actions to realize those choices are often directed by how they see themselves. Thus, how individuals identify themselves serves as a meaning-making guide to motivate action and regulate behavior (Oyserman et al., 2014). As identity development is a powerful motivational guide, scholars have theorized that developing an engineering identity has important implications for students' continued progression along their engineering pathway (Stevens et al., 2008; Tonso, 2006). Recent scholarship has empirically validated the effect of seeing oneself as an engineer as a heuristic toward (a) supporting students' choice to pursue an engineering major (Godwin & Kirn, 2020; Verdín & Godwin, 2021); (b) engendering positive perceptions about persisting in engineering (Jones et al., 2013; Patrick et al., 2018; Verdín, 2021); and (c) supporting students' identity congruence by fostering a sense of belonging in the discipline (Verdín et al., 2018; Verdín, 2021; Wilson et al., 2010).

Identifying as an engineer can be thought of as a construction process: that is, "thinking about oneself as an engineer, performing an engineer self, and ultimately being thought of as an engineer" (Tonso, 2006, pp. 273–274). While identities can be constructed within the flow of activity, they are also "caught in the tensions between past histories that have settled in them and the present discourses and images that attract them or somehow impinge upon them" (Holland et al., 1998, p. 4). When understood through an asset-based perspective, the past histories that have settled on Latinx students are their historically accumulated and culturally developed bodies of knowledge—obtained through daily activities, experiences, and practices (i.e., funds of knowledge). Esteban-Guitart and colleagues (2021; 2014) theorize that funds of knowledge are resources and repositories that people can access and that help define who they are (i.e., identity development). They postulate that funds of knowledge help define one's identity when actively leveraged (Esteban-Guitart, 2021; Esteban-Guitart & Moll, 2014), thus suggesting a relationship between engineering role identity development and students' funds of knowledge.

The main focus of this chapter is the premise that students' funds of knowledge can be leveraged to help explain how they come to define themselves. Specifically, I examine *how* Latinx engineering students establish their engineering identity by understanding how their funds of knowledge are used as resources in their development process. Additionally, this chapter examines how Latinx students' funds of knowledge and engineering identity inform their beliefs about graduating with an engineering degree, obtaining their desired engineering job, and succeeding in their chosen career path (conceptualized as career certainty).

THEORETICAL FRAMEWORKS

In this section, I discuss the funds of knowledge framework and its shift to the higher education space, particularly in the engineering context. I will also introduce the funds of knowledge survey instrument, developed using ethnographic interviews with low-income, first-generation college students (see Verdín et al., 2021b): this is the first survey instrument of its kind. A discussion of these frameworks is warranted, as the results presented in this chapter offer empirical evidence of how Latinx students' funds of knowledge support their engineering identity development and career certainty.

Funds of Knowledge in Higher Education

Funds of knowledge are conceptualized as the "historically accumulated and culturally developed bodies of knowledge and skills essential for household or individual functioning and well-being" (Moll et al., 1992, p. 133). The funds of knowledge framework acknowledges that working-class households and individuals have rich experiences, cultural practices, resources, and bodies of knowledge as a result of their daily activities (González et al., 2005; Moll et al., 1992). Foregrounding individuals' lived experiences as rich with funds of knowledge dismisses the presumption that students' households and communities can be reduced to being poor economically and in the quality of their experiences (Moll et al., 1992). Instead, students' lived experiences are treated as sources of knowledge, and a family's knowledge, social networks, and resourcefulness are emphasized as assets (González et al., 2005)—assets that can be leveraged to learn engineering. A core tenet of the funds of knowledge framework is to disrupt hegemonic views of working-class families by recognizing and celebrating the utility their knowledge systems can have in students' education (González et al., 2005).

Traditionally, the funds of knowledge framework has served as a heuristic for primary and secondary educators, enabling them to use students' household and community knowledge to support their learning. Through that approach, educators identified students' funds of knowledge and brought them into the classroom to support student learning. However, adults (i.e., college students) also have accumulated bodies of knowledge, practices, experiences, and skill sets through their participation in the workforce and by supporting their family households. Oughton (2010) acknowledges a shift from only viewing children as drawing from their household funds of knowledge to acknowledging that adults also draw from their *own* funds of knowledge. Within this shift, there has been an emphasis on widening the conceptual understanding of funds of knowledge from practical skills to include interpersonal and meta-cognitive skills (Oughton, 2010). More recently, Marquez Kiyama and Rios-Aguilar (2017) have further motivated a call to view adult learners (i.e., college students) as transmitters of their historically accumulated bodies of knowledge. Specifically, they recognize that college students' funds of knowledge are internalized and transformed into capital when they leverage their practices, experience, and skill sets in their adult lives to support their college transition (Rios-Aguilar & Kiyama, 2012).

Smith and Lucena (2016) studied first-generation college students in an engineering context and found that when these students were able to use their knowledge from home or work-related experiences to support their engineering learning, they subsequently felt a sense of belonging in engineering. To further understand the effects of students' funds of knowledge on their learning dispositions, Verdín, together with Smith and Lucena (2021b), categorized ethnographic student data into discrete and community knowledge, practices, and experiences that deliberately supported their engineering coursework and engineering-related work experiences to create a funds of knowledge survey instrument. Some of the funds of knowledge identified in prior work and of particular interest to this chapter are categorized as tinkering knowledge from home, perspective taking, community networks from college friends, and connecting experiences (Verdín et al., 2021b).

These funds of knowledge are not exhaustive, as scholars working in this area have identified multiple funds of knowledge held by minoritized students (see Wilson-Lopez et al., 2016). Nevertheless, they provide the building blocks for investigating how some funds of knowledge may support Latinx engineering students' engineering identity development. Definitions and example survey items for each fund of knowledge construct can be found in table 5.1.

Prior studies have detailed how the hands-on household skill sets students gained as adolescents—their tinkering knowledge from home—served as sources of knowledge in their current engineering coursework or internships (Smith & Lucena, 2016; Verdín et al., 2021b, 2021a). "Perspective taking" as a construct was

TABLE 5.1 Funds of Knowledge Survey Constructs

Funds of knowledge	Definitions	Survey items	Survey scale
Connecting experiences	Students' ability to draw from hobbies or home environment activities to scaffold what they are currently learning in engineering.	Q3a. I see connections between my hobbies and what I am learning in my engineering courses (e.g., design projects, homework, exams, presentations). Q3b. I see connections between experiences at home and what I am learning in my engineering courses. Q3c. I draw on my previous experiences at home when little instruction is given on how to solve an engineering task. Q3d. I draw on my previous experiences from my hobbies when little instruction is given on how to solve an engineering task.	Anchored numeric rating scale of 0 ("completely disagree") to 6 ("completely agree")
Tinkering knowledge: home	Tinkering knowledge from home relates to activities (i.e., repairing, assembling, or building) that students have engaged with in their home environment.	Q4a. At home, I learned to use tools to build things. Q4b. At home, I worked with machines and appliances (considered broadly, e.g., gym equipment, sewing machines, lawn mower, bikes, etc.). Q4c. I learned to fix things around the house (considered broadly, e.g., plumbing, furniture, electrical wiring, etc.) Q4e. At home, I learned to assemble and disassemble things	Anchored numeric rating scale of 0 ("completely disagree") to 6 ("completely agree")
Perspective taking	The capacity to examine a situation or examine another person's experience.	Q6a. I am open to listen to the point of view of others. Q6b. I consider other people's point of view in discussions. Q6c. I like to ask people questions about their experiences. Q6d. I like to view both sides of an issue.	Anchored numeric rating scale of 0 ("very inaccurately") to 6 ("very accurately")
Community network: college friends	Friends made while in college provided advice, resources (i.e., material or non-material), and/or support to aid in their engineering coursework.	Q1d. Friend(s) in my current school have given me advice that helped me in my engineering coursework (e.g., design projects, homework, exams, presentations). Q1e. Friend(s) in my current school have given me resources that helped me in my engineering coursework. Q1f. Friends(s) in my current school have given me emotional support that helped me continue my engineering coursework.	Anchored numeric rating scale of 0 ("not at all true") to 6 ("very true")

developed to capture the cognitive work that occurs as a result of navigating between worlds (i.e., students' households and their lives as adults outside their homes). In Verdín et al.'s (2012a) study, students' funds of knowledge associated with perspective taking and tinkering knowledge from home both supported their choice of pursuing an engineering major. The construct "community network from college friends" was born from the shared experiences of first-generation college students who used their expanded network of college peers as supportive agents throughout their studies.

The construct "connecting experiences" aligns with the recent shift that focuses on students' funds of knowledge when transitioning from secondary to higher education (e.g., Marquez Kiyama & Rios-Aguilar, 2017; Oughton, 2010). Integral to the funds of knowledge framework is the idea that students enter classrooms with lived experiences that have led to knowledge gains (González et al., 2005). Therefore, the construct "connecting experiences" was developed to understand whether students explicitly recognized and applied their accumulated bodies of knowledge in their engineering coursework. Recent work using a sample of minoritized, first-year engineering students found that students' funds of knowledge of connecting experiences supported their choice of pursuing an engineering major (Verdín et al., 2021a). Specifically, when these students could connect their experiences from home to their engineering learning, that connection reinforced their self-efficacy beliefs and interest, which solidified their choice to pursue engineering (Verdín et al., 2021a).

Engineering Role Identity

I will take some time in this section to conceptually discuss the framing of identity used throughout this chapter (i.e., engineering role identity). Role identity is defined as an individual's internalized meaning of the role in which they see themselves (i.e., "I see myself as an engineer"), expectations associated with the role, and attitudes and behaviors adopted to help the individual define themselves in a context (i.e., engineering; Burke & Stets, 2009; Gee, 2001). Roles are the expectations tied to a social position within social structures that guide an individual's attitudes and behaviors (Burke & Stets, 2009; Stryker & Burke, 2000). An individual learns the expectations tied to a role through interactions with others in that community of practice (Burke & Stets, 2009) and through the past histories and present discourses defining what it means to occupy that role (Holland et al., 1998).

Authoring an engineering identity or seeing oneself as an engineer requires that students adopt roles, attitudes, and behaviors that accompany the specific context. For example, a Latinx student who takes on an engineering role identity adopts the meanings and expectations accompanying that particular role and then acts to embody and preserve those meanings and expectations. Since the student sample in this study consisted of undergraduates, the meaning and expectations associated with the engineering role identity were contextualized with regard to their life stage: that is, students in postsecondary education instead of those in a professional environment.

From a student's standpoint, seeing oneself as an engineer requires building a repertoire of knowledge, skills, and confidence in their abilities to perform within the domain (Carlone & Johnson, 2007; Verdín et al., 2018). Beliefs about performing well and understanding content material (i.e., performance and competence beliefs) overlap with the concept of self-efficacy. Performance/competence beliefs can be thought of as general or global attitudinal views of an individual's capabilities—while the attitude that accompanies an individual who sees themselves as an engineer is their *interest* in engaging in activities, topics, or learning within the domain. Interest can be understood as a learner's predisposition to participate in disciplinary content over time and is an essential motivating behavior for persistence and engagement (Hidi & Renninger, 2006; Renninger, 2009; Renninger et al., 1992).

Considering one's interest foregrounds the importance of developing a nonprofessional, STEM-related role identity instead of treating it as an implicit component of how someone comes to define themselves (Hazari et al., 2010). "Recognition" is both an external manifestation and an internal state, both of which are required for identity development (Carlone & Johnson, 2007; Potvin & Hazari, 2013). Gee (2001) notes that the contextual identity a student seeks to take on is further reinforced when "they are recognized by [themselves] or others" (p. 102). Therefore, defining oneself in a given context (i.e., engineering) requires that individuals be "recognized as a certain 'kind of person,' in a given context" based on one's performance (Gee, 2001, p. 99) and based on others' perceptions of what it means to be in that role. Altogether, external recognition, being interested in the subject domain, and believing one can perform well and understand the course material helps develop an engineering role identity.

Researcher Positionality

I am an engineering faculty member who identifies as a Latina of Mexican heritage. I hold multiple degrees in engineering, meaning I have lived through the tensions and struggles of trying to develop an engineering identity amid the masculine hegemonic culture of engineering. My own lived experiences have made me acutely aware of how societal and institutional structures constrain and facilitate how I came to define myself as an engineer. As a result, one of my goals is to reconstruct the discourse around what it means to identify as an engineer and to facilitate the development of an engineering identity for marginalized students.

METHODOLOGY

Sample Demographics

Data were drawn from a cross-sectional survey administered to engineering students enrolled in nine four-year institutions in the West, South, and Mountain regions of the United States. Among the participating institutions, five are classi-fied as Hispanic-Serving Institutions. The analysis for this chapter focuses exclu-sively on students who identified as Latinx, $n = 226$ out of 816. Of these, 181 (80 percent) were first-generation college students (i.e., their parent[s] had less than a bachelor's degree). In addition, 149 Latinx students (66 percent) received or had received a Pell Grant in the past, a proxy typically used to identify students who come from low-income households. The majority of our sample (172 [76 percent]) indicated that they were non-transfer students, while 54 (24 percent) had transferred to their current institution. Most students (139 [62 percent]) reported that English was not the primary language spoken at home. The majority of the sample (146 [65 percent]) was male, while 80 (35 percent) identified as female. There was a broad range in years of enrollment: 41 (18 percent) were enrolled in their first year, 37 (16 percent) in their second year, 67 (30 percent) in their third year, and 81 (36 percent) in their fourth year or higher. Sixty-one percent were enrolled in a Hispanic-Serving Institution.

Survey Measures

The engineering role identity measures used in this analysis focus on interest, rec-ognition, and performance/competence beliefs and have been extensively used with engineering undergraduate students (Godwin & Kirn, 2020; Verdín, 2021; Verdín & Godwin, 2021). Performance/competence beliefs were initially theo-rized as two separate constructs intended to capture students' perceptions of their performance and knowledge (i.e., competence) in a subject area (see Carlone & Johnson, 2007). However, repeated analyses focused on understanding the factor structure of the constructs concluded that both constructs were conceptualized similarly by students (i.e., loaded onto one factor; Cribbs et al., 2015; Godwin, 2016; Hazari et al., 2010). The Cronbach's alpha value for the engineering identity constructs were between .78 and .88. The four funds of knowledge constructs used in this study have been validated using a large sample of undergraduate first-generation college students studying engineering. The process of gathering evi-dence to validate the four constructs used in this study is extensively discussed in prior work (Verdín et al., 2021b).

Table 5.1, shown earlier, provides the definitions, survey items, and anchored numeric rating scales used in the four funds of knowledge constructs (i.e., con-necting experiences, tinkering knowledge from home, perspective taking, and community network from college friends). The Cronbach's alpha value for the four funds of knowledge constructs were between .76 and .90., indicating accept-able internal consistency (Cronbach, 1951).

Career certainty refers to the "degree to which individuals feel confident, or decided, about their occupational plans" (Hartung, 1995, p. 1). This construct is a holistic measure of career certainty, capturing students' confidence in graduating with a degree, obtaining a job in their respective discipline, and having confidence in their occupational knowledge and career success. The construct "career certainty" consists of the following survey items: "I feel certain about graduating with an engineering degree," "I feel certain about getting a job in my desired engineering discipline," "I am certain I have the knowledge needed to be successful in my career," and "I am certain about being successful in my chosen career path." All response measures used in the analyses were measured on a seven-point anchored numeric scale based on students' level of agreement (i.e., strongly disagree to strongly agree). The Cronbach's alpha value for career certainty was .85.

Data Analysis Procedure

Structural equation modeling is a technique that helps researchers test and model complex networks of relationships. In this study, I examined the relationship between Latinx students' funds of knowledge, engineering identity, and their career certainty. Preliminary steps were taken to ensure the quality of the final model. Specifically, assumptions of univariate normality were examined and found to be within acceptable ranges (i.e., ± 2.0 for skewness and ± 7.0 for kurtosis; West et al., 1995). However, these data did violate assumptions of multivariate normality, as examined using Mardia's test (Korkmaz et al., 2014). Given the violation of multivariate normality, a Satorra–Bentler ($SB\chi^2$) mean-adjusted test statistic was used (Satorra & Bentler, 2010). All analyses were conducted in R statistical programming software, version 3.5.3 (R Core Team, 2019), through the lavaan package (Rosseel, 2012).

Measurement Model and Structural Model Fit

A structural equation model has two components: the measurement model focuses on the constructs' factor loadings, and the structural model focuses on the regression paths. Table 5.2 provides a summary of the parameters associated with the measurement model. The goodness of fit using a Satorra–Bentler adjusted chi-square test for the measurement model was $SB\chi^2 = 511.87$, $df = 349$, $p < .001$. The fit indices suggest a good overall model fit: a comparative fit index (CFI) of 0.94, Tucker–Lewis index (TLI) of 0.93, root mean square error of approximation (RMSEA) of 0.045 with a 90 percent CI [confidence interval; 0.038, 0.053], and standardized root mean squared residual (SRMR) of 0.059. The factor loadings were all above the recommended range of 0.45 (Tabachnick & Fidell, 2013); for most of the items, their reliabilities were above .50 (Brown, 2015), and the average variance extracted (AVE) were all above 0.50 (Kline, 2016).

The overall measure of goodness of fit for the structural model was $SB\chi^2 = 564.84$, $df = 332$, $p < .001$. The fit indices of the final model were as follows: a CFI of 0.94, TLI of 0.93, RMSEA of 0.045 with 90 percent CI [0.038, 0.052], and SRMR of 0.06. All the fit indicators listed (i.e., CFI, TLI, RMSEA, and SRMR) provide evidence to support the claim that the final model is a good fit. This chapter will not detail the overall model fit parameters, but they are stated to assure the reader that the final model presented is adequate and specified through best practice metrics (Brown, 2015; Kline, 2016; Satorra & Bentler, 2001).

TABLE 5.2   Unstandardized Loadings (Standard Errors), Standardized Loadings, and Significance Levels for Each Parameter in the Structural Equation Model (n= 226)

Parameter estimate	Unstandardized (std. error)	Standardized	p	Item reliability	AVE
Factor Loadings					
Exp → Q3a	0.91(0.11)	0.62	< .001	0.38	.51
Exp → Q3b	1.17(0.10)	0.68	< .001	0.46	
Exp → Q3c	1.19(0.10)	0.75	< .001	0.56	
Exp → Q3d	1.33(0.10)	0.78	< .001	0.60	
TinkH → Q4a	1.54(0.08)	0.86	< .001	0.74	.70
TinkH → Q4b	1.43(0.10)	0.84	< .001	0.71	
TinkH → Q4c	1.59(0.08)	0.85	< .001	0.72	
TinkH → Q4e	1.40(0.08)	0.85	< .001	0.72	
Persp → Q6a	0.76(0.07)	0.78	< .001	0.61	.53
Persp → Q6b	0.78(0.05)	0.86	< .001	0.74	
Persp → Q6d	0.76(0.06)	0.78	< .001	0.61	
Persp → Q6e	0.65(0.08)	0.55	< .001	0.30	
NetFri → Q1d	1.03(0.11)	0.78	< .001	0.61	.52
NetFri → Q1e	1.05(0.10)	0.76	< .001	0.58	
NetFri → Q1f	1.06(0.12)	0.65	< .001	0.42	
Recognition → Q10d	0.75(0.11)	0.54	< .001	0.30	.58
Recognition → Q10e	1.27(0.09)	0.85	< .001	0.72	
Recognition → Q10f	1.22(0.09)	0.84	< .001	0.71	
Int → Q10h	0.58(0.11)	0.86	< .001	0.74	.77
Int → Q10i	1.03(0.11)	0.98	< .001	0.81	
Int → Q10j	1.00(0.12)	0.89	< .001	0.96	
PC → Q10k	1.30(0.09)	0.90	< .001	0.81	.65
PC → Q10l	1.14(0.10)	0.82	< .001	0.67	
PC → Q10m	1.11(0.10)	0.70	< .001	0.49	
PC → Q10n	1.08(0.08)	0.82	< .001	0.67	
Certain → Q12a	0.80(0.14)	0.64	< .001	0.41	.61
Certain → Q12b	1.14(0.12)	0.78	< .001	0.61	
Certain → Q12d	1.10(0.08)	0.81	< .001	0.66	
Certain → Q12e	1.12(0.10)	0.87	< .001	0.76	

NOTE. Exp = connecting experiences; TinkH = tinkering knowledge from home; Persp = perspective taking; NetFri = network from friends at school; Recognition = engineering external recognition; Int = engineering interest; PC = engineering performance/competence beliefs; Certain = career certainty

FINDINGS

The final model depicted in figure 5.1 helps bridge two important theoretical concepts, funds of knowledge and engineering identity development, to understand how they inform perceptions of career success. In a structural equation model, relationships are read from left to right, and scholars typically discuss their results through

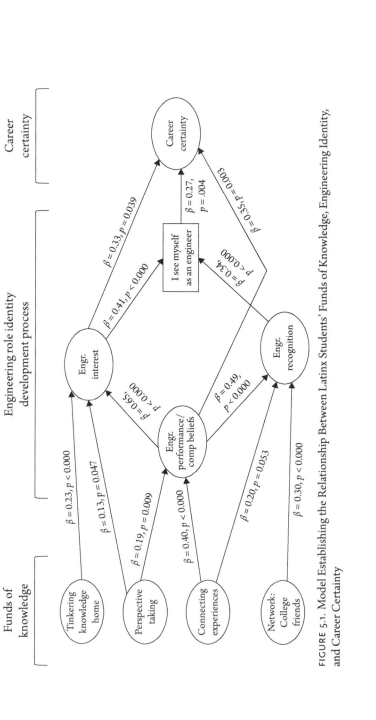

FIGURE 5.1. Model Establishing the Relationship Between Latinx Students' Funds of Knowledge, Engineering Identity, and Career Certainty

the same approach. However, I have organized the discussion of my findings differently. I will start by discussing the empirical relationships between the components that support engineering role identity development (i.e., interest, recognition, and performance/competence beliefs). Prior work has not empirically established these relationships with Latinx engineering students. Nevertheless, examining whether the same claims about identity development hold for this specific minoritized student group is important. I then discuss the factors within the engineering identity development process that inform Latinx students' career certainty. I also draw awareness to the connection between students' funds of knowledge and career certainty when considering engineering role identity as a mediational process.

The shapes and directional arrows in figure 5.1 are purposeful. The arrows represent relationships analogous to simple regression: the latent construct or variable receiving the arrowhead is predicted by the construct or variable transmitting the arrow. Latent variables are represented as ovals, and each latent construct has two or more observable measures. The observable measures associated with each latent construct are not shown in the figure. Squares represent practical measures, that is, survey questions that were directly asked to participants. The model shows only one observed measure (i.e., "I see myself as an engineer"). All values reported in the final model are in standardized form.

Performance/Competence Beliefs, Interests, and Recognition Inform Latinx Students' Engineering Role Identity Development Process

The process that helps promote Latinx students' engineering identity development can be understood as occurring through an interplay between three consequential components. That is, when Latinx students feel they are confident in their abilities to perform and understand engineering concepts, that performance/competence belief in turn reinforces their interest ($\beta = 0.63$, $p < .000$) and prompts recognition by others (i.e., instructors, parents, peers; $\beta = 0.50$, $p < .000$). In turn, through students' interest in engineering and external recognition, they come to see themselves as engineers ($\beta = 0.41$, $p < .000$, and $\beta = 0.34$, $p < .000$, respectively).

Latinx students' beliefs about performing well in engineering and understanding engineering content (labeled "engr. performance/competence beliefs") do not directly influence their perceptions of themselves as engineers (i.e., "I see myself as an engineer"). This draws awareness to the reality that it is not enough that Latinx students feel efficacious in their performance/competence abilities: their intrinsic interest in engineering needs to exist, and their bids for recognition need to be validated. Notably, the engineering identity finding in figure 1 challenges prior assertions that external recognition is the strongest predictor of engineering identity (e.g., Godwin & Kirn, 2020), an assertion made with an engineering sample composed predominantly of white males. Overall, the outcome of this model reinforces our understanding of how the identity development process—that is, performance/competence beliefs, interest, and external and internal recognition—interplay specifically for Latinx engineering students. This finding underscores that it is *equally* important to continue fostering Latinx students' engineering interests and affirming their view of themselves as engineers through external recognition, as these both inform their self-definition.

Engineering Role Identity Development Process and Its Effect on Career Certainty

Empirical findings reported in this chapter reinforce the importance of developing an engineering role identity, as it bolsters Latinx students' certainty of being successful in their chosen career path. Specifically, Latinx students' internalized views of themselves as engineers (i.e., "I see myself as an engineer") significantly supported their certainty of being successful in their chosen career path ($\beta = 0.27$, $p = .004$). Two of the components that promote identity development, interest and performance/competence beliefs also helped inform students' career certainty ($\beta = 0.33$, $p = .039$ and $\beta = 0.35$, $p = .003$, respectively). Interestingly, receiving recognition from influential others did not directly impact students' career certainty. Findings from this model and prior work underscore the importance of considering the effect of the individual mechanisms that inform identity development, not just the overall outcome of identifying as an engineer (i.e., "I see myself as an engineer").

When zooming out and examining the model holistically, one can see that career certainty is also impacted by four funds of knowledge, albeit indirectly through the mediational effect of the engineering identity development process (i.e., funds of knowledge → identity development process → career certainty). Students' funds of knowledge support the development of an engineering role identity. In turn, the process of identifying oneself as an engineer supports students' certainty of being successful in their careers. The model sheds light on an indirect relationship between students' funds of knowledge and career certainty when mediated by the engineering identity development process. Altogether, 80 percent ($R^2 = 0.797$) of the proportion of the variance for career certainty can be explained by the engineering identity development process.

Funds of Knowledge That Help Inform Engineering Identity Development

An important contribution to the engineering identity literature is the understanding of the drivers that help inform (a) how minoritized engineering students come to develop their performance/competence beliefs and interest and (b) how they receive external recognition. For example, the performance/competence beliefs construct gives a general view of students' capabilities to perform well in their coursework or in engineering-related areas, yet it does not elucidate *how* Latinx students came to develop this belief. The engineering recognition construct captures students' beliefs that their instructors, parents, and peers see them as engineers. While performance/competence beliefs inform engineering recognition, this is only a partial view of how students come to be recognized as engineers. Similarly, concerning engineering interest, while it is partially informed by Latinx students' performance/competence beliefs, more needs to be understood about how they came to develop their general interest or enjoyment in learning engineering. Performance/competence beliefs, interest, and recognition can be cultivated by the available experiences and how learners perceive, understand, and represent those experiences. The findings presented in this chapter point to four funds of knowledge that inform the engineering identity development process. These funds of knowledge are a constellation of interpersonal skills (i.e., perspective taking); practices and skill sets acquired from home (i.e., tinkering knowledge from home and connecting experiences); and the strategic use of community resources (i.e., network from college friends).

Latinx students' interpersonal perspective-taking skills and their ability to connect their experiences from home both support their beliefs about performing well and understanding relevant engineering knowledge ($\beta = 0.19$, $p = .009$, and $\beta = 0.40$, $p < .000$, respectively). Notably, in this model, Latinx students' ability to (a) see connections between their home experiences and what they are learning in engineering and (b) leverage experiences from home to support their learning (i.e., connecting experiences) was more than twice as influential to their performance/competence beliefs. The proportion of the variance explained for performance/competence beliefs was 24 percent ($R^2 = 0.24$). The fund of knowledge of perspective taking and connecting experiences mildly helped explain how Latinx students came to believe they could perform well and understand engineering coursework. Nevertheless, this finding is a significant contribution, as studies examining how students' funds of knowledge inform their performance beliefs are sparse.

Interest in engineering was cultivated through Latinx students' home experience of using tools, working with machines, and assembling/disassembling things at home (i.e., tinkering knowledge from home) and through their perspective-taking capability ($\beta = .25$, $p < .000$ and $\beta = .13$, $p < .047$, respectively). The two funds of knowledge (i.e., perspective taking and tinkering knowledge from home) helped explain 65 percent ($R^2 = 0.65$) of the variance for the variable engineering interest. My findings further reinforce how Latinx students' everyday home practices of tinkering with household items and the interpersonal skill of considering other people's situations or life circumstances promote their engineering interest.

When trying to understand how Latinx students' funds of knowledge inform how they are recognized as engineers, I found that the advice, support, and resources provided by their school peers (i.e., network from college friends) afforded them positive external recognition as engineers ($\beta = .30$, $p < .000$). Students' use of their home knowledge to scaffold their learning (i.e., connecting experiences) allowed them to be externally recognized as engineers ($\beta = .20$, $p = .053$). The variance explained for students' engineering recognition was largely informed, at 52 percent ($R^2 = 0.519$), by two funds of knowledge: connecting experiences and network with college friends.

DISCUSSION

How individuals define themselves in a context has important implications for their behaviors, attitudes, and ways of being. Specifically, people look to their identities or how they define themselves in a given context (e.g., engineering) to make choices and to provide a meaning-making lens to their actions (Oyserman et al., 2014). The engineering education community has broadly adopted the understanding that establishing an identity as an engineer is instrumental for students' persistence in the field. Moreover, recent scholarship has confirmed that one's identity as an engineer supports students' engineering career aspirations (Rohde et al., 2019) and persistence beliefs (Verdín, 2021; Verdín & Godwin, 2021). While prior work has informed us that students' identity is developed through a process that includes the mechanisms of performance/competence beliefs, interest, and external recognition (Carlone & Johnson, 2007; Cribbs et al., 2016; Godwin, 2016; Hazari et al., 2010), there is insufficient empirical evidence documenting how the identity development process and its corresponding survey items behave with a sample of Latinx engineering students. Latinx students enter engineering with unique knowledge and lived experiences, and this repertoire is tied to how they define and position themselves and are positioned by others. Thus, understanding how Latinx students develop an engineering role identity requires an examination of the "historically accumulated, culturally developed, and socially distributed" practices, beliefs, and knowledge that students use to help author their self-definitions (Esteban-Guitart & Moll, 2014, p. 31).

A novel contribution from this chapter is a further understanding of how Latinx students' lived experiences, understood as their funds of knowledge, help inform the identity development process. Findings from this chapter point to four distinct funds of knowledge constructs: tinkering knowledge from home, perspective taking, connecting experiences, and networking with college friends. Perspective taking is essential in an engineer's work, as it is directly connected to empathetic reasoning and practice (Hess et al., 2016). The empathic practice produced through perspective taking is a necessary skill of the engineers' work, helping them to define and generate solutions to societal problems (Leydens & Lucena, 2017; Zoltowski et al., 2012). Allowing Latinx engineering students the opportunity to capitalize on this fund of knowledge and apply it in a classroom will subsequently help affirm their performance/competence beliefs and solidify their interest in engineering.

Students can more readily be afforded or denied recognition through interaction with members of the engineering community of practice. For example, college friends play a vital role in supporting engineering students' navigation and persistence through engineering (Denton & Borrego, 2021; Simmons & Martin, 2014). My model further builds on those findings by underscoring how the strategic use of college friends positively supports Latinx students' bids for recognition. Specifically, college friends were a resource that Latinx students drew on to support them in their engineering coursework. This network gave Latinx students the necessary advice, resources, and emotional support to help them with difficult course material.

Additionally, findings from my model show that when Latinx students are able to connect and capitalize on their home experiences to scaffold their learning, that connection and capitalization can increase their confidence in performing well in their engineering coursework and support their bids for recognition. Learning engineering content and gaining confidence in one's abilities to understand course material should not be restricted to classroom environments. Students' lived experiences and their home and work environments should be leveraged to help bridge engineering ways of knowing and doing. In engineering, students have also been enculturated into believing there is one way of knowing and doing engineering. How a profession bounds its epistemologies has implications for how students see themselves in relation to that profession. Scholars have long documented how engineering knowledge shifted to rigidly encompass only the scientific, analytical, or knowledge obtained through "scholarly means" (as synthesized in Kant & Kerr, 2019), leaving no room for practical knowledge learned at home or as working adults. However, findings from this chapter point to the importance of elevating practical knowledge obtained through hands-on experiences or from home as equally valuable as theory-based knowledge.

The final contribution I will emphasize is the effect that students' engineering identity has on their career certainty. Not much is understood about how an engineering identity impacts Latinx students' career outcome expectations. In this chapter, career outcome was conceptualized as certainty about one's chosen career path; it goes beyond beliefs about graduating with an engineering degree to incorporate Latinx students' conviction to obtain their desired engineering job, hold the necessary knowledge to perform that job, and, broadly, succeed in their desired career pathway. Latinx students' interest, performance/competence beliefs, and perceptions of themselves in relation to the role of an engineer all closely influence their certainty about their chosen career path. Therefore, it is essential that educators focus on promoting Latinx students' continued interest in the field, nurture confidence in their performance abilities, and reinforce their internalized views of themselves as engineers.

Implications for Practice

The deliberate use of Latinx students' funds of knowledge has implications for how they come to define themselves as engineers. Current opportunities for Latinx students to connect their funds of knowledge with their engineering coursework may be limited by dominant epistemologies defining what "counts" as engineering knowledge. The messages students receive about the types of learning experiences, or ways of knowing, that are valued in engineering can affect their engagement in engineering and how they come to define themselves as engineers. Engineering is recognized as the application of scientific knowledge, and this understanding of engineering leaves little room for the application of home knowledge to inform scientific practices. However, overlooking the value and utility of Latinx students' funds of knowledge in their engineering learning undercuts the opportunity to support their engineering identity development and, ultimately, their career certainty.

First-year engineering design courses, found in most institutions with engineering programs, are an ideal avenue for infusing students' funds of knowledge into their design thinking and problem-solving approach. A conversation could be had with this group of eager first-year engineering students about how engineering ways of knowing and doing not only rigidly encompass calculus, physics, or other theory-based knowledge but also include the everyday practices learned from home experiences. Instructors can be more intentional in asking students to reflect on how they apply hands-on and interpersonal strategies and practices from home to solve their design challenges. These early messages that transmit the value and utility of their funds of knowledge can reinforce how they see themselves fit within the engineering community of practice.

CONCLUSION

For Latinx students to see themselves as engineers, it is essential to understand the necessary factors within the broader process of developing an engineering role identity. When unpacking the first step in their identity development process, we can see that Latinx students need to feel confident and capable of performing well in their engineering coursework. The positive beliefs about their performance and the competence they acquire, in turn, reinforce their interest in engineering and afford them recognition by others as individuals who can be engineers. Interest in engineering and receiving recognition by others subsequently helps explain how Latinx students come to define themselves as engineers.

Nevertheless, *how* Latinx students obtained confidence in their performance beliefs was an area that needed further exploration. In addition, performance/competence beliefs alone do not wholly explain *how* Latinx students foster their interest in the field or *how* they come to be granted recognition by peers, family, and instructors. The novel contribution from this chapter connected the *how* component of the identity development process to students' funds of knowledge. The funds of knowledge framework is an asset-based perspective that underscores how minoritized students come from communities rich with knowledge, skills, and practices. Moreover, the knowledge, skills, and practices acquired from home or through students' participation in the world as adults can help inform how they come to see themselves as engineers. Collectively, the findings from this chapter suggest that funds of knowledge can be a basis for the construction and continuous development of an engineering identity among Latinx students. Capitalizing on students' funds of knowledge to learn engineering content offers a student-centered approach that supports the engineering identity development process and, indirectly, the certainty of their chosen career path. Engineering courses remain the site where most engineering problem-solving and epistemologies are learned, which contribute to engineering students' self-definitions as problem solvers (Lucena & Leydens, 2015)—thus presenting a unique and as yet untapped opportunity for the integration of students' funds of knowledge with course content.

REFERENCES

Brown, T. A. (2015). *Confirmatory factor analysis for applied research* (2nd ed.). The Guilford Press.

Burke, P. J., & Stets, J. E. (2009). *Identity theory*. Oxford University Press.

Carlone, H. B., & Johnson, A. (2007). Understanding the science experiences of successful women of color: Science identity as an analytic lens. *Journal of Research in Science Teaching, 44*(8), 1187–1218. https://doi.org/10.1002/tea.20237

Cribbs, J. D., Cass, C., Hazari, Z., Sadler, P. M., & Sonnert, G. (2016). Mathematics identity and student persistence in engineering. *International Journal of Engineering Education, 32*(1), 163–171.

Cribbs, J. D., Hazari, Z., Sonnert, G., & Sadler, P. M. (2015). Establishing an explanatory model for mathematics identity. *Child Development, 86*(4), 1048–1062. https://doi.org/10.1111/cdev.12363

Cronbach, L. J. (1951). Coefficient alpha and the internal structure of tests. *Psychometrika, 16*(3), 297–334. https://doi.org/10.1007/BF02310555

Denton, M., & Borrego, M. (2021). Navigational capital of Latinx engineering transfer students: A qualitative study. *Journal of Women and Minorities in Science and Engineering, 27*(2), 61–86. https://doi.org/10.1615/JWomenMinorScienEng.2020035154

Esteban-Guitart, Moises. (2021). Advancing the funds of identity theory: A critical and unfinished dialogue. *Mind, Culture, and Activity, 28*(2), 169–179. https://doi.org/10.1080/10749039.2021.1913751

Esteban-Guitart, Moisès, & Moll, L. C. (2014). Funds of Identity: A new concept based on the Funds of Knowledge approach. *Culture and Psychology, 20*(1), 31–48. https://doi.org/10.1177/1354067X13515934

Gee, J. P. (2001). Identity as an Analytic Lens for Research in Education. In W. G. Secada (Ed.), *Review of Research in Education* (Vol. 25, pp. 99–126). American Educational Research Association. https://doi.org/10.3102/0091732X025001099

Godwin, A. (2016, June 26–29). *The development of a measure of engineering identity.* [Conference presentation]. American Society for Engineering Education Annual Conference & Exposition, New Orleans, LA, United States. https://doi.org/10.18260/p.26122

Godwin, A., & Kirn, A. (2020). Identity-based motivation: Connections between first-year students' engineering role identities and future-time perspectives. *Journal of Engineering Education, 109*(3), 362–383. https://doi.org/10.1002/jee.20324

González, N., Moll, L. C., & Amanti, C. (2005). *Funds of knowledge: Theorizing practices in households, communities, and classrooms.* Routledge.

Hartung, P. J. (1995). Assessing career certainty and choice status. *ERIC Digests.* https://www.counseling.org/resources/library/eric%20digests/95-19.pdf

Hazari, Z., Sonnert, G., Sadler, P. M., & Shanahan, M. C. (2010). Connecting high school physics experiences, outcome expectations, physics identity, and physics career choice: A gender study. *Journal of Research in Science Teaching, 47*(8), 978–1003. https://doi.org/10.1002/tea.20363

Hess, J. L., Fila, N. D., & Purzer, S. (2016). The relationship between empathic and innovative tendencies among engineering students. *International Journal of Engineering Education, 32*(3), 1236–1249.

Hidi, S., & Renninger, K. A. (2006). The four-phase model of interest development. *Educational Psychologist, 41*(2), 87–98. https://doi.org/10.1207/s15326985ep4102_4

Holland, D., Lachicotte, W., Skinner, D., & Cain, C. (1998). *Identity and Agency in Cultural Worlds.* Harvard University Press..

Jones, B. D., Ruff, C., & Paretti, M. C. (2013). The impact of engineering identification and stereotypes on undergraduate women's achievement and persistence in engineering. *Social Psychology of Education, 16*(3), 471–493. https://doi.org/10.1007/s11218-013-9222-x

Kant, V., & Kerr, E. (2019). Taking stock of engineering epistemology: Multidisciplinary perspectives. *Philosophy and Technology, 32*(4), 685–726. https://doi.org/10.1007/s13347-018-0331-5

Kline, R. B. (2016). *Principles and practices of structural equation modeling* (4th ed.). The Guilford Press.

Korkmaz, S., Goksuluk, D., & Zararsiz, G. (2014). MVN: An R package for assessing multivariate normality. *R Journal, 6*(2), 151–162. https://doi.org/10.32614/rj-2014-031

Leydens, J. A., & Lucena, J. C. (2017). *Engineering justice: Transforming engineering education and practice.* John Wiley & Sons.

Lucena, J. C., & Leydens, J. A. (2015, June 14–17). *From sacred cow to dairy cow: Challenges and opportunities in integrating of social justice in engineering science courses* [Conference presentation]. American Society for Engineering Education Annual Conference & Exposition, Seattle, WA. https://doi.org/10.18260/p.24143

Marquez Kiyama, J., & Rios-Aguilar, C. (2017). *Funds of Knowledge as a Culturally Responsive Pedagogy in Higher Education.* Routledge.

Moll, L. C., Amanti, C., Neff, D., & Gonzalez, N. (1992). Funds of knowledge for teaching: Using a qualitative approach to connect homes and classrooms. *Theory into Practice, 31*(2), 132–141. https://doi.org/10.1080/00405849209543534

Oughton, H. (2010). Funds of knowledge—A conceptual critique. *Studies in the Education of Adults, 42*(1), 63–78. https://doi.org/10.1080/02660830.2010.11661589

Oyserman, D., Elmore, K., & Smith, G. (2014). Self, self-concept, and identity. In M. R. Leary & J. Prince Tangney (Eds.), *Handbook of self and identity* (2nd ed., pp. 69–104). The Guilford Press.

Patrick, A. D., Borrego, M., & Prybutok, A. N. (2018). Predicting persistence in engineering through an engineering identity scale. *International Journal of Engineering Education, 34*(2), 351–363.

Potvin, G., & Hazari, Z. (2013). The development and measurement of identity across the physical sciences. *2013 Physics Education Research Conference (PERC) Proceedings,* 281–284. American Association of Physics Teachers. https://doi.org/10.1119/perc.2013.pr.058

R Core Team. (2019). *R: A language and environment for statistical computing.* R Foundation for Statistical Computing, Vienna, Austria. https://www.R-project.org/

Renninger, K. A. (2009). Interest and identity development in instruction: An inductive model. In *Educational Psychologist* (Vol. 44, Issue 2, pp. 105–118). https://doi.org/10.1080/004615 20902832392

Renninger, K. A., Hidi, S., & Krapp, A. (1992). *The role of interest in learning and development.* Psychology Press.

Rios-Aguilar, C., & Kiyama, J. M. (2012). Funds of knowledge: An approach to studying Latina(o) students' transition to college. *Journal of Latinos and Education, 11*(1), 2–16. https://doi.org/10.1080/15348431.2012.631430

Rohde, J. A., Verdin, D., Doyle, J., Godwin, A., Kirn, A., Benson, L., & Potvin, G. (2019). Investigating the intersection of career aspirations and engineering beliefs in first year engineering students. *2018 IEEE Frontiers in Education Conference (FIE)*, San Jose, CA, USA, 2018, pp. 1–8. https://doi.org/10.1109/FIE.2018.8659311.

Rosseel, Y. (2012). lavaan: An R package for structural equation. *Journal of Statistical Software, 48*(2), 1–36. https://doi.org/10.18637/jss.v048.i02

Satorra, A, & Bentler, P. M. (2010). Ensuring positiveness of the scaled chi-square test statistic. *Psychometrika, 75*(2), 243–248. doi: 10.1007/s11336-009-9135-y.

Satorra, Albert, & Bentler, P. (2001). A scaled difference chi-square test statisti. *Psychometrika, 66*(4), 507–514. https://link.springer.com/content/pdf/10.1007%2FBF02296192.pdf

Simmons, D. R., & Martin, J. P. (2014). Developing effective engineering fictive kin to support undergraduate first-generation college students. *Journal of Women and Minorities in Science and Engineering, 20*(3), 279–292. https://doi.org/10.1615/JWomenMinorScienEng.2014010979

Smith, J. M., & Lucena, J. C. (2016). Invisible innovators: How low-income, first- generation students use their funds of knowledge to belong in engineering. *Engineering Studies, 8*(1), 1–26. https:doi.org/10.1080/19378629.2016.1155593

Stevens, R., O'Connor, K., Garrison, L., Jocuns, A., & Amos, D. M. (2008). Becoming an engineer: Toward a three dimensional view of engineering learning. *Journal of Engineering Education, 97*(3), 355–368. https://doi.org/10.1002/j.2168-9830.2008.tb00984.x

Stryker, S., & Burke, P. J. (2000). The past, present, and future of an identity theory. *Social Psychology Quarterly, 63*(4), 284–297. https://doi.org/10.2307/2695840

Tabachnick, B. G., & Fidell, L. S. (2013). *Using multivariate statistics* (6th ed.). Pearson Education Inc.

Tonso, K. L. (2006). Student engineers and engineer identity: Campus engineer identities as figured world. *Cultural Studies of Science Education, 1*(2), 273–307. https://doi.org/10.1007/s11422-005-9009-2

Verdín, D. (2021). The power of interest: Minoritized women's interest in engineering fosters persistence beliefs beyond belongingness and engineering identity. *International Journal of STEM Education, 8*(33), 1–19. https://doi.org/10.1186/s40594-021-00292-1

Verdín, D., & Godwin, A. (2021). Confidence in pursuing engineering: How first-generation college students' subject-related role identities support their major choice. *2021 IEEE Frontiers in Education Conference (FIE)*, Lincoln, NE, USA, 2021, pp. 1–9. http://doi.org/10.1109/FIE49875.2021.9637157.

Verdín, D., Godwin, A., Kirn, A., Benson, L., & Potvin, G. (2018, April 29–May 2). *Understanding how engineering identity and belongingness predict grit for first-generation college students* [Conference paper]. 2018 CoNECD—The Collaborative Network for Engineering and Computing Diversity Conference, Crystal City, VA. https://peer.asee.org/29589

Verdín, D., Godwin, A., & Ross, M. (2018). STEM roles: How students' ontological perspectives facilitate STEM identities. *Journal of Pre-College Engineering Education Research (J-PEER), 8*(2), 4. https://doi.org/10.7771/2157-9288.1167

Verdín, D., Smith, J. M., & Lucena, J. C. (2021a). Funds of knowledge as pre-college experiences that promote minoritized students' interest, self-efficacy beliefs, and choice of majoring in engineering. *Journal of Pre-College Engineering Education Research, 11*(1), Article 11. https://doi.org/10.7771/2157-9288.1281

Verdín, D., Smith, J. M., & Lucena, J. C. (2021b). Recognizing the funds of knowledge of first-generation college students in engineering: An instrument development. *Journal of Engineering Education, 110*(3), 671–699. https://doi.org/10.1002/jee.20410

West, S. G., Finch, J. F., & Curran, P. J. (1995). Structural equation models with nonnormal variables: Problems and remedies. In R. H. Hoyle (Ed.), *Structural Equation Modeling: Concepts, Issues, and Applications* (pp. 56–75). Sage Publications, Inc.

Wilson, D. M., Bell, P., Jones, D., & Hansen, L. (2010). A cross-sectional study of belonging in engineering communities. *International Journal of Engineering Education, 26*(3), 687–698.

Wilson-Lopez, A., Mejia, J. A., Hasbún, I. M., & Kasun, G. S. (2016). Latina/o adolescents'
    funds of knowledge related to engineering. *Journal of Engineering Education,* 105(2), 278-311.
    https://doi.org/10.1002/jee.20117
Zoltowski, C. B., Oakes, W. C., & Cardella, M. E. (2012). Students' ways of experiencing human-
    centered design. *Journal of Engineering Education, 101*(1), 28–59. https://doi.org/10.1002/j
    .2168-9830.2012.tb00040.x

6 · HOW LATINX STUDENTS ENGAGE SOCIAL AND NAVIGATIONAL CAPITAL TO RESIST EXCLUSIONARY ENGINEERING EDUCATION

RENATA A. REVELO AND JANICE MEJÍA

Engineering as a field and an academic discipline has long been described as innovative and life improving. Yet, it has also been characterized as rigorous, challenging, and a place of exclusion (Camacho & Lord, 2013; Cross, 2020), privileging individualism and meritocracy (Baber, 2015; Pawley et al., 2018), and harboring a "chilly climate" (Hall & Sandler, 1982). Engineering tends to attract Latinx students in high numbers, and Latinx students tend to have high aspirations for pursuing engineering. [1] However, graduation and retention numbers continue to be inequitable for these students in US higher education across the board (National Center for Science and Engineering Statistics [NCSES], 2023). As a result, engineering institutions have not historically served Latinx students equally, leaving students to find ways of navigating institutional structures that were not created for them. Such navigation led to the formation of organizations like the Society of Hispanic Professional Engineers (SHPE), established in 1974 in Los Angeles. Students found ways to persist in their degrees, despite institutional barriers and real and perceived obstacles. This navigation through engineering academic programs at Historically White Institutions is a form of resistance, perhaps driven by survival (McGee et al., 2019), as are other ways of navigation that students have been forced to employ. This chapter reviews ways that Latinx undergraduate students resist these exclusive environments, navigating their engineering education and maintaining their sense of engineering identity in the process.

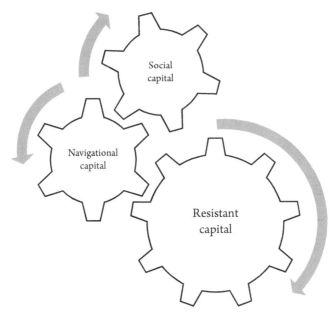

FIGURE 6.1. Interactions between Navigational and Social Capital
That Activate Resistant Capital

To start, we review how we perceive the interplay of resistant capital, navigational capital, and social capital in the stories of engineering students. This
dynamic between the various forms of capital is not linear, is always complex, is
unique for each individual, and is also discussed in greater detail later in the chapter, as shown in figure 6.1.

RESISTANCE IN ENGINEERING (RESISTANT CAPITAL
AND NAVIGATIONAL CAPITAL)

Perhaps unsurprisingly to the reader, the topic of resistance is quite sparse in the
engineering literature. Activism and advocacy are rarely welcomed within the
field of engineering in the United States, nor are they seen as aligning with a field
that is intended to be value-neutral or objective (Riley, 2008). The incongruence
between engineering and aspects of social justice may lie in engineering's military
roots, as it was used as a tool for national defense against foreign enemies (Lucena,
2011; Riley, 2008). The pursuit of such a goal, coupled with military culture, fosters
a tendency to conform to the norms and directives of those in power, instead of
challenging them to create political and social change (Baillie, 2006). While the
history of militarism in engineering is entrenched, its presence in modern engineering is less apparent or visible and more covert or subtle. However, certain engineering fields still uphold the notion that the innovation and glory of engineering
stem from objectivity—as if engineering is undertaken in a vacuum (e.g., apolitical

and acultural). Stripping these aspects of humanity from the doing of engineering and its solutions creates a space wherein it is much harder for engineers to operate from a place of advocacy or activism. Nonetheless, in recent years, there has been a growing body of literature within engineering education that uses resistance concepts or frameworks (e.g., Holly Jr, 2021; McGee, 2021; Rincón et al., 2020).

In this section, we synthesize scholarship that has investigated resistance as it pertains to engineering students (including those aggregated with the larger STEM population) and their marginalization in and navigation through engineering. We focus on the intersection of engineering with marginalization because Latinx students have been a minoritized group of people throughout the professional existence of engineering in the United States. This intersection frames the study and conversation in this chapter. When referring to "resistance," we include "resistant capital" from the Community Cultural Wealth (CCW) framework (Yosso, 2005), as well as other resistance theories and related concepts (e.g., the resilience and oppositional behavior that students use to navigate engineering education; Giroux, 1983; Solorzano & Bernal, 2001).

While resistance as a framework has been underutilized and resistant capital has been understudied, many scholars have discussed resistance concepts—such as resilience in the field and oppositional behavior (McGee, 2021). Before delving into the literature on resistance, we summarize a common theme across the studies: that while individuals' resistance, as a form of capital, is an asset and should be nurtured, supported, and celebrated, it is not a sustainable form of institutional change. As people who resist forces of oppression or marginalization, students use psychological and physiological resources to enact their resistant capital—resources that could be used elsewhere if these forces were eliminated by the institution. In other words, it should not be the students' job to dismantle these forces of oppression or marginalization; rather, the institution should be undertaking systemic work to do so, informed by students' experiences and ways of knowing.

Communities of color often utilize resistant capital to counter exclusionary culture in engineering education. While some of the ways students do this may not have been conceptualized via a resistant capital lens in the literature, they nevertheless exemplify this concept. For example, in their study of Latinx students in STEM, Rincón et al. (2020) found that students engage in "moves" of disengagement and negotiation to resist within and navigate STEM education. These moves were initiated by instances of incongruence between students' values (often based on community and collectivistic goals) and institutional values (often based on meritocratic and individualistic goals). Similarly, studies have also looked at how faculty of color employ resistant capital. Black engineering faculty use strategic moves to engage in voluntary service and address racial inequities within their disciplines: sometimes they do this on their terms instead of serving in normative ways (e.g., diversity, equity, and inclusion committees; McGee et al., 2022). These strategic moves of engagement and disengagement can be viewed as a way to implement their own practices that may have an immediate and direct effect on students of color, and to resist normative practices that may be slow to produce results. This strategic behavior on behalf of students and engineering faculty exemplifies ways to navigate institutions by opposing norms that have not historically benefited communities of color.

Unfortunately, the exclusionary nature of engineering is such that, for some groups, mere survival can be a form of resistance. Moore III et al. (2003) found that African American males "became even more determined and committed to engineering when they perceived that their intellectual capability in engineering was doubted or slighted" (p. 70). "Proving them wrong" became a mantra or a driving force for completing their engineering education and excelling in it; this kind of resistance strategy has also been shown to be used in mathematics (McGee & Martin, 2011) and other STEM fields, including engineering (McGee, 2021). As argued by McGee (2015), such a strategy may represent a fragile professional (e.g., engineering) identity rather than a robust one. In other words, although acting in an oppositional way to prove someone wrong may still lead to persistence and retention, it may come at the high cost of professional identity development. Energy spent on oppositional behavior could instead be devoted to integrated engineering identity development. Nonetheless, in their studies, McGee and others remind us that communities of color—in particular, Black students— possess the resistant capital to utilize and navigate these chilly environments. While these studies provide an insight about the experiences of Black students, more work around resistance and Latinx students in engineering is therefore needed.

Another form of resistant behavior can be found in the ways that communities of color navigate institutions that were not built with them in mind. In doing so, they are resisting existing structures and norms for success. Rincón and Rodriguez (2021), as an example, found that students enact navigational capital to persist in engineering and maintain their success in the field. The authors argue that, instead of the onus being on the student to know how to navigate these systems, it should be on the institution to know its impact on students and to understand how to adapt to the student body—thus, changing the institutional norms for success instead of changing the student. Similarly, Rodriguez et al. (2020) found that Latinas used their persistence as a form of resistance. In their study of Latinx engineering transfer students, Denton and Borrego (2021) report that students found opportunities to navigate engineering and redefine success through their own persistence strategies. In their study of the navigational capital of African American engineering students at Predominantly White Institutions, Damas and Benson (2021) found that students enacted navigational capital in similar ways as those described by Denton and Borrego (2021). However, none of the authors emphasize, studied, or discuss how this form of navigation (wherein the students change their behavior to move through engineering) maintains the institution's status quo for success.

SOCIAL CAPITAL AND LATINX ENGINEERING STUDENTS

"Social capital" consists of an individual's social networks and the resources that can be accessed within those networks (Lin, 2001). The social group or class to which an individual belongs determines the size and strength of their networks as well as their capacity to acquire and access social capital. Within education, parents are a fundamental source of social capital for students since they provide information and resources to navigate educational systems (Coleman, 1988). While these studies are within the fields of sociology and education, social capital has also been studied in engineering education. For example, Mobley, Shealy, and Brawner (2013) found that engineering students whose parents do not have a college degree or experience in the US higher education system may be unable to access this resource as readily as students with college-educated parents; they may therefore rely on other forms of capital to navigate the educational system.

Postsecondary education scholars have examined how access to social capital may reproduce class and gender inequality in mathematics and science (Holland & Eisenhart, 1990). Within engineering education, minoritized women students may be treated as outsiders in social networks, since white males have traditionally dominated the field (Tate & Linn, 2005). These students are less likely to have access to social relationships within these networks and the individuals who are considered "insiders," capable of sharing information and informal resources that lead to academic success in engineering and other STEM fields (Stevens et al., 2008). It is important to understand the social networks within engineering education and the types and forms of social capital that can be both accessed and activated within these spaces, by whom, and why.

Prior research has conceptualized social capital as the informal interactions that students have with faculty and other institutional agents, including administrators and staff members (Dika & Singh, 2007; Perna & Titus, 2005; Stanton-Salazar, 2001). An institutional agent is "an individual who occupies one or more hierarchical positions of relatively high-status and authority" (Stanton-Salazar, 2011, p. 1067); this individual may transmit, negotiate, or provide institutional resources, services, or support. In a study completed in Puerto Rico, Dika (2012) found that college students' perceptions of the quality of relations with faculty were more important than the frequency or quantity of those interactions. Other studies have found that the quality and nature of student interactions with faculty may vary based on student demographics. For example, minoritized, first-generation, and working-class students have reported feeling hesitant to initiate contact with white faculty and may have less positive and frequent interactions (Noel & Smith, 2006; Moschetti & Hadley, 2008). Students may be more likely to initiate communication with professors that are seen as supportive and accessible. When faculty take the lead on cultivating these relationships and creating welcoming and supportive spaces, it helps remove the onus placed on minoritized engineering students to learn how to navigate the engineering academic climate: this is imperative for academic persistence.

For Latinx students in engineering, studies have found that there are multiple sources of social capital. Faculty are an important and primary source, since they may serve as role models—having successfully navigated the educational system—and can provide both formal and informal resources and guidance (Leslie et al., 1998; Martin et al., 2013). Leslie et al. (1998) found that students who completed engineering and science degrees also tended to emphasize the contributions of a faculty member to their success. Furthermore, Cole and Espinoza (2008) note that faculty support and encouragement may be considered one of the most important college experiences impacting the academic performance of Latinx students in STEM. Other studies have found that graduate students who serve as teaching assistants may also provide social capital by sharing information about graduate school and careers in engineering as well as informal resources that help students feel more confident about their undergraduate experience (Martin et al., 2012). Professional engineering organizations, including the National Society of Black Engineers (NSBE) and SHPE, also provide social capital by helping students transition to higher education in a supportive and welcoming environment within cocurricular and extracurricular contexts (Daily et al., 2007; Ross & McGrande, 2016; Revelo & Baber, 2018). Next, we introduce and discuss a more comprehensive framework that includes and extends beyond social capital, and is designed for and by communities of color. We note that access to social capital enables navigational capital and resistant capital within engineering education contexts for Latinx students.

FORMS OF CAPITAL AND WEALTH

In this section, we describe CCW as the foundational framework for resistant capital and the various forms of capital that communities of color possess and utilize. Yosso (2005) conceptualized CCW into a framework that incorporates six forms of capital (i.e., resistant, familial, aspirational, navigational, social, and linguistic). Driven by critical race theory, CCW was developed to challenge racism and "reveal" cultural wealth. In the educational setting, this framework helps to oppose notions of deficit about students of color, and instead to promote understanding of the wealth that students of color bring to educational institutions. Within engineering education, there is a recent wave of work that specifically promotes the use of critical frameworks such as CCW (Dietz et al., 2022) to better approach equity and justice in the field via praxis (Mejia et al., 2018) and justice-oriented approaches (Holly Jr & Quigley, 2022).

The various forms of capital that are part of CCW, while presented here individually, can interact and overlap with one another. Yosso (2005) states that "these various forms of capital are not mutually exclusive or static, but rather are dynamic processes that build on one another as part of community cultural wealth" (p. 77).

Within the context of engineering, a Latinx student at a Predominantly White Institution may use their social capital to build a network of peers and support to create community and navigate a chilly or icy climate, meanwhile enacting their aspirational capital when faced with obstacles in their engineering journey. Similarly, a Latinx student at a Minority-Serving Institution may use their aspirational capital to serve as a role model for others in the institution and create a network of people who want to effect change within their engineering departments.

In the next section, we provide a brief overview of each of the forms of capital conceptualized by Yosso, with an example of how it may be viewed within the engineering context. Note that the example is provided only for context—it is not meant to limit the definition of the particular form of capital in any way.

Resistant Capital

As defined by Yosso (2005), resistant capital entails "knowledges and skills fostered through oppositional behavior that challenges inequality" (p. 80). This definition is grounded in literature that asserts communities of color as those that are capable, and often must oppose normative structures in various ways. Focused on higher education and students, resistant capital may be used by students of color to navigate institutional systems that were not established to serve them. Engineering is not the exception. Latinx students continue to be a minoritized group within engineering education, and resistance capital can be seen in the way that they resist normative ways of conceptualizing engineering identity.

Social Capital

Social capital refers to the "networks of people and community of resources" (Yosso, 2005, p. 79). The social capital to which an individual has access lies within the network of their social relationships (Bourdieu, 1986). Studies have shown that social capital has been linked to increased involvement in campus life (Ellison et al., 2007; Trice, 2004). College social networks include but are not limited to academic programs and student organizations. These networks provide the exchange of socioemotional goods that include information, support, advice, and validation (Glass & Gesing, 2018). When students view themselves as members of these networks and exchange socioemotional goods, they create a sense of community on campus. For underrepresented students, ethnic or cultural student organizations also provide a way to experience and examine naturally occurring social networks that exist within and beyond the university (McFaul, 2016; Rose-Redwood & Rose-Redwood, 2013) and in similar ways to their familial capital (Yosso, 2005).

Familial Capital

Familial capital "refers to those cultural knowledges nurtured among *familia* (kin) that carry a sense of community history, memory, and cultural intuition" (Yosso, 2005, p. 79). This form of capital includes family as a central and supportive component of a student's college trajectory, but it expands the definition of family to include extended family and the community. For Latinx engineering students, ties to family are critical to their success in their engineering journey, as are the creation of family-like bonds with others at their institution. For example, in a separate strand of this project, we found that Latinx students developed family-like bonds with peers and advisors as a way to build community and navigate the field (Revelo Alonso, 2015). As part of community well-being, Latinx engineering students may be deeply interested in the direct applications of their degrees to uplifting communities and individuals (Revelo Alonso, 2015).

Linguistic Capital

Linguistic capital "includes the intellectual and social skills attained through communication experiences in more than one language or style" (Yosso, 2005, p. 78). For Latinx engineering students, such linguistic capital may entail bilingualism, including Spanish and Indigenous languages. Communication skills such as storytelling, joke telling, and communication via the use of proverbs (*dichos*) may be used by Latinx engineering students, especially in the face of adversity (e.g., chilly or icy climates in engineering) and as a way to persist.

Aspirational Capital

Aspirational capital "refers to the ability to maintain hopes and dreams for the future, even in the face of real and perceived barriers" (Yosso, 2005, p. 77). As an example, in another strand of this work, it was found that Latinx students used aspirational capital to serve as role models for other engineers and to sustain their success and persistence in the field of engineering (Revelo Alonso, 2015).

Navigational Capital

Navigational capital "refers to skills of maneuvering through social institutions" (Yosso, 2005, p. 80). Specifically, this form of capital is enacted in social institutions that may be racially hostile, such as higher education institutions. Given engineering's exclusionary history, navigation capital can be crucial for students' success in this context. While exclusionary practices may not be overt, they can be in place in covert but powerful ways, designed to exclude students of color from engineering. These can include recruitment practices, retention or probation policies, resource allocation to programs that support students of color, and advising practices.

Current Study

In this chapter, we primarily focus on navigational, resistant, and social capital to address the following research question: What are the ways that students use their social capital in order to enact their resistant capital to navigate engineering education? We present all forms of capital, as conceptualized in the CCW framework, to provide the reader with an overview of the framework that drove this work.

STUDY DETAILS AND METHODS

The research presented in this chapter is drawn from a broader study that addressed a research question about bridging engineering culture,[2] cultural wealth, and engineering identity for Latinx undergraduate students. In this chapter, we focus on discussing resistance, resistant capital, and social capital as it pertains to the larger research study. Specifically, we address the ways in which students use their social capital to enact their resistant capital to navigate engineering education. Thus, we provide a summary of the details of the larger study from which the data are drawn. The full methodological and study details can be found in Revelo Alonso (2015). To address the research question in this chapter, we used the data from the qualitative portion of the larger mixed methods study.

In addition to CCW, the science identity model (Carlone & Johnson, 2007) and engineering identity concepts were used to drive instrument development (including the interview protocol) and data analysis. Methods and findings about the engineering identity aspect of this work can be found in Revelo Alonso (2015) and Revelo (2015). For the data shared in this chapter, the authors used an established codebook to answer the research question and identify themes around resistance that was enabled by social capital. Some of these results overlap with those shared in Revelo and Baber (2018) and were reanalyzed and examined through other forms of capital captured by the CCW framework, to address the research question for this chapter.

Participants

Participants were undergraduate students from across the United States, recruited from SHPE—a national professional organization in the United States. Recruitment commenced in 2013. The first author attended the SHPE's national conference and recruited students there, with the approval of and assistance from the SHPE national headquarters. Eligible study participants included students who identified as Latina/o, Chicana/o, or Hispanic and those who had been members of SHPE for at least one year. From the national recruitment, purposeful sampling was used to select interview participants based on diversity of major, gender, and institution type (Hispanic-Serving Institution vs. non-Hispanic-Serving Institution).

There were twenty participants in this study, all of whom were undergraduates at various institutions in the United States at the time of the interviews. All the students identified as Latina/o* and traced their heritage to Latin American countries, the majority being Mexico, Puerto Rico, the Dominican Republic, and Cuba. Most of the students identified their families, both immediate and extended, as their primary support system throughout their engineering education. Thirteen out of twenty were first-generation college students, defined as students whose parents did not complete a baccalaureate degree, growing up in bilingual or primarily Spanish-speaking households.[3]

Their stories, as an aggregate, represent the stories of young adults excited to pursue a "prestigious"[4] engineering degree while navigating institutional and life barriers along the way. They were intrinsically motivated to pursue engineering; math and science were often their strength courses in high school, regardless of whether their high schools prepared them for a "rigorous"[5] engineering curriculum. At the same time, they understood the social mobility that came with a career in engineering, as well as the responsibility of achieving a degree not just for themselves, but also for their families and community. Most of the participants (fifteen out of twenty) attended institutions that were not built for them (e.g., Predominantly White Institutions, private institutions); those who attended Hispanic-Serving Institutions tended to have more supportive institutional agents (e.g., faculty, mentors), yet they still had some barriers that they were navigating to achieve their engineering degree. Eight of the twenty students were transfer students. Table 6.1 presents the participants' demographic information.

TABLE 6.1 Study Participant Demographics

Pseudonym	Gender	Year in university	Engineering major	University location	HSI status	First-generation college status	Transfer status
Karina	Female	4th	Mechanical	New York	Non-HSI	No	No
Emily	Female	2nd	Electrical	Missouri	Non-HSI	No	No
Cosmo	Male	4th	Mechanical	Florida	HSI	Yes	Yes
Cesar	Male	4th	Computer science	Illinois	Non-HSI	No	No
Edgar*	Male	5th	Architectural	Kansas	Non-HSI	Yes	Yes
Linda	Female	4th	Material science	Illinois	Non-HSI	Yes	No
Bob	Male	5th	Civil	Illinois	Non-HSI	Yes	No
Joaquin*	Male	4th	Mechanical	Ohio	Non-HSI	No	No
Isabel*	Female	1st	Civil	New Jersey	Non-HSI	Yes	No
Fernanda*	Female	4th	Computer	New York	Non-HSI	Yes	Yes
Mike	Male	3rd	Electrical	Illinois	Non-HSI	Yes	Yes
Hector	Male	4th	Mechanical	Colorado	Non-HSI	Yes	Yes
Anthony	Male	4th	Mechanical	Colorado	HSI	Yes	Yes
Carol	Female	3rd	Mechanical	Pennsylvania	Non-HSI	No	Yes
Robby	Male	6th	Electrical	California	Non-HSI	Yes	Yes
Manolo	Male	6th	Mechanical	California	Non-HSI	No	No
Ivan	Male	3rd	Chemical	Puerto Rico	HSI	No	No
Jacob	Male	6th	Mechanical	California	HSI	Yes	No
Luis	Male	3rd	Mechanical	California	Non-HSI	Yes	No
Mike II	Male	2nd year	Electrical	Texas	HSI	Yes	No

* Pseudonyms were chosen by the first author

Data Collection

Twenty participants were interviewed either in person or via video conference. A semi-structured interview protocol was followed, which included questions about engineering identity, the educational journey, and cultural perspectives of engineering identity. The development of the interview protocol was guided by the CCW framework. The interviews were on average fifty minutes, transcribed verbatim by the first author, and analyzed using HyperRESEARCH software. Inductive and deductive data analysis was used in line with the theoretical frameworks guiding the work. The first author performed individual coding, followed by expert checking with two engineering education scholars, alongside member checking— all the study participants were presented with a set of overarching themes for the broader study and had the opportunity to provide feedback and choose a pseudonym for their quotes. Feedback from the expert and member checking was used to strengthen the codebook, align it with the theory being applied, and enhance the study's trustworthiness (Lincoln & Guba, 1986).

Data Analysis

The interview data were individually coded by the first author, involving two cycles of coding (Saldaña, 2015) and the development of a codebook. Inductive coding (Miles & Huberman, 1994) entailed coding the data for dimensions and aspects of the CCW framework that addressed the research question for the larger study. A list of the codes analyzed for this chapter is included in table 6.2. Note that this is only a partial view of the codebook, given that the larger study addressed a broader research question. Once coding was completed, the first author engaged in theme development, guided by the research question addressed in this chapter: specifically, the first author reviewed all the data for instances coded with the codes summarized in table 6.2. Partial inclusion of codes in this study was done only with instances that addressed the research question.

REFLEXIVITY

As mentioned earlier, this work was driven by the need to promote institutional (rather than individual) change in engineering education. Both authors are engineering educators and have been working in various higher-education roles, having held multiple responsibilities (e.g., administrating at program and institutional levels; teaching college preparatory, undergraduate, and graduate courses; overseeing and designing engineering education curricula for multiple departments; developing extra curricula; recruiting students; and providing student support through academic advising) for a total of twenty-two years, collectively. Both authors acknowledge their role in promoting institutional change while being part of multiple higher education public and private institutions. Moreover, both have long had a vested interest in conducting research that benefits marginalized

TABLE 6.2   Partial Codebook Addressing the Research Question

Code/subcode	Description	Included in this study
CCW/resistant capital	Instance of enacting, acknowledging, or describing ways of resisting institutional practices as it relates to undergraduate engineering experience	Yes
CCW/navigational capital	Instance of enacting, acknowledging, or describing ways of navigating engineering education	Partial
SHPE Experience/"not all about you" (in vivo)	External motivation for engineering education beyond oneself	Yes
CCW/social capital	Access to resources and people as it relates to undergraduate engineering experience	Yes
Journey as an engineer/ challenging or difficult	Recounting of challenging or difficult experiences or moments as they relate to undergraduate engineering experience	Yes
SHPE experience/help community or giving back	Instance of helping the community or giving back to the community as it relates to undergraduate engineering experience or engineering identity	Yes
CCW/familial capital	Instance of enacting, acknowledging, or describing familial capital	Partial

engineering students, in particular Latinx students, with an understanding that such research often benefits all students. Both authors have had marginalizing experiences while attending Predominantly White Institutions and feel empowered by being able to bring an insider perspective to this type of research; at the same time, we feel the responsibility of being agents of change in our respective institutions.

FINDINGS

In this section, we highlight results that exemplify the use of resistant capital, enabled by social capital that facilitated navigational capital. We decided to analyze the data from this perspective, to begin exploring the overlaps or connections between the different forms of CCW. We found that students used their social capital to access and activate resistant capital, in order to ensure their success in engineering and navigate engineering education. Students exemplified the following forms of resistance, which were enabled by social capital in their navigation of exclusive engineering spaces: survival as a form of resistance, community as a form of resistance, and peers as a form of resistance. These uses of resistant capital were enabled by social capital and, while they were not necessarily acts of activism or transformational resistance (Solorzano & Bernal, 2001), they were uses of CCW necessary to resist exclusionary culture and develop a strong sense of identification with the field of engineering.

Survival (Not Just for Me) as a Form of Resistance

Students were used to overcoming obstacles and being resilient in the face of real barriers that could have prevented them from pursuing or completing their engineering degrees. This was evident in their precollege journeys as well as their time at higher-education institutions. Surviving could in part be encapsulated by persistence, but these are not synonyms in that students should not have to use survival skills to persist in engineering. As such, we view survival as a form of resistance because these students saw their existing and persisting in engineering as a responsibility to achieve, not just as another achievement in life. This responsibility to achieve was often connected to a responsibility to others, most saliently family, and also to their membership in a larger group (e.g., being Mexican American). In the quote below, Bob, a civil engineering student at a Predominantly White Institution, recounts his motivation for pursuing and persisting in engineering as one driven by breaking stereotypes. He also shares that he, and people like him, must overcome numerous obstacles. When queried further, Bob clarified that he was referring to his Mexican American community in the city where he grew up: "I look at myself as opposed to other students or other people around me, it's a lot of um overcoming obstacles, may they be directly implementing by other people or just indirectly, or in an indirect fashion either through an assumption or through a stereotype."

Similarly, Hector, a first-generation mechanical engineering transfer student at a Predominantly White Institution shares that being part of an ethnic student organization and "passing the torch to the younger generations" was an accomplishment for both the group and for him (as a leader of the group). However, the focus of the achievement was not on the achievement itself, but on the meaning for Latinx people: "You know, another great accomplishment for the [student group] chapter and to kind of, what I like to say too, is pass the torch to the younger generations, you know, that STEM is important, but you don't have to work in the field, you know, there are opportunities there. I think that was definitely a big moment and one that was really heartfelt as well."

Survival, as a form of persistence and perseverance, was also fueled by family. Family was a motivation for students to enroll in college and persist, even in the face of barriers. Survival was not just for the individual but also for the family. Edgar, a first-generation architectural engineering student, recounts being motivated to achieve in engineering because of his parents and siblings.

Well, growing up my parents always pushed my siblings and I to do the best that we can and to take advantage of all the opportunities presented to us and so my parents were my motivation for my journey to college because I wanted, because I've always wanted to do well for them and other people that have played an important part of my journey to college have been my siblings. I've always wanted to do well for my siblings because I wanted to be the . . . I wanted to be the person that my siblings look up to, I wanted to be a positive example for my siblings and so both my parents and my siblings have been an important motivation throughout my journey in college and to graduate from college.

Social capital, in the form of access to people (e.g., peer groups, large group membership, family), was the fuel necessary to use and enact resistant capital in the form of survival. Students conceptualized their need to survive their engineering journeys as something that they were doing not just for themselves but for others. Other people were key players in the students' activation and use of resistant capital to persist in engineering.

Community as a Form of Resistance

Unfortunately, for the majority of participants (excluding some attending Hispanic-Serving Institutions), the climate within their engineering schools (colleges and departments) required the creation of community spaces that incorporated Latinx culture as a critical survival strategy for resisting normative values in and navigating engineering. This creation of community included adopting their peers as *familia*—brothers and sisters—to take on the engineering journey together, as exemplified in the quote below: "And those are the people that, well we stuck with each other almost the entire year. Of course, everyone starts to go off their own way later on in the college careers, but they were like my brothers and sisters at the time so . . . we took care of each other."

The community also included purposeful outreach events with middle and high schools to increase the number of Latinx students in engineering. Undergraduate students provided examples of many volunteer outreach events being held exclusively by them—some with help from their institution—to recruit for engineering. This activity was not just motivated by altruism but also by the drive to change the face of engineering. Many of these students, but primarily those at Predominantly White Institutions, were "the only ones" in their classrooms, nonethnic student organizations, research labs, and other engineering spaces. They felt the need to change that for other students via community efforts. They bore the responsibility to give back to others and to "pay it forward," in their continuation of a communal effort to succeed in engineering. Success was thus not just an individual achievement but rather a group achievement that required a group effort to achieve.

In the quotation below, Emily, a second-year electrical engineering student at a Predominantly White Institution, discussed how because there were not many women or Latinx students in her classes, she felt the need to be part of SHPE to have a group "of people who are more like [her]." When asked what she meant by people who are more like her, she added: "Uh-huh. I don't really know how to explain it. Mmm. The way I was raised, it's just very different than most of the people here. I'm not actually from this area. I'm from Tennessee and I was also raised with a more Hispanic, different household and I see that a lot more with the SHPE members than I do with some of the other students on campus." Community as a form of resistance is enabled by social capital. Specifically, students created and used peer networks to nurture success in engineering. They also worked as a group, connected in family-like relationships, to run outreach events with middle and high school students that could lead to changing the face of engineering. While acknowledging the support and resources that they received from older students in the professional engineering organization (SHPE) of which they were part, they also continued to share information, support, and social capital with younger students and communicate how to access and navigate undergraduate engineering at their institution.

Peer Role Models as a Form of Resistance

In previous work (Revelo & Baber, 2018), the authors discussed how students' engaging in successive role modeling exemplified their use of resistant capital. Successive role modeling entails having access to a role model and then becoming a role model as a way to "give back." Here, we analyze role modeling from a social capital perspective. While the majority of participants did not have direct, non-peer engineering role models in their engineering journeys, they did have access to peer role models through a Latinx engineering student organization. Having access to peer role models was groundbreaking for some of the students. As an example, Bob (the civil engineering student) recounts how it felt to find other "Latinos" in engineering who were achieving in the way he imagined his own achievements.

> When I came [to this university] and I met members from SHPE, I was like "Man, these people are almost like gods because they look like me, they spoke like me, but yet they didn't have–they basically–they shattered that stereotype of Latinos not being able to achieve whatever it is to achieve or at least Latinos over-exceeding the expectations." And that was to me, for me at least, it was a single person's journey because that's what I was doing. They always considered me *the* smart Latino even though I'm also a person–I'm not just categorized by my ethnicity.

The majority of the students did not have engineering role models before or during college, but they found the most accessible role models to be other Latinx students. We see this as an example of students carving their path in engineering to navigate the educational system, being led by peer Latinx engineering students, and resisting alongside them.

SO WHAT? A TARGETED DISCUSSION OF OUR FINDINGS

In answering the research question, we found that social capital and navigational capital were driving forces for students to enact resistant capital. As shown in figure 1, for resistant capital to be utilized or activated, students must access social capital in order to understand their navigational capital options. The relationship between these three forms of capital is not linear and may be influenced by other forms of capital within the CCW framework in a nuanced way for each student: based on their access to, activation of, and experience with each form of cultural wealth. In this way, a student can use navigational capital to enact social capital as a means to resist. For example, a student navigating a chilly climate in their engineering lab may seek guidance from their equity and inclusion program to address the issue and promote change in the lab. A student could also use social capital to enact navigational capital as a means to resist: for example, enlisting peers in the major to sign a petition to hire more than one faculty member to teach a required course, if the only instructor has a history of weeding out (Suresh, 2006), pushing out, or simply failing students via this course as a way to maintain their imagined standards for the major.

These uses of social and navigational capital are not permanent, as different forms of capital can interact in multiple ways. The way that students enable resistant capital through the interaction of social and navigational capital can also be situational and temporal. In the example of "survival as a form of resistance," students utilized support from family and peer groups (social capital) to navigate chilly environments (navigational capital) as a form of resistance in engineering. In this case, social capital enabled navigational capital in order to enact resistant capital. Conversely, in the "peer role models as a form of resistance" example, students found peer role models (social capital) while navigating chilly environments (navigation capital) and then served as such to help others do the same (social capital), drawing on their personal experiences and what they had learned from their older peers. The interactions of social and navigational capital to drive resistant capital can be simultaneous, successive, or iterative. Nonetheless, such interaction was integral to students' activation and use of resistant capital in their pursuit of an engineering education.

While our analysis focused on the presence of resistant capital, we posit that students can have an interaction of social capital and navigational capital without activating resistant capital. This aspect is not reflected in our findings because we did not address it as a research question; however, from our review of the relevant codes shown in table 6.2, we note examples where students discussed interactions of social and navigational capital without activating resistant capital. We argue that the absence of resistant capital in these cases may be due to the need that students feel to acculturate. Acculturation has been conceptualized as a process of socialization into the field of engineering, but the negative side effects (e.g., attrition) of acculturation have not necessarily been studied (Joseph, 2014). Some recent calls for funding explicitly welcome studies on acculturation into the engineering profession. In cases where the interaction of social and navigational capital does not activate resistant capital, a student may stop their navigation through engineering education at acculturation—using social and navigational capital to further the status quo in engineering and not necessarily to enact change within. We are not suggesting that it is a student's responsibility to change the institution, but instead that without the activation of resistant capital, the result may be acculturation into the field.

In agreement with the first author's previous work on resistant capital (Revelo & Baber, 2018), we found that Latinx engineering students use and enact their resistant capital in their pursuit of an engineering education. As discussed at length in our previous work (Revelo & Baber, 2018), while this form of resistant capital may not be transformational resistance, it is a form of resistance that can be activated within the bounds of engineering. In answering the research question addressed in this chapter, we found that social and navigational capital are key activators of resistant capital for Latinx engineering students. This analysis also allowed us to further incorporate the institutional environment as an enveloping force, wherein social and navigational capital interact. More specifically, the interactions of navigational and social capital discussed in our findings are there as a result of a marginalizing, alienating, and exclusionary environment and culture in engineering.

Connected to previous work (Revelo & Baber, 2018), we found that students may not be poised to reach transformative resistance in such institutions or microinstitutional environments (e.g., colleges or departments); such a culture is so pervasive that going against it is a task that extends beyond a student's engineering trajectory. In other words, Latinx students—especially those who are part of an ethnic organization promoting seemingly positive aspects of acculturation (e.g., professional and leadership development)—may find it easier to resist by participating in a professional engineering organization similar to SHPE so that they have a supportive space to persist and graduate. Engineering students may not know how to question the specific space that they are trying to learn from and actively participate in as students. We posit that all engineering students hope to learn the norms of the engineering profession, but not all have the same experience based on their access to social networks and capital. Therefore, those students need to carve out new ways to survive and resist engineering education norms that were not designed to include their experiences. Different academic outcomes are linked to disparate and unequal sources of social capital. As engineers, we can and should redesign these systems to ensure that all students have access to the same resources, formal and informal, to succeed.

WHAT CAN I DO? IMPLICATIONS FOR ENGINEERING EDUCATORS

Two of the main questions that came to mind during this work, and that were often asked when presenting it, were "How do we put this into practice? What can we do with these findings?" The work was born from a necessity to address something deeper than answering a research question or moving forward with a framework (i.e., engineering identity, as in an earlier manuscript; Revelo Alonso, 2015). While these were indeed driving forces, the work has always been about what can be done on the ground from an institutional perspective. In this section, we want to provide some meaningful takeaways for institutional agents such as staff, faculty, and administrators, and all institutional individuals who influence students' livelihood and success in engineering. Our goal with these takeaways is to affect systematic, institutional change and to place the onus of change on the institution rather than the student.

- Resistance capital is not a commodity; it is an asset that engineering students, primarily students of color, are often forced to engage or enact to be successful. As an asset, it is important to acknowledge and nurture it but not necessarily to expect its use. The mere presence and use of resistance capital indicate that there is something to be resisted. Instead of attempting an exhaustive list of action items, we provide some implications below as they relate to one of the themes in this chapter. In light of the differences in context across institutions, we urge the reader to focus on understanding what is being resisted and why, considering students' perspectives and developing solutions alongside all stakeholders. In the example provided earlier of the theme "community as a form of resistance," we see that students felt the need to create a community among themselves because the existing institutional climate (e.g., department, college) was not welcoming of them. Administrators, in particular, could (continue to) devote time and energy to ensuring community across and not just within particular spaces. A community for Latinx students can entail meaningful curriculum connections to projects in the community (Mejia & Revelo, 2022); inclusion and acknowledgment of family as a visible and critical component of students' success and motivation in college (Revelo Alonso, 2015); and curricular time to acknowledge culture as it relates to learning and doing engineering (Castaneda & Mejia, 2018; Mejia & Wilson-Lopez, 2016).
- As in the theme "peer role models as a form of resistance," students who do not see themselves reflected in the curriculum, student body, faculty or instructors, programming, or success practices figure out their own path to be successful in the field. In the findings presented, this path entailed seeking role models from their peers, then becoming peer role models themselves. Institutional agents such as faculty and administrators could (continue to) devote time and energy to dismantling acculturation into the field and promoting integration into the field. In this way, students would not only see themselves reflected in engineering education in the ways mentioned earlier but they would also see the various ways that people—who may look or act like them—are successful in the field.
- As captured by the theme "survival (not just for me) as a form of resistance," students' values connected to motivation and persistence in engineering

(e.g., motivated by family and community to work in the field) may not be in sync with institutional values to ensure student success. Students draw from collectivistic rather than individualistic values and culture. The latter is often associated with the field of engineering, and such incongruence may lead students to feel disconnected from their field. Institutional agents could devote time to (re)constructing educational components (e.g., curriculum, programs) to address this incongruence.

CONCLUSION

To conclude this chapter, instead of providing concluding words to summarize our work, we ask the reader to take the time to reflect on and write down action items with which they can engage locally, regionally, and nationally, whether in a group or at a personal level, to effect positive change for Latinx students in engineering education.

With regard to reflection on what students are resisting and why, engineering educators can focus energy on what they can do locally: as bell hooks, in her brilliant series of public dialogues hosted by The New School in New York City (see Swan, 2014), reminds us, "Start where you are"—quoting the work of Chodron (2021). To change the culture that pushes people away from engineering, ask the following questions: What can you change as an engineering educator to foster community in the classroom? To allow students to grow as professionals and also as individuals, without stripping their identity to receive an engineering degree? To take care of each other? Instead of telling students, "Look to your right; look to your left; only one of you will make it"—a far-too-common practice in engineering schools to communicate a culture of competition, meritocracy, and individualism—could we instead tell them, as Lin-Manuel Miranda told a group of young singers in 2021 on the set of "Encanto at the Hollywood Bowl," "Look at the person next to you on one side, look at the person on the other side, and say 'I got your back.'" What can you, as an engineering educator, do (as in, change) to remind students that they can support each other in a harmonious journey to become a full engineer, one who grows in all aspects of one's profession?

NOTES

1. At the time of the interview, "Latinx" was not used as an option for identification and the binary ("Latina/o") was presented to them as an option, as was "Hispanic" and "Chicana/o."
2. When we discuss "engineering culture," we refer to the existing norms, practices, and ways of operating and being (Jensen & Cross, 2021), as well as mindsets (Riley, 2008). As Jensen and Cross (2021) remind us, engineering culture also encompasses the historical roots of engineering in the military, the myth of meritocracy, and a "culture of suffering and rigor" (p. 372).
3. As defined by TRIO (*Higher Education Act of 1965, 1998 Higher Education Act Amendments Subpart 2—Federal Early Outreach and Student Services Programs*, Chapter 1—Federal *TRIO* programs, Sec. 402A. 20 U.S.C. 1070a–11, n.d.)

4. The term "prestigious" is written with quotation marks to denote that, while students described engineering degrees and professions as prestigious, the authors challenge such a description due to the connotation that engineering is superior to others or that it is only for some people.

5. The term "rigorous" is written with quotation marks to denote that, while many engineering institutions (e.g., schools, colleges, departments) pride themselves on being rigorous or offering a rigorous curriculum, the authors contest, as have others (Riley, 2017; Slaton, 2010), that rigor is not necessary to obtain an engineering degree; in fact, "rigor" is usually a placeholder for exclusionary practices and limiting access to an engineering education.

REFERENCES

Baber, L. D. (2015). Considering the interest-convergence dilemma in STEM education. *The Review of Higher Education, 38*(2), 251–270. https://doi.org/10.1353/rhe.2015.0004.

Baillie, C. (2006). Engineers within a local and global society. *Synthesis Lectures on Engineering, Technology and Society, 1*(1), 1–76. https://doi.org/10.1007/978-3-031-79934-1

Bourdieu, P. (1986). The forms of capital. In J. Richardson (Ed.), *Handbook of Theory and Research for the Sociology of Education.* Greenwood Press.

Camacho, M. M., & Lord, S. M. (2013). Latinos and the exclusionary space of engineering education. *Latino Studies, 11*(1), 103–112. https://doi.org/10.1057/lst.2012.57

Carlone, H. B., & Johnson, A. (2007). Understanding the science experiences of successful women of color: Science identity as an analytic lens. *Journal of Research in Science Teaching: The Official Journal of the National Association for Research in Science Teaching, 44*(8), 1187–1218. https://doi.org/10.1002/tea.20237

Castaneda, D. I., & Mejia, J. A. (2018). Culturally relevant pedagogy: An approach to foster critical consciousness in civil engineering. *Journal of Professional Issues in Engineering Education and Practice, 144*(2), 02518002. https://doi.org/10.1061/(ASCE)EI.1943-5541.0000361

Chodron, P. (2021). *Start where you are: A guide to compassionate living.* Shambhala Publications.

Cole, D., & Espinoza, A. (2008). Examining the academic success of Latino students in science technology engineering and mathematics (STEM) majors. *Journal of College Student Development, 49*(4), 285–300.

Coleman, J. S. (1988). Social capital in the creation of human capital. *American Journal of Sociology, 94,* S95–S120.

Cross, K. J. (2020). Racism is the manifestation of White supremacy and antiracism is the answer. *Journal of Engineering Education, 109*(4), 625–628. https://doi.org/10.1002/jee.20362

Daily, S. B., Eugene, W., and Prewitt, A. D. (2007). The development of social capital in engineering education to improve student retention. *Proceeding of the 2007 ASEE Southeast Section Conference,* Louisville, KY. https://doi.org/10.2514/6. 2007-5256

Damas, S. A., & Benson, L. C. (2021, October 13–16). Navigational capital of African American students in engineering at a predominantly White institution. *2021 IEEE Frontiers in Education Conference (FIE),* 1–5. https://doi.org/10.1109/FIE49875.2021.9637394

Denton, M., & Borrego, M. (2021). Navigational capital of Latinx engineering transfer students: A qualitative study. *Journal of Women and Minorities in Science and Engineering, 27*(2), 61–86. https://doi.org/10.1615/JWomenMinorScienEng.2020035154

Dietz, G. A., Douglas, E. P., McCray, E. D., Mejia, J. A., Pawley, A. L., & Revelo, R. A. (2022). Learning from antiracist theories to reframe engineering education research on race. *Journal of Women and Minorities in Science and Engineering, 28.* https://doi.org/10.1615/JWomen MinorScienEng.2022036609

Dika, S. L. (2012). Relations with faculty as social capital for college students: Evidence from Puerto Rico. *Journal of College Student Development, 53*(4), 596–610. doi:10.1353/csd.2012.0051.

Dika, S. L., & Singh, K. (2007, April 9–13). *Differentiating potential and activated social capital: A measurement model* [Conference paper]. Annual Conference of the American Educational Research Association, Chicago, Illinois.

Ellison, N. B., Steinfield, C., & Lampe, C. (2007). The benefits of Facebook "friends": Social capital and college students' use of online social network sites. *Journal of Computer-Mediated Communication, 12*(4), 1143–1168.

Giroux, H. (1983). Theories of reproduction and resistance in the new sociology of education: A critical analysis. *Harvard Educational Review, 53*(3), 257–293. https://doi.org/10.17763/haer.53.3.a67x4u33g7682734

Glass, C. R., & Gesing, P. (2018). The development of social capital through international students' involvement in campus organizations. *Journal of International Students, 8*(3), 1274–1292. https://doi.org/10.32674/jis.v8i3.52

Hall, R. M., & Sandler, B. R. (1982). *The classroom climate: A chilly one for women?* Association of American Colleges, Project on the Status and Education of Women. https://eric.ed.gov/?id=ED215628

Harper, S. R., & Hurtado, S. (2007). Nine themes in campus racial climates and implications for institutional transformation. *New Directions for Student Services, 2007*(120), 7–24. https://doi.org/10.1002/ss.254

Holland, D., & Eisenhart, M. (1990). *Educated in romance: Women, achievement, and campus culture.* University of Chicago Press.

Holly Jr, J. (2021). Criticality is crucial: Fidelity in what we say and what we do. *Studies in Engineering Education, 2*(2), 46–53. https://doi.org/10.21061/see.78s

Holly Jr, J., & Quigley, L. T. (2022). Reckoning with the harm of anti-Blackness in engineering education: A reparatory justice research approach. *Journal of Women and Minorities in Science and Engineering, 28*(2). 95–110. https://www.dl.begellhouse.com/journals/00551c876cc2f027,219bdde45bcaaa20,2e0165c41ce5952f.html

Jensen, K. J., & Cross, K. J. (2021). Engineering stress culture: Relationships among mental health, engineering identity, and sense of inclusion. *Journal of Engineering Education, 110*(2), 371–392. https://doi.org/10.1002/jee.20391

Joseph, J. (2014). Acculturation, not socialization, for African American females in the STEM fields. *Sociological Research Online, 19*(2), 1-8. https://doi.org/10.5153/sro.3313

Lane, T. B. (2016). Research environments as counterspaces? Examining spaces that inhibit and support science identity development for Black students in STEM. *Urban Education Research & Policy Annuals, 4*(1), 160-169.

Leslie, L. L., McClure, G. T., & Oaxaca, R. L. (1998). Women and minorities in science and engineering: A life sequence analysis. *Journal of Higher Education, 69*(3), 239–276. https://doi.org/10.1080/00221546.1998.11775134

Lin, N. (2001). *Social capital: A theory of social structure and action.* Cambridge University Press.

Lincoln, Y. S., & Guba, E. G. (1986). But is it rigorous? Trustworthiness and authenticity in naturalistic evaluation. *New Directions for Program Evaluation, 1986*(30), 73–84. https://doi.org/10.1002/ev.1427

López, E. J., Basile, V., Landa-Posas, M., Ortega, K., & Ramirez, A. (2019). Latinx students' sense of familismo in undergraduate science and engineering. *The Review of Higher Education, 43*(1), 85–111. https://doi.org/10.1353/rhe.2019.0091

Lucena, J. (2011). What is engineering for? A search for engineering beyond militarism and free-markets. In G. Downey & K. Beddoes (Eds.), *What is global engineering education for? The making of international educators* (pp. 361–383). Springer.

Martin, J. P., Miller, M. K., & Kennedy, M. S. (2012, June 10–13). *Graduate students: Influential agents of social capital for engineering undergraduate researchers* [Conference paper]. American Society for Engineering Education Annual Conference & Exposition, San Antonio, Texas. https://doi.org/10.18260/1-2--21436

Martin, J. P., Revelo, R. A., Stefl, S. K., Garrett, S. D., & Adams, S. G. (2016, June 26–29). *Ethnic student organizations in engineering: Implications for practice from two studies* [Conference paper]. American Society for Engineering Annual Conference & Exposition, New Orleans, Louisiana. https://doi.org/10.18260/p.26744

McFaul, S. (2016). International students' social network: Network mapping to gauge friendship formation and student engagement on campus. *Journal of International Students, 6*(1), 1–13. https://doi.org/10.32674/jis.v6i1.393

McGee, E. O. (2015). Robust and fragile mathematical identities: A framework for exploring racialized experiences and high achievement among black college students. *Journal for Research in Mathematics Education, 46*(5), 599–625. https://doi.org/10.5951/jresematheduc.46.5.0599

McGee, E. O. (2021). *Black, brown, bruised: How racialized STEM education stifles innovation.* Harvard Education Press.

McGee, E. O., Griffith, D. M., & Houston, S. L. (2019). "I know I have to work twice as hard and hope that makes me good enough": Exploring the stress and strain of Black doctoral students in engineering and computing. *Teachers College Record, 121*(4), 1–38. https://doi.org/10.1177/016146811912100407

McGee, E. O., & Martin, D. B. (2011). "You would not believe what I have to go through to prove my intellectual value!" Stereotype management among academically successful Black mathematics and engineering students. *American Educational Research Journal, 48*(6), 1347–1389. https://doi.org/10.3102/000283121142372

McGee, E. O., Naphan-Kingery, D., Miles, M. L., & Joseph, O. (2022). How Black engineering and computing faculty exercise an equity ethic to racially fortify and enrich Black students. *The Journal of Higher Education,* 1–33. https://doi.org/10.1080/00221546.2022.2031704

Mejia, J. A., Revelo, R. A., Villanueva, I., & Mejia, J. (2018). Critical theoretical frameworks in engineering education: An anti-deficit and liberative approach. *Education Sciences, 8*(4), 158. https://doi.org/10.3390/educsci8040158

Mejia, J. A., & Wilson-Lopez, A. (2016). Sociocultural analysis of engineering design: Latino high-school students' funds of knowledge and implications for culturally responsive engineering education. In S. Marx (Ed.), *Qualitative research in STEM* (pp. 68–90). Routledge.

Mejia, Joel Alejandro, & Revelo, Renata A. (2022). Literacies of design: Studies of equity and imagination in engineering and making. In A. Wilson-Lopez, E. Tucker-Raymond, A. Esquinca, & J. A. Mejia (Eds.), *Literacies of design: Studies of equity and imagination in engineering and making.* Purdue University Press.

Miles, M. B., & Huberman, A. M. (1994). *Qualitative data analysis: An expanded sourcebook.* Sage.

Mobley, C., Shealy, E. G., & Brawner, C. E. (2013). First-generation engineering transfer students: A qualitative study of social and cultural capital. *IEEE Frontiers in Education Conference (FIE),* Oklahoma City, Oklahoma, 1651–1653, https://doi.org/10.1109/FIE.2013.6685118

Moore III, J. L., Madison-Colmore, O., & Smith, D. M. (2003). The prove-them-wrong syndrome: Voices from unheard African-American males in engineering disciplines. *The Journal of Men's Studies, 12*(1), 61–73. https://doi.org/10.3149/jms.1201.61

Moschetti, R., & Hadley, C. (2008). Measuring social capital among first-generation and non-first-generation working-class White males. *Journal of College Admission, 198,* 25–30.

Noel, R. C., & Smith, S. E. (1996). Self-disclosure of college students to faculty: The influence of ethnicity. *Journal of College Student Development, 37,* 88–94.

Nuñez, A.-M. (2011). Counterspaces and connections in college transitions: First-generation Latino students' perspectives on Chicano studies. *Journal of College Student Development,* 52(6), 639–655. https://doi.org/10.1353/csd.2011.0077

Ong, M., Smith, J. M., & Ko, L. T. (2018). Counterspaces for women of color in STEM higher education: Marginal and central spaces for persistence and success. *Journal of Research in Science Teaching,* 55(2), 206–245. https://doi.org/10.1002/tea.21417

Pawley, A. L., Mejia, J. A., & Revelo, R. A. (2018, June 24–27). *Translating theory on color-blind racism to an engineering education context: Illustrations from the field of engineering education* [Conference paper]. American Society for Engineering Education Annual Conference & Exposition Proceedings, Salt Lake City, Utah. https://doi.org/ 10.18260/1-2-31161

Perna, L. W., & Titus, M. A. (2005). The relationship between parental involvement as social capital and college enrollment: An examination of racial/ethnic group differences. *Journal of Higher Education, 76,* 485–518.

Revelo Alonso, R. A. (2015). *Engineering familia: The role of a professional organization in the development of engineering identities of Latina/o undergraduates* [PhD Thesis]. University of Illinois at Urbana-Champaign.

Revelo, R. A. (2015). Culturally situated survey of engineering identity for Latina/o undergraduates. *IEEE Frontiers in Education Conference (FIE), 2015.* 32614 2015, 1–5. https://doi.org /10.1109/FIE.2015.7344394

Revelo, R. A., & Baber, L. D. (2018). Engineering resistors: Engineering Latina/o students and emerging resistant capital. *Journal of Hispanic Higher Education, 17*(3), 249–269. https://doi .org/10.1177/1538192717719132

Riley, D. (2008). Engineering and social justice. *Synthesis Lectures on Engineers, Technology, and Society, 3*(1), 1–152. https://doi.org/10.1007/978-3-031-79940-2

Riley, D. (2017). Rigor/us: Building boundaries and disciplining diversity with standards of merit. *Engineering Studies, 9*(3), 249–265. https://doi.org/10.1080/19378629.2017.1408631

Rincón, B. E., Fernández, É., & Dueñas, M. C. (2020). Anchoring comunidad: How first- and continuing-generation Latinx students in STEM engage community cultural wealth. *International Journal of Qualitative Studies in Education, 33*(8), 840–854. https://doi.org /10.1080/09518398.2020.1735567

Rincón, B. E., & Rodriguez, S. (2021). Latinx students charting their own STEM pathways: How community cultural wealth informs their STEM identities. *Journal of Hispanic Higher Education, 20*(2), 149–163. https://doi.org/10.1177/1538192720968276

Rodriguez, S. L., Bukoski, B. E., Cunningham, K. J., & Jones, A. (2020). Critiquing oppression and desiring social justice: How undergraduate Latina students in STEM engage in acts of resistance. *Journal of Women and Gender in Higher Education, 13*(3), 251–267.

Rodriguez, S. L., Doran, E. E., Sissel, M., & Estes, N. (2019). Becoming la ingeniera: Examining the engineering identity development of undergraduate Latina students. *Journal of Latinos and Education, 21*(2), 181–200. https://doi.org/10.1080/15348431.2019.1648269

Rose-Redwood, C., & Rose-Redwood, R. (2013). Self-segregation or global mixing? Social interactions and the international student experience. *Journal of College Student Development, 54*(4), 413–429. https://doi.org/10.1353/csd.2013.0062

Ross, M. S., & McGrade, S. (2016, June 26–29). *An exploration into the impacts of the National Society of Black Engineers (NSBE) on student persistence* [Conference paper]. American Society for Engineering Education Annual Conference & Exhibition, New Orleans, Louisiana.

Saldaña, J. (2015). *The coding manual for qualitative researchers.* Sage.

Slaton, A. E. (2010). *Race, rigor, and selectivity in US engineering: The history of an occupational color line.* Harvard University Press.

Solórzano, D., Ceja, M., & Yosso, T. (2000). Critical race theory, racial microaggressions, and campus racial climate: The experiences of African American college students. *Journal of Negro Education*, 60–73.

Solórzano, D. G., & Bernal, D. D. (2001). Examining transformational resistance through a critical race and LatCrit theory framework Chicana and Chicano students in an urban context. *Urban Education*, 36(3), 308–342. https://doi.org/10.1177/0042085901363002

Stanton-Salazar, R. D. (2001). *Manufacturing hope and despair: The school and kin support networks of U.S.-Mexican youth*. Teachers College.

Stanton-Salazar, R. D. (2011). A social capital framework for the study of institutional agents and their role in the empowerment of low-status students and youth. *Youth & Society*, 43(3), 1066–1109. https://doi.org/10.1177/0044118X10382877

Stevens, R., O'Connor, K., Garrison, L., Jocuns, A., & Amos, D. M. (2008). Becoming an engineer: Toward a three dimensional view of engineering learning. *Journal of Engineering Education*, 97(3), 355–368.

Suresh, R. (2006). The relationship between barrier courses and persistence in engineering. *Journal of College Student Retention: Research, Theory & Practice*, 8(2), 215–239. https://doi.org/10.2190/3QTU-6EEL-HQHF-XYF0

Swan, S. C. (2014). She came, she saw, she transgressed. *The New School Free Press*.

Tate, E. E., & Linn, M. C. (2005). How does identity shape the experiences of women of color engineering students? *Journal of Science Education Technology*, 14(5–6), 483–493.

Trice, A.G. (2004). Mixing it up: International graduate students' social interactions with American students. *Journal of College Student Development*, 45(6), 671–687.

Yosso, T. J. (2005). Whose culture has capital? A critical race theory discussion of community cultural wealth. *Race Ethnicity and Education*, 8(1), 69–91. https://doi.org/10.1080/13613320 52000341006

Yosso, T., Smith, W., Ceja, M., & Solórzano, D. (2009). Critical race theory, racial microaggressions, and campus racial climate for Latina/o undergraduates. *Harvard Educational Review*, 79(4), 659–691. https://doi.org/10.17763/haer.79.4.m6867014157m707l

PART 2

STRUCTURAL ISSUES AND INTERSECTIONALITY FOR LATIN* STUDENTS IN ENGINEERING

7 · A CRITICAL MIXED METHODS ANALYSIS OF LATIN* ENGINEERING STUDENTS IN DIVERSE CONTEXTS

LARA PEREZ-FELKNER, CIERA FLUKER, AND DA'SHAY TEMPLETON

The growing population of engineering students who identify as Latin*[1] have diverse identities and backgrounds, and these students are pursuing engineering at a range of institutional types across and beyond the United States. Institutions and societies reflect and enact systemic injustices. Therefore, it is imperative to directly study how Latin* students experience engineering through a lens that considers the structural challenges they encounter and how the contexts in which they are situated can shape their pathways to and through engineering degrees.

This chapter examines the diversity of Latin* engineering students' educational contexts to illuminate considerations and potential avenues for meaningful intervention in these patterns. We employ a critical race perspective, recognizing that disparities in Latin* students' representation and success in these fields neither occur naturally nor because of innate differences in motivation or skill. Rather, the historical and continued underrepresentation of Latin* students arise out of systemic and compounding differences in students' exposure, opportunities, and support for engineering study and careers. As such, interventions aiming to reduce these disparities must attend to structural as well as individual solutions.

This chapter integrates quantitative and qualitative approaches to examine Latin* engineering students' counter-stories in context from a critical social justice perspective (see McGee, 2021). We report on variation among a nationally representative sample of Latin* students from high school through college who are interested in engineering. Next, we narrow our focus to a Minority-Serving Institution context that is not specifically focused on (albeit does have high enrollment of)

Latin* engineering students: a Historically Black University. Here, we build on research in this volume that attends to the role of institutional (e.g., Hispanic-Serving and Historically White Institutions) and social–relational supports for Latin* secondary and postsecondary students in engineering. In this chapter, we investigate the following research questions: (1) What are characteristics of US Latin* students who enroll in postsecondary engineering programs, in relation to their outcomes in these and other fields, their intersectional identities, and knowledge? and (2) How do social contexts shape intersectionally diverse Latin* engineering students' pathways to and through experiential engineering knowledge development? Below, we offer a theoretical overview to ground our analyses, followed by a brief summary of our methodological approach, we then review our analyses and discuss implications.

THEORIZING LATIN* ENGINEERING STUDENTS' CONTEXTS AND EXPERIENCES

We employ a critical framing for this chapter to capture the experiences of Latin* students who are too often seen as being outside of or anomalous in engineering labs and academic environments. Below, we identify and define our *critical race mixed methods* (CRMM) approach, wherein we attend to intersectional differences among a Latin* engineering student sample. Next, we align a theoretical tool complementary to our CRMM approach.

Critical Race Mixed Methods Framework

Critical race theory (CRT) in education draws on legal frameworks (e.g., Bell, 2008) to explain how racial disparities arise not simply from individuals' behavior but mainly through social structures that facilitate the reproduction of the racial status quo (Ladson-Billings & Tate, 2016). CRMM is a theoretical framework that pulls from CRT and mixed methodology to form one method of inquiry. Following DeCuir-Gunby et al. (2019), we employ a CRMM through an explanatory sequential mixed methods design that focuses explicitly on Latin* students' racialized and gendered experiences (quant→QUAL) in diverse contexts. Our CRMM theoretical framework draws from three key CRT tenets—*the centrality of race and racism* (Bell, 2008; Omi & Winant, 2014), *intersectionality* (Collins, 2000; Crenshaw, 1991), and *the centrality of experiential knowledge* (Delgado, 1989)—which we cover briefly next.

The Centrality of Race and Racism

Guided by CRMM, we examine how society enacts racialization across educational settings and over time (Cabrera, 2018; Tate, 1997; Yosso et al., 2009). Systematic disenfranchisement in education based on race and other identities occurs in institutions, fields, and social groups, even among those who consider themselves to "not see" social distinctions that nonetheless continue to be axes of inequality across generations. STEM fields such as engineering may be especially relevant for a critical examination of how field norms and practices may maintain, reproduce, and exacerbate racial and gender inequality (Carter et al., 2019).

Too often, as established in prior chapters in this text and elsewhere (see especially chapters 5, 6, and 9, in this volume), Latin* students can be made to feel marginalized and/or anomalous in engineering labs and academic environments by their classmates and faculty. With respect to systemic and persistent racial disparities regarding who enrolls in and completes engineering undergraduate programs, these patterns manifest in the persistent disparities often cited in the literature—as well as in this book. The patterns are aggregates of institutional and departmental patterns that often reflect differential opportunities afforded to students of color, within and across gender identities.

For this reason, we examine patterns at the national level as well as within institutions and labs. We also investigate how students process and relate to these experiences, which are often both racialized and gendered. Over time, additional frameworks using a critical perspective have emerged. These include LatCrit (which centers critical studies on the experiences of the Latin* community), critical feminist studies, and other critical lenses that attend to how the salience of individuals' identities can condition their experience and opportunities (e.g., Bernal, 2002; Hernández, 2016; Solórzano & Villalpando, 1998). Accordingly, we now turn our attention to how identities—such as race, ethnicity, and gender—can be intrinsically connected in students' pathways through engineering education.

Intersectionality

Individuals from marginalized groups may experience discrimination and oppression associated with multiple intersecting identities, including race and gender, rendering a double bind for those pursuing STEM degrees who can encounter discrimination associated with two or more identities (Gaston Gayles & Smith, 2018; Malcom & Malcom, 2011; Ong et al., 2011). Intersectionality theory recognizes how social structures can disempower individuals on multiple axes, including but not limited to representation (Cho et al., 2013; Collins, 2015; Crenshaw, 1991). Indeed, intersectionality can and has been leveraged as both a theoretical and a methodological tool to examine such phenomena (Haynes et al., 2020; Ireland et al., 2018).

In research and practice, it is important to "shift the default" in our categories of who we consider the "norm" in engineering to facilitate meaningful opportunities to broaden participation (Pawley, 2019, 2020). An intersectional lens can reveal unexpected findings regarding disparities in engineering education, but too few intersectional studies of Latin* engineering students have been published (for exceptions, see Fluker et al., 2022; Rodriguez et al., 2022). This chapter and book aim to inspire future research in this area.

The Current Study: U.S. National Sample and Latin* Engineering Students
at Historically Black Colleges and Universities

We begin with an examination of the US national context, describing Latin* engineering students' diversity of backgrounds, identities, and educational experiences. We then turn to an original study to examine their experiences in greater
depth in a setting intentionally focused on racially minoritized populations and
too infrequently considered in the larger literature on Latin* students—a racially
heterogenous population. Notably, institutions that have historically been organized around serving underrepresented students have been found to supply more
supportive contexts by providing students with structured opportunities to build
and sustain confidence and resilience in engineering (Perna et al., 2009). As other
chapters in this volume explain, this can be found at Hispanic-Serving Institutions
(Madsen Camacho & Lord, 2011) as well as Historically Black Institutions
(Buzzetto-More et al., 2010; Lundy-Wagner, 2013; Mack et al., 2010). Engineering
is understood to be "hard," requiring long hours of study, with high barriers to
entry into gateway courses and research/mentoring opportunities. To address this
perceived and often real inaccessibility for racially marginalized engineering students, it is essential to critically examine the contextual and lived experiences of
Latin* college students aspiring to become engineers.

CRITICAL RACE MIXED METHODS AND MATERIALS

This study captures counter-stories of Latin* engineering students in diverse contexts to contest the historical and contemporary majoritarian experiences of
White cis-gender male engineers at Predominately White Institutions. Instead of
viewing racially and ethnically marginalized students as culturally disadvantaged,
CRT's centrality of experiential knowledge recognizes the knowledge and unique
capital that students of color bring to campus, including aspirational, linguistic,
familial, social, navigational, and resistant capital (Delgado, 1989; Yosso, 2005).[2]
We center the experiential knowledge of engineering-interested Latin* undergraduates at Historically Black colleges and universities (HBCUs) because their
counter-stories provide critical minoritarian perspectives, which provide stakeholders with considerations and avenues for interventions.

Our analysis of Latin* engineering students' educational contexts begins with a
national picture, drawing on the nationally representative High School Longitudinal Study of 2009/2016. We then report on an intervention research study in an
HBCU setting, in which neither whiteness nor Latinidad are the dominant norms
that they would be at Predominantly White Institutions (PWIs) and Hispanic-
Serving Institutions (HSIs). At such campuses, Latin* students are not necessarily
a majority identity among students or faculty and staff but rather comprise a minimum of 25 percent of the student body (Crisp et al., 2009; Núñez et al., 2016).
Accordingly, "Hispanic-Serving" Institutions may be growing in the United States

but do not necessarily serve Latin* students as their core mission and function (Garcia, 2019; Garcia et al., 2019), especially in disciplines (like engineering) with a historically low representation of Latin* students.

Our study was set in a Historically Black University with over 20 percent of its engineering students identifying as Latin*. Unlike PWIs, HBCUs structure opportunities for racially minoritized students to build and sustain their confidence and resilience in engineering. Indeed, we found this to be the case in prior research drawing on this site—with careful attention and structured support around research mentoring for undergraduate research students (Perez-Felkner et al., 2020). This institutional context allows for an examination of racialization with less expectation that one of these identities may be preferred and privileged. Essentially, Latin* students can consider and value the complexities of their identities more fully when, for example, Afro-Latin*, multiracial, and other marginalized identities are acknowledged and valued.

Positionality

The chapter authors are three proud Puerto Rican women researchers who hold diverse positionalities and backgrounds. Lara Perez-Felkner is a Puerto Rican woman researcher and former STEM undergraduate who recognizes the multiplicity of identities that students may hold and is increasingly engaging in critical and equity-focused scholarship. She regularly engages with engineering faculty and researchers on projects to equitably expand opportunity and build capacity in the field. Ciera Fluker is an Afro-Latina, cisgender researcher whose experiences navigating harmful K–20 STEM environments have encouraged her to problematize structural inequities that influence racially marginalized students' access and participation in higher education and engineering. Da'Shay Templeton identifies as an Afro-Latina, Indigenous, and Puerto Rican statistician, and a bisexual, disabled mother of three Brown boys. Her experience watching her *tío* and *papa* build and fix cars, houses, and other structures with incredible ingenuity has shaped her definition of engineering and the true measure of an engineer.

Critical Mixed Methods Rationale and Integration

This critical mixed methods analysis follows an iterative sequential explanatory approach, such that we begin with a quantitative analysis followed by a qualitative approach, serving as the focus of our quant→QUAL analysis and reporting (Creswell & Creswell, 2017; Mayoh & Onwuegbuzie, 2015; Onwuegbuzie et al., 2013). Both methodological approaches are connected to larger lines of inquiry which were pursued in tandem. The quantitative analysis of the national data draws on past and current research led by the first author on reducing gender and racial disparities in STEM fields (see, e.g., Zhao & Perez-Felkner, 2022). Here, a focused subsample was drawn for a novel analysis of backgrounds and experiences of Latin* engineering students that would contextualize their experiences. The qualitative analysis followed, drawing on a five-year engineering evaluation research study that includes sixty undergraduate research participants over five cohorts. As with the quantitative design, we drew on a subsample of Latin*-identified students to investigate and explain the academic, family, and socioeconomic patterns identified and reported on in the quantitative analysis. We frame our study as applying a CRMM approach because CRT guides our explanatory sequential mixed methodology.

Quantitative Methodology

To ground our qualitative study, we leverage the most recent nationally representative US cohort of secondary school youth, from the High School Longitudinal Study of 2009/2016 (HSLS), which commenced when the students were in ninth grade (in 2009) and continued with four waves of survey and administrative information. A panel of over 25,200 students were followed from high school through college.[3] Of these, 23,300 students self-identified with one or more racial/ethnic groups and formed our initial analytic sample. To our knowledge, this chapter presents the first analysis of Latin* engineering-aspiring students from this national sample. We focus on demographic characteristics and a select series of educational measures to illustrate the variation in Latin* engineering students' characteristics before and after entering college. Latin* students comprised 17.2 percent of the HSLS sample ($n = 4{,}000$). Rather than comparing Latin* students to other racial/ethnic groups such as white students (for similar design and rationale, see, e.g., Strayhorn, 2010), we draw on critical methodological paradigms to intentionally center minoritized populations understudied and marginalized in the literature; our intent here is to leverage this subsample to tell stories through the data in order to establish a context for the qualitative case study data that follow. We use the same approach—drawing on a subsample of Latin* engineers from a larger study—for the qualitative dimension described next.

Qualitative Methodology

In this study, we analyzed data from a larger engineering evaluation research study based at a Historically Black University in the southeastern United States. Specifically, we examined the experiences of students who participated in a paid summer research internship in materials science and engineering and attended varied institutions across the country ($n = 60$). Our analytic sample in this chapter consists of the seven undergraduate students who identified as Hispanic or Latina/o/x, some of whom presented and identified as Afro-Latin*. Four of these students identified as men, and three identified as women. We focus on student self-reported data from a confidence-building intervention training (including an affirmation writing exercise and survey questions) and semi-structured interviews conducted by the first and second authors. Our data analysis consisted of analyzing students' self-reported demographic information as well as deductively coding and analyzing the transcribed interviews.

FINDINGS

Quantitative: Contextualizing Latin* Engineering Students

Latin students' enrollment and completion postsecondary outcomes in engineer-ing.* Table 7.1 reports on the distribution of Latin*-identified students across post-secondary majors. The first column lists the number and percentage share of students who were considering this major on entry, drawing on the source variable

TABLE 7.1   Latin* Students' Engineering Initial Major and Degree Field of Study

Field of study	Initial major considered	Change direction	Degree major
Engineering and engineering technologies	140 (7.9%)	–	130 (7.5%)
Computer and information sciences	0 (3.3%)	–	50 (2.9%)
Biological and physical sciences, science technologies, and mathematics	177 (9.8%)	–	150 (8.5%)
Social sciences	150 (8.4%)	+	170 (9.9%)
Health fields	350 (19.3%)	–	270 (16.1%)
Education	80 (4.2%)	–	70 (4.0%)
Business	240 (13.0%)	+	240 (14.3%)
Other applied	370 (20.1%)	+	340 (21.1%)
Arts and humanities	90 (5.1%)	+	100 (5.8%)
General studies	40 (2.4%)	–	40 (2.2%)
Don't know/undeclared	120 (6.5%)	+	130 (7.7%)
Valid Total *N* per indicator	**1,810 (100.0%)**		**1,700 (100.0%)**
Missing or not applicable (e.g., did not earn degree)	*2,190 (54.7%)*		*2,300 (57.6%)*

SOURCE. Authors' analysis of High School Longitudinal Study of 2009/2016 Restricted-Use Data Files. N = 4,000. Institute for Education Sciences restricted-use secure data policy mandates rounding to the nearest 10 for individual counts and nearest 10th decimal for descriptive figures. Figures may not total exactly because of rounding. We include a reporting line for missing data that refers to the number and share of Latin* students without valid data on this indicator, because they did not respond generally or were not considered eligible for these questions (i.e., they did not enroll in postsecond-ary education). Given the critical orientation of the chapter, we go against convention and note here the share of initial ninth graders identifying as Latin* who are "missing" from these data, to represent their potential representation if granted greater exposure and opportunities to gain entry to the field. NOTE. Because categories vary across the variables used, some fields of study are aggregated to allow for comparison across indicators, particularly for non-STEM fields. The variables used to generate these analyses are derived from the following source variables: *x4entrymaj4y* (major considered upon postsecondary entry) and *x4rfdgmj14y* (fourth-wave reference degree major), where "change direction" refers to the direction of change for the percentage of total students in that category, from initial to degree major.

x4entrymaj4y. Next, we assess how many Latin* students completed degrees in these same fields to date, using the source variable *x4rfdgmj14y*. We also note the direction of change between initial and completed majors for Latin* students. Of primary interest are students who report interest in "engineering and engineering technologies." One hundred forty students in the study were initially interested in this field, or 7.9 percent of respondents. The share of Latin* students reporting completion of degrees is slightly lower, representing 7.5 percent ($n = 130$). Compared to the other STEM fields listed in table 7.1, this is a similarly moderate drop as that experienced in computer and information sciences, which enrolls fewer than half as many Latin* students and a smaller drop than that experienced in the larger STEM category of natural sciences and mathematics. Latin* students experienced some attrition from health fields and were more likely to shift toward social sciences and applied fields, such as business. Overall, there is a robust and steady share of students who enter and complete engineering and related degrees across their disciplinary specialties.

Intersectionality: Gender. Table 7.2 begins an intersectional turn to more closely examine the degree fields of Latin* students by gender.[4] Latino men comprise a larger share of those aspiring to engineering degrees (83.2 percent) as compared to Latina women (16.8 percent). Perhaps it is for this reason that we see more variation among the men than we do the women, with respect to their ultimate postsecondary degree field. Interestingly, while Latinas represent a smaller share of engineering-interested students, they have higher levels of representation among undergraduate degree holders in engineering than do Latino men. We now look more closely at the characteristics of students who initially considered engineering majors. We constrained the analytic sample to those 140 (rounded) students who reported intending engineering majors.

Intersectionality: Socioeconomic Experiences and Racialization. The findings reported in table 7.3 describe the share of Latin* students who aspire to engineering majors, with a focus on family background and personal academic characteristics. First, we see that nearly 40 percent of those Latin* students interested in engineering have parents who completed a bachelor's degree or higher. Further, 65 percent of students' fathers and 15 percent of their mothers are employed in STEM fields. This is a high percentage, suggesting that, nationally, Latin* students who aspire to engineering fields are more socioeconomically advantaged than those who do not aspire to these fields. Most students also attended four-year colleges as the first postsecondary institution they attended (not shown in table). Interestingly, Latin* engineering students vary by immigrant generational status, whereby the majority identifies as a first- or second-generation immigrant to the United States. This may partially explain the low number and share of students reporting a racial identity other than Latino/Hispanic/Other when asked about their race rather than their (Hispanic) ethnicity ($n = 1$, who chose "White, non-Hispanic").

STEM Knowledge Perceptions among Future Latin* Engineering Students. Finally, we examined Latin* students' social psychological relationship

TABLE 7.2   Latin* Engineering-Aspiring Students' Degree Major by
Biological Sex/Gender[a]

Degree major*	Latino men %	Latina women %
Did not earn degree/not applicable	4.2	0.0
Don't know/undecided	4.2	*
Computer and information sciences and support services (11)	3.4	0.0
Engineering (14)	61.3	75.0
Engineering technologies/technicians (15)	18.5	*
Biological and biomedical sciences (26)	0.0	*
Mathematics and statistics (27)	0.0	*
Science technologies/technicians (41)	*	0.0
Social sciences (45)	*	0.0
Visual and performing arts (50)	*	0.0
Health professions and related clinical sciences (51)	3.4	0.0
Business, management, marketing, and related support services (52)	3.4	*
Total of Latin* engineering students	83.2 (N = 120)	16.8 (N = 20)

SOURCE. Authors' analysis of High School Longitudinal Study of 2009/2016 Restricted-Use Data Files. N = 1,810. Institute for Education Sciences restricted-use secure data mandates rounding to the nearest 10 for individual counts and nearest 10th decimal for descriptive figures. Figures may not total exactly due to rounding. Because of small individual cell sizes in this analysis, some N totals may not appear to match the percentage associated, which is derived from the non-rounded indicator (e.g., N = 1 corresponds to 0.84% of the men's sample).

NOTE. X4rfdgmj12 is the source variable used here for degree major, using the two-digit CIP (classification of instructional program) codes. [a] We constructed a gender non-binary category for students who did not select male or female, or who changed gender over the course of the study. However, none of the ~140 students who reported an interest in engineering and a Latin* identity responded to the sex/gender questions in categories other than male/female. Although this is a longitudinal study starting in ninth grade, our focus on engineering is centered more strongly in postsecondary education when students are adults. For this reason, we use the terms "men" and "women" to signal the social category for gender rather than the biological category of sex. * Institute for Education Sciences restricted-use data mandates constrain reporting for cell sizes based on raw numbers lower than n = 3. Therefore, in those cases, we place an asterisk to signify that students did earn degrees in this field, but do not report the specific percentage share.

with engineering, and its related fields in K–12 schooling and earlier on in college to understand potential opportunities to intentionally broaden opportunities in high school. Notably, the path toward engineering postsecondary degrees begins with coursework and extracurricular activities in middle and high school (Erichsen et al., under review). Specifically, we assessed perceived ability and interest in science and mathematics in high school (ninth grade) during the study's baseline.

TABLE 7.3   Characteristics of Latin* Engineering-Aspiring Students

Student family and individual characteristics	N	%
Socioeconomic status: parental education		
Earned baccalaureate degree or higher	50	39.1
Socioeconomic status: parental occupation		
Father/male guardian works in STEM field	90	65.0
Mother/female guardian works in STEM field	20	15.0
Socioeconomic status: family income		
$ 35,000 or lower	39	33.9
$ 35,001–55,000	30	21.0
$ 55,001–95,000	18	15.4
$ 95,001–135,000	22	19.1
$ 135,001 or higher	18	15.4
Immigration generational status		
First-generation immigrant	40	29.2
Second-generation immigrant	50	44.2
Third-generation immigrant or higher	30	26.7
Psychological factors (baseline; ninth grade)	**Mean**	**SD**
Science interest (scale); alpha = 0.76	3.0	0.6
Perceived science ability (scale); alpha = 0.84	3.0	0.5
Mathematics interest (scale); alpha = 0.85	3.0	0.7
Perceived mathematics ability (scale); alpha = 0.85	3.2	0.5
Mathematics identity (NCES-generated scale)	0.5 (range: −1.7 to 1.7)	1.0
Science identity (NCES-generated scale)	0.3 (range: −1.6 to 2.2)	0.9

SOURCE. Authors' analysis of High School Longitudinal Study of 2009/2016 Restricted-Use Data Files.
NOTE. $N = 143$. Institute for Education Sciences restricted-use secure data mandates rounding to the nearest 10 for individual counts and nearest 10th decimal for descriptive figures. Figures may not total exactly because of rounding.

The indicators we developed have strong reliability, as noted by their alpha scores (reported in table 7.3).

Overall, Latin* students report positive self-ratings of their interest in mathematics and science fields (no engineering-specific questions were offered), which suggests a positive trajectory, from high school and beyond, for those students who aspire to engineering degrees at the start of college. With respect to mathematics and science identity, they hover near the mean with respect to science identity and slightly higher with respect to mathematics identity. It may be due to Latin* students' exposure to mathematics, science, and engineering at the start of high school being limited and developing further later in high school and college. We investigate this further with our qualitative data in the next section. Having reviewed the national context of Latin* engineering students, we now turn our attention to the experiences of Latin* students participating in engineering labs during their undergraduate years.

Qualitative: Experiences of Latin* Engineering Students

Intersectional identities and their variation among Latin engineering students.* Findings from our qualitative data analysis set in a Historically Black University revealed that students had a range of social identities that shaped and influenced their college experiences. Additionally, the salience students placed on their racial and ethnic identity differed. For example, Michelle was an international engineering student from Panama who described finding community and support with other Panamanian students on her campus. She stated, "I think a big part of that is the Panamanian students that have a lot of Panamanian friends. I don't know if it's just like an industrial [engineering] thing, but there's a lot of them, which is really cool because—so, they come from Panama, and they come here to study, and I'm really good friends with them. So that, like, helps out the community." Meanwhile, Darrell did not feel a strong connection with his Mexican heritage and expressed identifying more as white: "Mainly White though because the only thing that's Hispanic about me is my last name and my skin color and my ancestors—I can't speak Spanish to save my life."

In addition to differences in race and ethnic identity salience, students had a range of intersecting social identities, including religious identity. Nelie listed her racial/ethnic identity as Black/African American and Hispanic on her program application but described her gender and racial/ethnic identity as being unimportant compared to being Christian: "For me, like, I guess perceived ideas of identity, like race, or like gender and different things like that don't really affect me as much, because, um, it's one of those things where I just kind of, I'm like 'Oh, yeah, I'm just me.'" Similarly, Alejandro identified as Black and Hispanic and thus sought community through both Latina/o/x and Black student organizations on campus. Alejandro also described the salience of his Christian religious identity and his decision to join a Christian student organization on his campus: "Just interacting with those people and building community—specifically for Christians—was the goal for me there." Christian and Alejandro's male and Christian social identities may align more with the cultural identity of their engineering programs and overall institutions, making them feel more comfortable and accepted at their institutions than students with other marginalized social identities (e.g., Afro-Latina and Muslim students; Smith, 2015). Their experiences highlight the ways Latin* students' multiple and intersecting social identities simultaneously marginalize them while also affording them some levels of power and privilege in engineering spaces.

While many students described finding social identity within student organizations, Michelle described exclusion on her campus among Latin* students from differing racial and ethnic groups as well as differing citizenship statuses. She explained, "I feel like there's like a little segregation between like the Panamanian students and then like the Hispanic students that are from the U.S. are, like, naturalized citizens and so on. Maybe they were born here but they come from His-panic or Latino backgrounds." Michelle's experience showcases the nuances of racialization in her use of two racial categories, "Hispanic or Latino," and her realiza-tion that critical differences exist between Latin* students who belong to the same racial group but have distinct experiences with immigration and naturalization.

Counter-Stories: Affordability and Experiential Knowledge. In addition to social identity, students' socioeconomic status seemed to shape their college experiences as well as their educational and career plans. They discussed relying on jobs, financial aid, and scholarships to cover the cost of their undergraduate education. For example, Michelle discussed juggling outside employment while taking advantage of research opportunities on campus: "I don't know how much time I have for the fall or spring but because it's like you're just volunteering or just doing the research, you know, not getting paid. For me, I kind of like having the job for an extra income." While unpaid research opportunities could expose Michelle to critical skills and knowledge, these opportunities created an addi-tional burden for her as she had to forgo extra income on which she normally relied. Similarly, prior to participating in the undergraduate engineering research program, Nelie had trouble juggling employment with her coursework and research: "[T]here wasn't a lot of time where I could just like, *breathe*. And so I think that just the main thing was just like, when you have to pay for your own schooling on top of being an engineer, and then also like "Oh well, I kind of want to be involved in this [non-science extracurricular activity]" because I know for me it was like, if I don't do anything on campus then I feel like there's part of my college at college experience that I wasted. So, juggling all three of those things was really hard." Balancing steady paid work opportunities with short-term funded and unfunded research opportunities is a struggle for many of the students we spoke with in the larger study, including the Latin* students.

Nelie's narrative underscores affordability challenges. When we first met, Nelie had just lost a prestigious paid position through undergraduate housing and found herself without a place to sleep after receiving a low (albeit contested) grade, bringing her GPA to below a 2.5. She explained:

That's the way that housing works: They were like, "Well, . . . unless you can fix your grade, in, like, two days you're fired," and I'm just like "Well, okay" [laughs].

I: That's pretty disruptive in your life.
R: Yes. [*laughs*] Yeah.
I: Because that comes with housing as well?
R: Yeah, yeah . . . So, that was fun. [*laughs*] But, it's, I don't know, I feel like it's a blessing in disguise, cause the, the last semester was awful. I like, slept here [in the lab] some days. And so, it was like I don't think I would have been able to put in the time, um, for, um, my job in here, and be able to get the grades that I want to get. . . . Trade-offs. [*Pause*] I feel like I probably wouldn't have even applied for this

internship, because, like, in the back of my mind, like "Oh, I need to get an internship," but like, more driving is like, "Oh, I need money to live." And so, if I knew that I would have a job over the summer anyways, then I might not have applied. So, it's, it worked out. I had to take out loans, but, you know, it's not the worst thing in the world.

This counter-story highlights the challenge for students to balance the demands of engineering studies (and the lower GPAs associated with these majors) and initial entry into volunteer internships as a pathway to their longer-term career plans.

Nelie went on to graduate studies in engineering. She had moral support from her family, and she had research funding resulting from her paid undergraduate summer internship: "My dad kind of wanted me to get a job to go to grad school. But now that he knows that they're going to pay for my education, and they'll be paying me, he's more on board . . . so they're all kind of excited about it." Funded support for students to complete their studies tends to matter, perhaps especially given the expectations that engineering students engage in extracurricular and often unpaid training to show their interest and commitment—in addition to managing strenuous course loads, the costs of attending college, and the desire to (as noted in Nelie's earlier quote) be a regular college student and engage in life beyond the lab.

Similarly, Samantha described leaving her job and relying on financial support from her mother, so she could focus on her studies. She explains: "My mom was like, 'I'd rather you have good grades and have your degree than do a job that you're not going to be working,' you know? So thankfully she helps me out with my rent and stuff although I know it's a lot for her, because it's just her by herself." Samantha went on to explain the positive impact of the stipend she received for the summer research program: "This internship helped a lot with the money that I get paid for and I get to help her a lot too with it." When such financial support is not available, it can be difficult to stay engaged in and on track with engineering studies.

Students' socioeconomic status also seemed to influence their academic and career goals. For example, Alejandro had concerns about how he would afford graduate school: "I don't know if I would want to do that if it would cost me, like, a significant amount of money because I can't really pay for undergrad out of pocket." He explained further that his state merit scholarship covered sufficiently covered his expenses at present: "So it's I'm fine as I am now but, that led me to not worry about much about money now." For Alejandro and other students, graduate school was a financially risky proposition, especially following the challenges they experienced with financial security as undergraduates.

On track to graduate with degrees in engineering, students might earn high salaries but are aware of the costs of addressing life's basic needs—housing, transportation, food—and spoke at times indirectly ("I'm fine as I am now but") and directly (e.g., Nelie's case) about how financial insecurity in their lived experience before and during college has shaped their career plans (regarding college student needs insecurity, see also Anderson et al., 2023; Martinez et al., 2021). This apprehension, perhaps also informed by their family experiences and perceived importance of steady, stable employment, appeared to shape their intentions concerning whether to continue in engineering beyond college graduation.

Findings Summary

Table 7.4 summarizes our results across methodologies. In our quantitative data analysis, we found a steady and robust share of students entering and completing engineering and related degrees across all disciplinary specialties. Students from Latin* backgrounds reported positive self-ratings of their interest in mathematics and science fields (no engineering-specific questions were asked), suggesting that students who hope to pursue engineering degrees in college will have a positive trajectory, from high school through college. In terms of intersecting identities, Latinos comprise a larger share of those aspiring to engineering degrees as compared to Latinas. Although Latinas make up a smaller share of engineering-interested students, they make up a larger proportion of undergraduate engineering degree holders than Latinos. Furthermore, the majority of Latin* students who aspire to engineering fields had family members—mostly men—who worked in STEM fields (but not necessarily in professional occupations). In addition, Latin* engineering students vary by immigrant generational status, with the majority identifying as first- or second-generation immigrants.

Our qualitative analysis of data from a public Historically Black University reveals that students have a range of social identities that shape and influence their college experiences. Furthermore, students' perceptions of their racial and ethnic identities differed. Students had a range of intersecting social identities, including gender and religious affiliation, as well as differences in race and ethnic identity salience. Additionally, their social positions in terms of citizenship and naturalization seemed to influence their sense of belonging. Lastly, students' socioeconomic status appeared to influence both their college experiences and their educational and career goals.

DISCUSSION: PRECARITY AND SUCCESS IN THE PURSUIT OF ENGINEERING CREDENTIALS

Across our analyses, we found heterogeneity in the backgrounds and experiences of intersectionally diverse Latin* engineering students. We also found (albeit with small Latin*-focused subsamples) a decentering of US notions of race and racial identity among the students we observed. Their counter-stories offer insights into how tensions and challenges around finances and opportunity—shaped indirectly by race—mattered in their sensemaking about current and potential future experiences in engineering.

Our related research—drawing on the engineering internship program studied here—underscores (a) the importance of undergraduate students' prior experiences in enhancing their engineering identity and successful navigation of engineering research (Fluker et al., 2022); and (b) the potential for mentored research in engineering to foster resilience and greater confidence in technical skills among racially minoritized undergraduate students (Perez-Felkner et al., 2020). This chapter attends especially to the socioeconomic and familial backgrounds of Latin* engineering students and centers this student population in its framing and analysis. We find that racially minoritized Latin* students bring their own rich experiences and knowledge to the pursuit of "hands-on" experiential engineering knowledge development via undergraduate research.

Recall the three tenets of CRT: the *centrality of race and racism, intersectionality,* and the *centrality of experiential knowledge.* Our CRMM analyses reflect how—even when race and racism is not directly visible in a given site—the intersectionality of race and other identities remains an important consideration. Students' counterstories serve as a powerful tool to understand why even funded programmatic investments in broadening participation in engineering are not enough (on their own) to undo the systemic disparities and challenges Latin* students encounter before, during, and after their undergraduate studies. We discuss in the next section and display in table 7.4 potential implications of our findings, each associated with recommendations for policy and practice, drawn from the literature.

Recommendations for Policy, Practice, and Future Research

First, with respect to racial identity, the demands of students' engineering coursework tend to limit their opportunities to engage in humanities and social science coursework that would foster greater understanding of race, racism, and their roles in higher education. As an associated recommendation for institutional and departmental policy and practice, it appears important for Latin* and other racially marginalized students to complete coursework that focuses on social justice and equity; doing so will offer them tools to more effectively navigate challenges with racism and other intersectional forms of marginalization, in and beyond their engineering studies (Salas Pujols, 2022).

A second implication with respect to the centrality of race in Latin* engineering students' lives and the power of reduced salience, it may be that students who chose to participate in a highly diverse engineering program at a Historically Black University were less tied to a specific conceptualization of race or racial identity. They may have sought to avoid the experience of being racially marginalized in pursuing their experiential knowledge training in an environment that more accurately represented the world than that which they might find at a Predominantly White University. Indeed, a Black undergraduate researcher woman interviewed in the larger study reported that she felt she could "relax" while working in the program's labs over the summer and at times "forgot" to think about race while she was there, compared to when she was in classes or working at her home campus.

If a Historically Black University setting can be similarly empowering for Latin* students—such that they can more easily let go of their racial awareness and

TABLE 7.4   Summary of Critical Race Mixed Methods Findings

Key findings by method	Quantitative	Qualitative	Combined
Centrality of race and racism	Most Latin* engineering students nationally were first- or second-generation immigrants.	Range in ethnic and racial identity and saliency	Connect threads—recency in the United States may explain why US-based race categories such as Black and Native American tended not to be selected. Stronger articulation of socioeconomic status and religion but does not mean race does not matter in their educational contexts.
Intersectionality	Latin* women reported lower math/science interest in high school than did men, but represented a greater share of completed engineering degrees (75.3% and 61.3%, respectively).	Influence of socioeconomic status on education and career Saliency of religious identities	Focus on gender and other throughlines. Religious and socioeconomic identities were salient to college students in engineering, but gender and even sometimes race were less important. While Latinas were more successful in degree completion than Latinos, both struggled with their racial identities in the engineering program.
Centrality of experiential knowledge	Engineering-aspiring Latin* students had family connections to STEM professions—despite most being first-generation college students.	Family	Family exposure to STEM led to positive beliefs about and interest in STEM fields, particularly engineering fields. Experiential knowledge of STEM may have led to students pursuing engineering degrees.

experiences with discrimination (a potential read not conclusively supported by these limited data)—this has potential for the success of other racially marginalized communities. Research on racial reappraisal theory—explained next—can serve as a useful frame to capture these processes, whereby Latin* engineering students who are aware of racialized stigma associated with their identity in their field may limit interaction with racialized, predominantly white contexts (i.e., buffering) while maintaining connection to and accessing resources associated with these environments (i.e., bridging; Abrica, 2022). There are potential research and practice implications of this meriting future study.

Third, with respect to intersectionality, Latin* students do not fit neatly into US racial categories, alternately being considered part of an ethnicity and a racial group (see Cuellar & Salinas, 2022; Zerquera et al., 2020). The quantitative results find few Latin* engineering students identified in the study's base year (ninth grade) with a specific racial category other than Hispanic or Latina/o/x; greater racial awareness may emerge as they enter adulthood. However, the qualitative results also tend to shift away from discussions of race or racism, even when talking with a Latina interviewer, which they all did (either the first or second author, depending on the cohort). In its place, they discussed a spiritual/religious identity that they valued more (see also Friedensen et al., 2020)—or, more often, they discussed their socioeconomic experiences in association with precarity. As a second recommendation, we therefore recommend that researchers (as well as engineering faculty and administrators) working with Latin* engineering students consider their intersectional identities holistically, with care and engagement as advisors, mentors, supervisors, and other key student-facing roles. These key actors can alternately serve as obstacles for students such as through racial microaggressions (e.g., García-Louis et al., 2020) or support the retention and success of students who have felt marginalized during their studies (e.g., Bensimon et al., 2019; Luedke, 2017; Madsen Camacho & Lord, 2011). Our findings indicate how Latin* engineering students bring lived experiences and insights into their development as scientists and engineers. Even if they are not directly attentive to the racialization of the Latin* community and racial stereotypes about who can be an engineer, these contextual signals and a range of supports to counter these forces may factor into their engineering identity and career ambitions.

Fourth and finally, we attend to the experiential knowledge Latin* students bring into their engineering studies and research endeavors. The qualitative and especially the quantitative findings indicate that, once enrolled in and pursuing these programs, Latin* students do well and successfully complete degrees. This is a narrative that needs to be told more clearly and investigated more deeply. The qualitative findings report on how Latin* undergraduate students perceive their interactions and challenges in engineering research environments, guiding our understandings of how to better broaden participation in engineering in a manner that is inclusive and can undo the racial hierarchies that have traditionally existed in these fields. Students, including those reported on here, repeatedly shared how their experiential knowledge shaped their interest in pursuing engineering and any apprehension they may have held about continuing in this field.

CONCLUSION

To enhance the success of Latin* students in engineering, it is important to support them structurally, not just individually. This necessitates systemic investment in and sustained efforts pertaining to (a) intersectionally diverse faculty and laboratories in which race is less salient, and they can experience a nurturing environment of excellence, (b) networks of skilled mentors effective in working with minoritized populations, and (c) stable and consistent financial support that allows them to focus on their studies and view engineering graduate studies and careers as realistic, viable ambitions.

Authors' Note

We have no conflicts of interest to disclose. This material is based on work supported by the National Science Foundation under grant nos. 1646897, 1735968, and 1920670. Any opinions, findings, conclusions, or recommendations expressed in this material are those of the authors and do not necessarily reflect the views of the National Science Foundation.

NOTES

1. This chapter employs "Latin*" as a term widely used among contemporary undergraduate students in the United States. Latinx and another recent term ("Latine") draw on Latin* to signify Latin American origin while not being gender-specific (see also Salinas, 2020). An older term, "Hispanic" (developed for government purposes), is still commonly used today, signifying ethnic origin in Spain or the Spanish-speaking former colonies of Spain in Latin America. Latino (male) and Latina (female) are also used to signify identities associated with specific biological sex and/or gender. While in Spanish grammar, Latino is inclusive of all identities, it still employs "male" as the default category. Given this chapter's positioning with critical theory, we choose Latin* as the primary term to refer to Latin American origin students' race/ethnicity, while recognizing that language is fluid and will continue to shift after the publication of this chapter. See also this book's introductory chapter and its associated references.
2. Using tenets of CRT, Yosso (2005) challenges traditional deficit approaches to studying cultural capital and the experiences of students of color. Instead of viewing students of color as culturally disadvantaged, community cultural wealth recognizes the unique capital that students of color bring to campus.
3. Our restricted-use licensed data from the National Center for Educational Statistics (Institute for Education Sciences) mandates rounding to the nearest ten for individual counts and nearest tenth decimal for descriptive figures.
4. Importantly, while we investigated gender beyond the binary (see also Perez-Felkner et al., 2024), we found that there were no Latin* engineering students who opted out of the male/female biological sex question in the base year (ninth grade), and only two later identified as transgender in a more inclusive, gender identity question asked in college (one identified earlier as male; one as female).

REFERENCES

Abrica, E. J. (2022). Revisiting the complexity of racial understandings and subjective experiences of race among students of color in STEM higher education: Toward a racial reappraisal framework. *Journal of Women and Minorities in Science and Engineering, 28*(2), 31–52. https://doi.org/10.1615/JWomenMinorScienEng.2022036573

Anderson, D.M. Broton, K. M., & Monaghan, D. B. (2023). Seeking STEM: The causal impact of need-based grant aid on undergraduates' field of study. *The Journal of Higher Education, 94*(7), 921–944. https://doi.org/10.1080/00221546.2023.2209003

Bell, D. A. (2008). *And we are not saved: The elusive quest for racial justice*. Basic Books.

Bensimon, E. M., Dowd, A. C., Stanton-Salazar, R., & Dávila, B. A. (2019). The role of institutional agents in providing institutional support to Latinx students in STEM. *Review of Higher Education, 42*(4), 1689–1721. https://doi.org/10.1353/rhe.2019.0080

Bernal, D. D. (2002). Critical race theory, Latino critical theory, and critical raced-gendered epistemologies: Recognizing students of color as holders and creators of knowledge. *Qualitative Inquiry, 8*(1), 105–126. https://doi.org/10.1177/107780040200800107

Buzzetto-More, N., Ukoha, O., & Rustagi, N. (2010). Unlocking the barriers to women and minorities in computer science and information systems studies: Results from a multi-methodological study conducted at two Minority Serving Institutions. *Journal of Information Technology Education, 9*, 115–131. https://doi.org/10.28945/1167

Cabrera, N. L. (2018). Where is the racial theory in critical race theory? A constructive criticism of the crits. *Review of Higher Education, 42*(1), 209–233. https://doi.org/10.1353/rhe.2018.0038

Carter, D. F., Razo Dueñas, J. E., & Mendoza, R. (2019). Critical examination of the role of STEM in propagating and maintaining race and gender disparities. In M. B. Paulsen & L. W. Perna (Eds.), *Higher education: Handbook of theory and research: Volume 34* (pp. 39–97). Springer International. https://doi.org/10.1007/978-3-030-03457-3_2

Cho, S., Crenshaw, K. W., & McCall, L. (2013). Toward a field of intersectionality studies: Theory, applications, and praxis. *Signs: Journal of Women in Culture and Society, 38*(4), 785–810. https://doi.org/10.1086/669608

Collins, P. H. (2000). *Black feminist thought: Knowledge, consciousness, and the politics of empowerment* (2nd ed.). Routledge.

Collins, P. H. (2015). Intersectionality's definitional dilemmas. *Annual Review of Sociology, 41*(1), 1–20. https://doi.org/10.1146/annurev-soc-073014-112142

Crenshaw, K. (1991). Mapping the margins: Intersectionality, identity politics, and violence against women of color. *Stanford Law Review, 43*(6), 1241–1299. https://doi.org/10.2307/1229039

Creswell, J. W., & Creswell, J. D. (2017). *Research design: Qualitative, quantitative, and mixed methods approaches*. Sage.

Crisp, G., Nora, A., & Taggart, A. (2009). Student characteristics, pre-college, college, and environmental factors as predictors of majoring in and earning a STEM degree: An analysis of students attending a Hispanic serving institution. *American Educational Research Journal, 46*(4), 924–942. https://doi.org/10.3102/0002831209349460

Cuellar, M. G., & Salinas, C. (2022). Latin* college students' diversity and intersectionality: Future directions for research and practice. In M. Bonous-Hammarth (Ed.), *Neighborhoods, communities, and urban marginality* (pp. 55–83). Springer Nature. https://doi.org/10.1007/978-981-16-8000-7_4

DeCuir-Gunby, J. T., Chapman, T., & Schutz, P. A. (2019). *Understanding critical race research methods and methodologies*. Routledge.

Delgado, R. (1989). Storytelling for oppositionists and others: A plea for narrative. *Michigan Law Review, 87*(8), 2411–2441. https://doi.org/10.2307/1289308

Erichsen, K., Perez-Felkner, L., Zhao, T., Hu, S., & Chen, J. (2020). *Longitudinal effects of sex segregation in STEM extracurricular activities on gender disparities in STEM efficacy, identity, and postsecondary majors.* American Educational Research Association, Annual Meeting, Virtual.

Espino, M. M. (2014). Exploring the role of community cultural wealth in graduate school access and persistence for Mexican American PhDs. *American Journal of Education, 120*(4), 545–574. https://doi.org/10.1086/676911

Fluker, C., Perez-Felkner, L., & McCoy, K. (2022). *Students' perceptions of their engineering identity development and REU summer internship program experiences: An equity-centered analysis.* American Society for Engineering Education Annual Conference [Peer proceedings]. Minneapolis, MN. https://peer.asee.org/40969

Friedensen, R. E., Rodriguez, S., & Doran, E. (2020). The making of "ideal" electrical and computer engineers: A departmental document analysis. *Engineering Studies, 12*(2), 104–126. https://doi.org/10.1080/19378629.2020.1795182

García-Louis, C., Sáenz, V. B., & Guida, T. (2020). How community college staff inflict pervasive microaggressions: The experiences of Latino men attending urban community colleges in Texas. *Urban Education,* 0042085920963594. https://doi.org/10.1177/0042085920963594

Garcia, G. A. (2019). *Becoming Hispanic-Serving Institutions: Opportunities for colleges and universities.* Johns Hopkins University Press. https://doi.org/10.1353/book.66167

Garcia, G. A., Núñez, A.-M., & Sansone, V. A. (2019). Toward a multidimensional conceptual framework for understanding "servingness" in Hispanic-Serving Institutions: A synthesis of the research. *Review of Educational Research, 89*(5), 745–784. https://doi.org/10.3102/0034654319864591

Gaston Gayles, J., & Smith, K. N. (2018). Advancing theoretical frameworks for intersectional research on women in STEM. *New Directions for Institutional Research, 2018*(179), 27–43. https://doi.org/10.1002/ir.20274

Haynes, C., Joseph, N. M., Patton, L. D., Stewart, S., & Allen, E. L. (2020). Toward an understanding of intersectionality methodology: A 30-year literature synthesis of Black women's experiences in higher education. *Review of Educational Research, 90*(6), 751–787. https://doi.org/10.3102/0034654320946822

Hernández, E. (2016). Utilizing critical race theory to examine race/ethnicity, racism, and power in student development theory and research. *Journal of College Student Development, 57*(2), 168–180. https://doi.org/10.1353/csd.2016.0020

Ireland, D. T., Freeman, K. E., Winston-Proctor, C. E., DeLaine, K. D., McDonald Lowe, S., & Woodson, K. M. (2018). (Un)hidden figures: A synthesis of research examining the intersectional experiences of Black women and girls in STEM education. *Review of Research in Education, 42*(1), 226–254. https://doi.org/10.3102/0091732x18759072

Ladson-Billings, G., & Tate, W. F. (2016). Toward a critical race theory of education. In A. D. Dixson, C. K. R. Anderson, & J. K. Donnor (Eds.), *Critical race theory in education* (pp. 10–31). Routledge.

Luedke, C. L. (2017). Person first, student second: Staff and administrators of color supporting students of color authentically in higher education. *Journal of College Student Development, 58*(1), 37–52. https://doi.org/10.1353/csd.2017.0002

Lundy-Wagner, V. (2013). Is it really a man's world? Black men in science, technology, engineering, and mathematics at historically Black colleges and universities. *Journal of Negro Education, 82*(2), 157–168. https://doi.org/10.7709/jnegroeducation.82.2.0157

Mack, K., Johnson, L. R., Woodson, K. M., Henkin, A., & Dee, J. R. (2010). Empowering women faculty in STEM fields: An examination of historically Black colleges and universities. *Journal of Women and Minorities in Science and Engineering, 16*(4), 319–341. https://doi.org/10.1615/JWomenMinorScienEng.v16.i4.30

Madsen Camacho, M., & Lord, S. M. (2011). *Quebrando fronteras*: Trends among Latino and Latina undergraduate engineers. *Journal of Hispanic Higher Education, 10*(2), 134–146. https://doi.org/10.1177/1538192711402354

Malcom, L., & Malcom, S. (2011). The double bind: The next generation. *Harvard Educational Review, 81*(2), 162–172. https://doi.org/10.17763/haer.81.2.a84201x508406327

Martinez, S. M., Esaryk, E. E., Moffat, L., & Ritchie, L. (2021). Redefining basic needs for higher education: It's more than minimal food and housing according to California university students. *American Journal of Health Promotion, 35*(6), 818–834. https://doi.org/10.1177/0890117121992295

Mayoh, J., & Onwuegbuzie, A. J. (2015). Toward a conceptualization of mixed methods phenomenological research. *Journal of Mixed Methods Research, 9*(1), 91–107. https://doi.org/10.1177/1558689813505358

McGee, E. O. (2021). *Black, Brown, bruised: How racialized STEM education stifles innovation.* Harvard Education Press.

Núñez, A.-M., Crisp, G., & Elizondo, D. (2016). Mapping Hispanic-serving institutions: A typology of institutional diversity. *Journal of Higher Education, 87*(1), 55–83. https://doi.org/10.1080/00221546.2016.11777394

Omi, M., & Winant, H. (2014). *Racial formation in the United States.* Taylor & Francis. https://books.google.com/books?id=T7LcAwAAQBAJ

Ong, M., Wright, C., Espinosa, L., & Orfield, G. (2011). Inside the double bind: A synthesis of empirical research on undergraduate and graduate women of color in science, technology, engineering, and mathematics. *Harvard Educational Review, 81*(2), 172–209. https://doi.org/10.17763/haer.81.2.t022245n7x4752v2

Onwuegbuzie, A. J., Collins, K. M. T., & Frels, R. K. (2013). Foreword. *International Journal of Multiple Research Approaches, 7*(1), 2–8. https://doi.org/10.5172/mra.2013.7.1.2

Pawley, A. L. (2019). Learning from small numbers: Studying ruling relations that gender and race the structure of U.S. engineering education. *Journal of Engineering Education, 108*(1), 13–31. https://doi.org/10.1002/jee.20247

Pawley, A. L. (2020). Shift the default in "broadening participation" in STEM equity research. *International Journal of Gender, Science and Technology, 11*(3), 9. http://genderandset.open.ac.uk/index.php/genderandset/article/view/668

Perez-Felkner, L., Ramirez Surmeier, L., & Rezaei, R. (2024). Gender and educational inequality. In M. Berends, B. Schneider, & S. Lamb (Eds.), *Handbook of Sociology of Education* (pp. 142–157). Sage.

Perez-Felkner, L. C., Shore, C., Dickens, T., & Yang, M. D. (2020, June 22–26). *Engineering resilience through research mentorship: Manufacturing pathways to careers* [Conference paper]. American Society for Engineering Education Virtual Annual Conference. https://doi.org/10.18260/1-2--34557

Perna, L. W., Lundy-Wagner, V., Drezner, N., Gasman, M., Yoon, S., Bose, E., & Gary, S. (2009). The contribution of HBCUs to the preparation of African American women for STEM careers: A case study. *Research in Higher Education, 50*(1), 1–23. https://doi.org/10.1007/s11162-008-9110-y

Rincón, B., Fernández, É., & Hinojosa, J. K. (2020). "I wanted to follow in her footsteps": Activating, nurturing, and extending community cultural wealth for students of color entering STEM pathways. *Teachers College Record, 122*(9). https://doi.org/10.1177/016146812012200903

Rincón, B. E., & Rodriguez, S. (2021). Latinx students charting their own STEM pathways: How community cultural wealth informs their STEM identities. *Journal of Hispanic Higher Education, 20*(2), 149–163. https://doi.org/10.1177/1538192720968276

Rodriguez, S. L., Doran, E. E., Sissel, M., & Estes, N. (2022). Becoming *la ingeniera*: Examining the engineering identity development of undergraduate Latina students. *Journal of Latinos and Education, 21*(2), 181–200. https://doi.org/10.1080/15348431.2019.1648269

Salas Pujols, J. (2022). "It's about the way I'm treated": Afro-Latina Black identity development in the third space. *Youth & Society, 54*(4), 593–610. https://doi.org/10.1177/0044118x20982314

Smith, D. G. (2015). *Diversity's promise for higher education: Making it work*. Johns Hopkins University Press. https://books.google.com/books?id=nfVoCQAAQBAJ

Solórzano, D. G., & Villalpando, O. (1998). Critical race theory, marginality, and the experience of students of color in higher education. In T. R. M. Carlos Alberto Torres (Ed.), *Sociology of education: Emerging perspectives* (pp. 211–222). State University of New York Press.

Strayhorn, T. L. (2010). When race and gender collide: Social and cultural capital's influence on the academic achievement of African American and Latino males. *Review of Higher Education, 33*(3), 307–332. https://doi.org/10.1353/rhe.0.0147

Tate, W. F. (1997). Critical race theory and education: History, theory, and implications. *Review of Research in Education, 22*, 195–247. https://journals.sagepub.com/doi/10.3102/0091732X022001195

Yosso, T., Smith, W., Ceja, M., & Solórzano, D. (2009). Critical race theory, racial microaggressions, and campus racial climate for Latina/o undergraduates. *Harvard Educational Review, 79*(4), 659–691. https://doi.org/10.17763/haer.79.4.m6867014157m707l

Yosso, T. J. (2005). Whose culture has capital? A critical race theory discussion of community cultural wealth. *Race Ethnicity and Education, 8*(1), 69–91. https://doi.org/10.1080/1361332052000341006

Zerquera, D., Haywood, J., & De Muchas Flores, M. (2020). More than nuance: Recognizing and serving the diversity of the Latinx community. In R.T. Teranishi, B. M. D. Nguyen, C. M. Alcantar, & E. R. Curammeng (Eds.), *Measuring race: Why disaggregating data matters for addressing educational equity* (pp. 154–169). Teachers College Press.

Zhao, T., & Perez-Felkner, L. (2022). Perceived abilities or academic interests? Longitudinal high school science and mathematics effects on postsecondary STEM outcomes by gender and race. *International Journal of STEM Education, 9*(1), 1–26. https://doi.org/10.1186/s40594-022-00356-w

8 · *SER MARICA ES PA' MACHOS [SER BICHA É PRA MACHO]*

Agency, Activism, and Coping While Engineering

HÉCTOR E. RODRÍGUEZ-SIMMONDS,
LEONARDO POLLETTINI MARCOS,
CRISTIÁN VARGAS-ORDÓÑEZ, AND
KEVIN JAY KAUFMAN-ORTIZ

HOW TO ENGAGE WITH THE AUTHORS' PURPOSE

Oye! Si te llama la atención el título de este capítulo, es por algo. Ya sea porque eres un educador que quiere entender a sus estudiantes y actuar para garantizar experiencias significativas de aprendizaje a estudiantes de ingeniería. Ya sea porque estudias ingeniería y quieres apoyar a tus colegas. Ya sea porque quieres hacer chisme de nuestras historias. En cualquier caso, para bolas que te queremos contar cómo hemos construido y experimentado nuestra identidad como educadores en ingeniería, siendo parte de la población LGBTIQA+ y de la comunidad Latina.

Olá leitor! Se o nome deste capítulo chama a sua atenção, é por alguma razão. Pode ser que você seja um educador que quer entender seus alunos e trabalhar para garantir experiências de aprendizagem significativas para estudantes de engenharia. Pode ser que você estude engenharia e queira apoiar os seus colegas. Pode ser que você queira fofocar sobre as nossas histórias. Então foque aqui, queremos te contar sobre como temos construído e vivenciado nossa identidade de educadores em engenharia como parte da população LGBTQIA+ e da comunidade latina.

Hey reader! If the title of this chapter catches your attention, it's for a reason. You may be an educator who wants to ensure meaningful learning experiences for your diverse engineering students. You may study engineering and want to support your colleagues. You may even want to gossip about our stories. Whatever your reason, we are here to tell you how we built and experienced our identity as Latino LGBTQ+ engineering educators.

We ask you to try to put yourself in our shoes. It will serve as an exercise to pave the way for a more empathetic engineering education. Throughout this process, we clashed with the English language several times. Because our native languages are Spanish and Portuguese, they more accurately depict deeper and complex feelings about our experiences. There is power associated with certain words that cannot be captured in English. We intentionally used Spanish and/or Portuguese throughout the text when we felt that English was insufficient.

Engineers come from many different places and backgrounds—they carry a multitude of identities, stories, and interpretations of their lived experiences. Intersecting identities and contextual factors dictate the way people interpret their experiences. In this chapter, we—the four gay Latinx male educator coauthors—demonstrate how our intersecting identities have simultaneously afforded us hostilities and agency throughout our engineering education journeys. We used a collaborative autoethnographic approach to elicit and portray our experiences (Chang et al., 2012). Villanueva Alarcón et al. (2022) suggest allowing participants to self-select the most appropriate identifier for themselves, as well as asking more nuanced demographic questions that capture the complex realities of participants. They urge scholars to think critically about the history and implications of terminology in their studies. In this chapter, we used the term "Latinx" but want to acknowledge the tension we felt when making this practical decision.

Latinx engineering students continue to be structurally excluded within the contiguous United States. Within this structural exclusion, however, individual experiences are varied, even when those experiencing exclusion share identity labels. People who have something in common feel more socially connected (Cwir et al., 2011). Thus, when we assembled our team, we initially connected through four shared identities that would form the basis of this chapter: "Latinx," "engineer," "gay," and educator." As we began our collaborative autoethnography, we realized that we had additional identities and interests that would help contextualize our stories for one another, which enabled deeper discussion and analysis.

Our intention is to show what the varied experiences of gay Latinx men can look like in engineering. As we analyzed and reflected on our data, we identified a progression in our stories: coping with the adversities that we faced, developing our agency, and enabling current and future forms of activism for ourselves. Based on that progression, we share actions that educators may take in their own spaces to be more inclusive of students in positions similar to ours.

INTRODUCTION TO THE AUTHORS

In this section, we discuss our positionality, our outlook, and how who we are informs how we see, experience, and interact with our world.

Héctor

Hello,

I was born in one Latinx place (Mérida, Yucatán, México) and immigrated to another, (Miami, Florida, the United States) when I was about four years old. My mother is Colombiana, my dad is Mexicano, and I grew up in Miami (pronounced "Mee-ah-mee"). I had the privilege of United States citizenship bestowed on me from my grandmother very early in my childhood. This privilege was nearly invisible to me (Rodríguez-Simmonds, 2022), but it had at least two side effects: (a) it freed me from fears of deportation, which were rampant on the nightly news, and (b) it labeled me as United Statesian, which as I grew up imposed United Statesian traits on me, such as individualism, capitalism, and automobile ownership.

My family raised me to value education above all else. My maternal grandparents, who I grew up with until I was eighteen, believed education was a path toward socioeconomic stability. I believe that, for them, moving from Colombia to the United States was part of acquiring stability and the chance for greater opportunities. When I was young, I figured that moving to the United States was solely about money. Money played a prominent role for me and my family in the United States. It was seemingly at the center of a "better life" and American notions of meritocracy. It was through the inquiry for this book chapter (in particular, conversations with Cristián) that I realized that I did not fully understand what having a better life, what immigrating to the United States, could mean to my family. Initially, I felt that money directly implied stability. Stability came from educational opportunities. Through this chapter, I realized that stability also meant health care, freedom from war and theft, freedom from corruption, and less socioeconomic and political volatility. Education was a means out of poverty, providing a better, safer, more consistent life. Having a profession, *siendo educado* [being educated], was a route toward stability and consistency. My understanding of this country, my family, my path to becoming an engineer, and my cultural identities is an ever-evolving landscape.

Leo

Oi,

I was born and raised in Brazil (in the state of São Paulo), a country that does not share Hispanic heritage with other Latin American countries (Bethell, 2010). I was raised in an economically privileged environment, where my family could always provide me with food, a comfortable home, and access to quality education. I had to move to a different city for my undergraduate studies in materials engineering, which allowed me to *conviver* [share daily living experiences with others] with people from different backgrounds than I was used to at that point in my life. This experience allowed me to understand other lived experiences and explore different parts of myself, including some of my societal beliefs, my sexuality, and what my reality up to that point meant. I constructed my identities in those environments, which broadened my perspectives. I studied engineering because the science behind it fascinated me, but I have always struggled to see myself working in a standard engineering environment for a living because my perspective on the engineering profession is not aligned with corporate interests. Starting my PhD in engineering education in the United States has led me to share experiences with people from different cultures and recontextualize some of the identities I constructed during my undergrad years. Sharing these experiences is helping me understand the elements that I share with other Latin American cultures, including a larger sense of community and proximity, a more welcoming attitude toward other people, and even food preparation. Education is highly valued in my family, with most of my relatives having an undergraduate degree. However, I am the first person in my immediate family to pursue a PhD.

Cristián

Hola,

The history of my country, Colombia, and my family have had the heaviest influences on my engineering identity. My mother and father migrated to Bogotá from a rural area (Tolima and Valle del Cauca, respectively) in the 1960s. They sought the opportunities that domestic violence and *el conflicto armado* [Colombian armed conflict] snatched from them. Bogotá was the promised land to survive the armed conflict; it was their top priority. However, in the early 1980s, my mom lost her husband due to a tragic, violent event, and—like in a Latin American soap opera—she met my dad during the corresponding legal process. Some years later, my sister and I were born (1988 and 1985, respectively).

Colombia is legally ordered by economic and social classes. We were in the lower-middle class in the city, which allowed us to receive a subsidy to reduce utility bills and facilitate access to some resources. It is tough to jump from one economic class to another in Colombia, and education is usually the means to acquire socioeconomic mobility (Uribe Mallarino & Ramirez Moreno, 2019). Essentially, if you have money, you can avoid violent experiences. My parents encouraged my sister and me to pursue higher education. Despite my desire, they did not let me become an artist because it meant being poor. I then turned to medicine, but that meant a substantial economic effort for my parents. Hence, engineering became my only option, since it was highly valued in my country. It would provide me with the social mobilization I needed.

I became a part of the middle class as a chemical engineer. This position let me see the overall effects of *el conflicto armado* and its effects on daily life when I started to work in different places around the country. My career gave me physical and socioeconomic safety which others did not have. Rising above *el conflicto armado* also helped me have more access to health care and, consequently, a better quality of life. If I were not a chemical engineer, I would have been bullied or killed when I discussed uncomfortable topics in some workplaces. These topics included my political views toward the armed conflict or the unethical behaviors of my colleagues or collaborators. Thankfully, I had the support and protection of the companies with which I worked as a chemical engineer. Nevertheless, outside Bogotá, I was twice threatened with death when I actively questioned the essence of the conflict in front of paramilitary commanders and corrupt collaborators. Being quiet when faced with injustice is something I learned from my family, and I followed those ideals. That is why I spoke, knowing that engineering would potentially help me out of a dangerous situation. I had the resources to leave those places. Being a chemical engineer afforded me privilege.

Because of the relative privilege that engineering brought me, I developed an interest in seeking peace instead of speaking out against violence. That is why I became an educator. I want to help stop the violence by fostering uncomfortable discussions in the engineering environment, which means challenging behaviors and discourses that might support civil or criminal violence. As I understood from my previous research on Colombian chemical engineers' perception of themselves, engineers are willing to help society through engineering. However, they do not know how to help—or helping is translated as generating a solution for a technical problem without considering other voices. Traditional engineering education supports this conception of engineering, which is the base of a hierarchical, privileged, and exclusive knowledge structure. Thus, my research is related to integrating the arts and engineering to create peace by challenging those hierarchies.

Kevin

Hey,

My brothers and I were born in 1996 in Brooklyn, New York, but we moved to Hormigueros, Puerto Rico (PR), when we were three months old. Our father grew up in Brooklyn, and our mother grew up in the Bronx before migrating to western PR. There were many factors that afforded me privilege in PR. Being in a somewhat rural area as one quarter white and three quarters Puerto Rican, my multiples and I stood out as being the triplets and "gringos" of the crowd. Others would look to us to clarify their English since they thought we spoke it perfectly, given our American accent. I navigate multiple identities that provide privilege while "othering" me, depending on where I am physically. For example, being gay in my smaller town could have quickly become the "talk of the town," but being gay at the university has afforded me the ability to integrate into LGBTQ+ spaces without repercussion, nor many questions asked. As another example, I am perceived as "better than the rest" by being white in Puerto Rico: an identity that is known to be the majority group in the mainland United States, an identity associated with success, money, and beauty. In other words, I'm white passing. But, I'm still a plantain! "Tengo la mancha de plátano de la isla" [a traditional saying that translates literally to "I have the stain of the plantain from the island"] is embedded in my culture and traditions. As a Puerto Rican, I can relate easily to other Latinx people, a privilege that the average white person does not have. From *aguinaldos*, cooking, language, tranquility, a sense of community, and a lack of voting rights, it is clear that if you get to know about me more deeply, I am not United Statesian.

METHODS

Héctor previously used autoethnography to investigate his own experiences in becoming a gay, Latinx engineer (Rodríguez-Simmonds, 2022). In this chapter, we used collaborative autoethnography to explore our experiences of becoming engineering educators at the intersection of our gay, Latinx, and engineering identities. This allowed us to understand ourselves in the company of others (Chang et al., 2012; Rodriguez-Simmonds & Hira, 2021) as we developed mutually trusting relationships.

We incorporated three data collection methods in our journey—having multiple sources of data is a common strategy in autoethnographic studies (Holly, 2018; Mejia, 2016). First, we created radar charts (following Smith, 2015) to visualize the magnitude with which we expressed our four similarly named identities (gay, Latinx, engineer, educator), comparing each other over different moments of our lives (see figures 1 and 2). Second, we created written reflection pieces that narrated the development of each of these identities, including meaningful experiences and our interpretations of them at the time. Finally, we used these reflections as the basis for probing and questioning each other, leading to additional notes and providing deeper insights into our lived experiences.

After gathering these data and becoming more knowledgeable about each other's experiences, we generated guiding questions to extract the parts of our stories that we felt would be most useful to fellow researchers and educators:

1. What aspects of their identities did these four gay "Latinx" engineers leverage as assets throughout their academic/educational pathways?
2. How did these four gay Latinx engineering educators cope with potentially hostile engineering spaces?
3. What might shifting engineering spaces look like for these four gay Latinx engineers?

    With those questions in mind, we pulled quotes from our own stories that we felt could answer them. However, given that the questions were simply thinking tools, we grouped and categorized the quotes in emerging themes. The grouping yielded three broad categories—coping, agency, and activism—which formed the basis of our findings.

Finally, we wanted to highlight that the process of building trust, understanding, and affection for each other carried throughout the entire research journey. We understood that vulnerability and openness were valuable for collaborative autoethnography, but did not explicitly discuss our boundaries at the outset.

The multiple instances of data collection were carried out in work sessions that we held in our own homes, which were safer and more welcoming spaces for sharing ourselves. Each time we shared more about ourselves, there was an implicit renegotiation of trust. Building trust was crucial to this project, as some of the experiences we shared were painful (Ellis et al., 2011).

FINDINGS

Now, we communicate what it means to be "gay, Latinx, engineering educators" in the United States by presenting three main sections: (a) perceived hostility, (b) coping with hostility, and (c) activism and external coping. Perceived hostility describes the non-institutional and institutional hostilities we perceived in our histories, which we must understand in order to contextualize how we coped. Then, we discuss the internal coping mechanisms we used to navigate the hostilities we experienced, leading to acknowledging our agency. Lastly, we discuss the external coping mechanisms we developed and leveraged to navigate hostilities while evolving into educators and starting to tap into activism.

PERCEIVED HOSTILITIES

Over the course of our lives, we have perceived hostilities from places, people, and systems specifically related to our gay identities. Some of these hostilities come from our families and friends in the Latin American culture we grew up in, which we refer to as "non-institutional hostility." Other hostilities are embodiments of the collective engineering culture and ingrained in larger educational or corporate structures that span ethnicity and race, which we discuss as "institutional hostility." We focus on our perceptions because of the nature of our data: we analyze our experiences from the perspective of an insider's reflections on the past. As a consequence of the epistemological stances we embody and the nature of qualitative inquiry (Trinchero, 2014), we make statements from within our lived, perceived, perspectives.

Non-Institutional Hostility

Homophobia within families, workplaces, and society are common factors that lead to hostile experiences (Montagno & Garrett-Walker, 2022). For us, this perceived hostility looked like explicit allusions of dislike toward diverse sexual orientations, leading to self-repressive attitudes. For instance, during college, one of Héctor's family members once referred to Ellen DeGeneres—the famous openly lesbian American comedian—as "Ellen Degenerate" when watching television, alluding to his family member's disapproval of diverse sexual orientations. Through this and other microaggressions, Héctor perceived homophobic tendencies and felt unsafe exploring his sexuality around family. Similarly, Cristián perceived family hostility when his parents threatened to remove financial support when they found out he was gay: "My mother and father were distraught and threatened me with interrupting my studies if I didn't convert [to a straight orientation]. My instinct was to say that I would try. Otherwise, I wouldn't have had the opportunity to finish my major by myself."

In a different way, Kevin implicitly experienced an oppressive "don't ask, don't tell" (Bilimoria & Stewart, 2009) environment at home that pushed him to self-restrict discussing and embodying traits associated with his gay sexual orientation. Indeed, parents' preference not to acknowledge their children's diverse sexual orientation at home is a homophobic mechanism (Baker, 2002). The previous examples show how homophobia appeared in moments where diverse sexual orientations could have become pertinent topics of conversation but were instead censored.

The homophobia each of us experienced within our families reverberated into and beyond university, reinforcing self-restrictive behaviors around engaging with our gay sexual orientations. Héctor decided to avoid talking or thinking about being gay for several years, trying to follow a heteronormative lifestyle as a way to repress his desires and thoughts. The oppression Kevin faced at home led him to consider moving out. For Cristián, his workplace was a hostile environment with negative consequences that led to him being fired because his boss learned about his sexuality. This situation led to anxiety and alcoholism. In Leo's case, he refrained from accepting his sexuality and coming out because of his perception that being gay did not fit with societal standards (i.e., being "perfect"). He explains: "I had to be 'perfect' to society's eyes when being gay did not meet those expectations. . . . I only told two people about it [being gay]: my best friend (who lived in a distant place), and one of my gay friends."

Our self-restricting behaviors intersected with our other identities. For example, Héctor and Kevin responded to the perceived hostility of being gay in their Latinx contexts by steering away from their Latinx identity and toward a white-passing identity. In parallel, passing as white—in the sense of how people perceived their language skills and their socioeconomic status—helped them navigate US spaces more easily. Héctor recounts feeling compelled to distance himself from seemingly unsuccessful Latinx professionals: "What I saw as common among these [educated, professional Latinx] individuals was that they did not speak English natively and were Hispanic/Latino. From those observations, I extrapolated that I needed to distance myself from these Hispanic/Latinos if I was to move upward socioeconomically and garner respect in the U.S. I also felt I needed to get away from Miami to whiter (more American) places."

In part, this restriction stopped Héctor from experiencing being gay in *Español*, keeping his gay and Latinx identities separated. The previous quote delineates how language is tied to culture. Eliason and Schope (2007) found that identity labels are ethnically and culturally bound as part of the LGB[1] identity formation process. Following this, it seems that Héctor was experiencing a cultural clash between a Latinx and LGB identity: a possible cause for him to want to separate the two, provoking unhealthy self-restricting behaviors.

Institutional Hostility

We also identified meaningful instances of hostility within formal education institutions, including our basic (K-12) and engineering education. As current engineering education PhD students and future engineering educators, it is important for us to reflect on these hostilities to consider how they may still affect us and impact our future work. We classified these instances of institutional hostility into three types: hostility related to perceived homophobia, hostility related to gatekeeping in engineering, and unintentional hostility.

In high school, Kevin was afraid of coming out to some of his social circles because he feared the social judgment he would experience within that educational institution. Therefore, he would selectively come out to people he considered trustworthy—those whom he felt would not harm him because of his sexuality. In contrast, Leo is sometimes still afraid of coming out to professors from traditional engineering departments: "[S]ometimes I repress this [gay] identity when I work with professors from more 'traditional' engineering disciplines, as I fear they would not receive this information very well."

These examples represent hostilities related to perceived homophobia. In those circumstances, we were/are afraid of the social judgment that could arise as a consequence of coming out. The fear of social judgment was/is even more pronounced in situations where the power dynamics did/do not favor us.

The second type of institutional hostility we identified was gatekeeping in engineering, which might be related to the elitism present in the field (Paul, 2022; Tonso, 2014). Héctor faced many adversities when pursuing his engineering degree, due to his struggles with an educational system that does not accommodate his unique needs (Héctor was diagnosed with ADHD while in his PhD program). As a consequence, he had to directly deal with the gatekeeping associated with the engineering educational path. According to Héctor,

> I went to [an engineering faculty member's] office with my dismissal letter to discuss if there was anything else that could be done so that I wouldn't be kicked out of the university for having a low (sub 2.0) GPA. Of the many things that hurt from the entire episode of slowly failing, lying to myself, friends, and family that everything with my academics was okay, and getting academically dismissed from [a large university in Florida] (aside from feeling like my life had ended), was that the [influential engineering faculty member] told me that engineering wasn't for everyone. That f-cking pissed me off!

Circumstances like these showed us that engineering was not very welcoming of students with diverse learning needs. Additionally, we witnessed classmates falling prey to the elitism in engineering by being held to different standards in classes due to their gender. As a consequence, we perceived engineering as an unwelcoming environment: one that leads to many instances of hostility toward different groups of people.

The last type of institutional hostility we identified was unintentional hostility. Such instances of hostility arise when well-intentioned people lack awareness of the context in which they are trying to intervene. This can happen due to a lack of preparation around handling different identities or to genuine ignorance about those identities. The following example from Cristián highlights this hostility: "In fifth grade, my teachers were concerned about [bullying] and made a special class where everyone could talk about all the bullying I was having. As it is possible to imagine, it didn't help me feel safe. It was the opposite. I was exposed in front of everyone. My immediate response was to seek refuge in the books."

In middle school, Cristián experienced bullying because of his peers' perceptions of his sexuality. Although his teachers were well-intentioned in trying to reduce or stop Cristián's suffering, their lack of understanding of how to handle the topic made him look for something else that could offer refuge from the consequences of the intervention. Experiencing many hostilities led to the need to generate mechanisms to cope with them, such as Cristián and Kevin's tendency to seek refuge in books.

MECHANISMS FOR COPING WITH HOSTILITIES

"El cuerpo no quiere doler. Y el ser no merece el sufrimiento." Writing this chapter helped us recall that, in various domains, we have experienced hostility and social exclusion—both of which physically hurt (Eisenberger et al., 2003). To deal effectively with this pain, some of us sought refuge in books, numbed ourselves through alcohol consumption, sought refuge in our other, less stigmatized identities, and realized our agency through supportive communities and individuals. To overcome the pain, we came to believe we had agency and control over those personal situations. Bandura (1989) refers to these as "self-efficacy beliefs." For example, Kevin's self-efficacy belief enabled him to be an individual who can be both gay and an engineer, simultaneously, and who can create change in his life. In that sense, the self-efficacy belief was a requirement for coping that we created for ourselves to move beyond the pain we were feeling. Next, we separately discuss internal and external coping mechanisms.

Internal Mechanisms

Throughout this section, we introduce subheadings to facilitate connection with our purpose. We believe that the use of language like "I do for I" will help you envision yourself in our shoes while you read this section.

I do for I. Internal coping mechanisms, which encompass shielding or escaping one's reality, manifested for us in two ways: altering our perception of reality (often with the help of substances) and escaping into books. Cristián's anxiety—

which resulted from coming out to his parents and being threatened with needing to move out and pay for college on his own—was something he had neither thought about nor dealt with. After a long, tense year of living with them, extensively crying, and losing a significant amount of weight, he graduated and acquired his first engineering industry job. In that job, he leveraged alcohol as a coping mechanism and avoided facing his anxiety: "A week after [graduation], I got my first job and became more independent. Alcohol became my best friend since I never had the opportunity to heal my anxiety, and I started to party a lot."

Books provided another form of escape for some of us and conferred power through knowledge. For example, being more knowledgeable, especially in engineering, helped Kevin prove himself to others *despite* his gay identity: "I put my face in the books to prove my worth [to my parents]. [I wanted to prove] that being gay did not define the fact that I won't be successful."

Seeking refuge in books and acquiring knowledge through them also helped Cristián and Kevin feel less stigmatized. Héctor, on the other hand, sought refuge in his work and "often escaped into engineering work to not think about other aspects of [his] life." Proving himself as an engineer helped him feel more powerful, competent, and valuable.

Feeling less stigmatized through the identity of a knowledgeable professional person (i.e., an engineer) was a common theme of coping—with other less stigmatized identities, such as masculinity and neurotypicality, also providing refuge. Through our reflections, we came to understand that our less stigmatized identities contributed to our self-efficacy beliefs by helping us escape the pain we felt: we could perceive ourselves as being worthy and useful for society because of our engineering knowledge. Leaning on our engineering identity allowed us to feel like we had a choice about what we could do.

I believe I can do and I accept what it is. We can only control what we believe we can control (Rotter, 1954/2017). In our narratives, we realized that there were certain things that fall outside of our control. Accepting that there are such things allowed us to deal with the pain/discomfort of our situations while also realizing that we have control. We wanted to change things for ourselves. However, sometimes we felt powerless to make relevant-enough choices.

As shown earlier, Kevin leveraged his engineering identity with his family to demonstrate that his gay identity did not define his entire persona. His family had an idea about what being gay meant: a failure on a professional and personal level. By accepting that he had no control over how his family perceived homosexuality, Kevin harnessed something that was within his control, to set a counterexample for his family. He relied on his developing engineering identity after his internship in Spain: "When I came back to Puerto Rico and lived at home, I felt more confident and bold to be out in more public settings. [...] when I was a laboratory instructor, I felt no shame in mentioning that I identified as part of the LGBTQ+ community." Kevin saw himself as a successful gay engineering student, professional, and instructor, earning him the respect he wanted at the intersection of those identities.

Similarly, Cristián decided to move away from engineering at one point in his professional life because he did not want to be associated with the engineering profession and its negative ethical impacts. Realizing that he had no power to change engineering practices at the time led him to work in other fields until he understood what he *could* change: how we educate our engineers and how they are taught to practice the profession.

In summary, the hostilities we faced were not solely negative life moments. They also helped us to change our perception of life and contributed to our ability to deal with adversity—and, subsequently, to effect change.

External Mechanisms

Others do for I. Although we had hostile experiences in different environments, other people were essential in providing us with support and helping us feel more comfortable with some of our identities. Their support was fundamental in getting us where we currently are and helping us understand that coping does not have to be a lonely and individual process. For Leo and Cristián, their gay identities benefited from the support of others, as this support gave them the strength to accept their identities and to endure the time that it took their family members to accept them:

> I only told two people about it [being gay]: my oldest friend (who lived in a distant place), and one of my gay friends. These two people helped me get through this process of understanding and accepting this part of myself, and after a few months I felt ready to come out to all my friends. (Leo)
>
> During these almost 11 years together [with my husband, Alejandro], people around us have changed too. My parents changed their negative perception of homosexuality and now support us. They talk with Alejandro and have visited us here in the U.S. My sister struggled significantly with managing the relationship between my parents and me but has been supportive and lovely. I am my niece's godfather. My brother-in-law and my sister listen to me when they need advice. She has been my friend for a long time. To be honest, with Alex, my best friend, my family is the most important, and I don't care what others think if I am well with them. (Cristián)

For Héctor, having engineers in his family helped him to navigate some of the challenges associated with becoming an engineer and also to form a better understanding of what an engineering identity was and could be. He explains: "[My engineer uncle] would sit with me in front of the whiteboard, and I would do my remedial algebra math problems with him, talking through all my steps. He would pick problems, and I would work my way through them. I hated it. It got results. My uncles served as powerful examples of engineers."

Therefore, we understand our community and familial support as an important source of our self-efficacy beliefs. Forming a stronger sense of self-efficacy was essential in making us build the confidence that we needed to affect the community around us, whether that was with the intention of promoting change or not.

I do for I and it affects others. "I do for I" has become a powerful mechanism to cope with difficult circumstances. Simultaneously, we are aware that our actions directly affect other people. We leverage our agency and decide to cope, regardless of how others are affected. It is a big step to make decisions that are best for us, even though we know others may disagree with those decisions. Through agency, we can push back, exerting ourselves against societal beliefs.

While Kevin was studying for his undergraduate degree in engineering, he lived at his parents' home. He made the difficult decision to move out, using all his savings, because he wanted to become more independent—regardless of the costs or what his parents thought of it. His decision came from sound reasoning: homophobia was a threat to Kevin in his household, and engineering was a way for him to escape. However, his parents were the ones who provided him with these savings. Kevin took a threatening tone against his parents when he mentioned that he would be using his savings to move out, indicating that he knew (or hoped) that it would impact them somehow. By moving out, he made a positive decision for his well-being ("I do for I"), stemming from the money his parents helped him save and cutting off their ability to see him every day ("it affects others").

Both Héctor and Leo found that they could cope with experienced hostilities by coming out. Leo sought to be his authentic self in the workplace, but if he hid parts of himself, he could not express the several identities he possessed. He felt that society had expectations of him to be perfect and perceived that "traditional" engineering professors would negatively perceive his gay identity. In contrast, Héctor felt that his research positively impacted the lives of other LGBTQ+ engineers. He saw that he could produce meaningful work by researching LGBTQ+ individuals as an LGBTQ+ individual. By conducting this kind of research, he came out to the audiences for whom he wrote, using it as a vehicle "to make myself come out of the closet (even if not overtly)." Consciously, when Héctor decided to write about LGBTQ+ individuals, he came out in the process, bringing out his best work when writing openly about these identities ("I do for I"). Héctor provided the reader with the perspective of an insider, writing about LGBTQ+ matters, as he disclosed in resistance to society ("it affects others").

I do for others. The previous section establishes how we made and still make decisions for ourselves that unintentionally impact the people around us. Conversely, as we progress in our careers and become more confident in our identities, we make increasingly deliberate decisions to impact the community around us. Our goal is for the engineering community to become more accepting of our shared identities through changemaking (Foster & Riley, 2019).

Our ability to engage in changemaking involves the spaces we are trying to change and the roles we have in those spaces. Considering our educator identities, changing the cultural aspects of an academic community is not a simple task: it requires the effort of many people; it requires time; and it requires the understanding of everyone in the community. We understand that being open about our identities and using them to empower our actions is a privilege that is granted to us societally, in contexts that privilege those identities—and that power shapes the extent of our influence and the changemaking efforts we can enact. To demonstrate how our influence changed during our lives and affected our ability to enact change in our communities, we present four instances of our intentional changemaking.

First, we have created safer working environments by bonding with our colleagues until we have felt comfortable coming out to them. For Leo, this happened when he was an intern at a small materials testing company. Initially, he had no understanding of how safe that space was for coming out. Coming out to his boss was a significant hurdle and Leo himself never came out to his boss while working there: "Through some interactions that [my queer friend and I] had with our boss, we knew that he was a very traditional person and did not seem to have a very positive view of LGBTQ people, as he would sometimes make fun of that topic. Therefore, we felt very motivated by having each other in the workplace."

Having a supportive colleague opened the space to talk freely about diverse sexual orientations, and it was through those efforts that Leo and his colleague were later able to come out to more people at the company—and his colleague was even able to come out to their boss, later on. As Leo describes:

> During most of my time there, [my queer friend and I] were the only people that knew about each other's sexualities, but during my last month there, we ended up telling most of the other people in the company (remember, this was a small company—we were less than 10 employees in total). We did not end up telling our boss while I was there. After my internship had already ended for a few months, this friend told me that she told our boss about her sexuality and even took her girlfriend at the time to a dinner the company hosted for employees.

By being in a supportive relationship with his queer colleague, Leo and she were able to make changes to their small company's environment, positively influencing their ability to come out to coworkers and eventually their boss.

Another instance of intentional changemaking involved Leo pioneering engineering education research at the Brazilian public university he had attended. This change later inspired engineering education research inside the traditional engineering environment. Leo communicated the importance of engineering education by aligning scholarly tools (e.g., grants and empirical research) with the values of the institution. The faculty also supported the ideal of changing the landscape for engineering education research: "[A]fter I started, more people started to demonstrate some interest in doing this type of research, and by the time I finished my master's, we had three other people doing that kind of research for their master's, one post-doc student and one undergraduate student. I like to think that I helped shift that space by making more people aware of engineering education research, and paving the way for future students to pursue similar studies" (Leo).

Shifting others' perceptions in traditional engineering education institutions has had an impact on our careers and paved the road for new ideas for those who come after us. On the other hand, changemaking has allowed us to advocate for change in engineering from different perspectives. Héctor focuses on the practical effect research can have on his students, while Cristián seeks to change the focus of engineering through his research. The research itself thus acts as a changemaking device. "My personal goal is to transform engineering from a utilitarian perspective into one based on care (big words!). I focus my research on integrating knowledge, specifically the arts and engineering. I found my way to integrating my concerns as a person and my interests in humanity" (Cristián).

Héctor uses his research to change engineering spaces. His research directly ties into his identities—notably, the gay identity—which is something that helps him navigate academic environments as a gay man: His research implicitly makes him come out. As he explains it: "Indeed, doing research in the area of sexual orientation at the intersection of other identities within engineering environments gives me plausibility to discuss sexual orientation and its relevance in engineering spaces."

When embodying the role of an instructor, Héctor introduces himself and his research. His research helps him be more attuned to microaggressions in the classroom, influencing day-to-day decisions about content choice, advocating for his students, and helps him serve as a role model. The consequences of this changemaking include him feeling better about being out in the classroom and showing others that you can be both gay and a successful engineer. "At this point, I have taught a class enough that coming out to students in these ways doesn't cost me credibility in the classroom. [...] At this point, I boldly walk in and, unapologetically, confidently stride into spaces. I've learned that people are less likely to bring things up if nothing seems amiss" (Héctor).

Coming out has the ancillary benefit of helping him connect to students because it shows them that he is a holistic person, not just a distant professor on a screen. It also helps him feel more comfortable, as he does not have to manage the cognitive load of pretending to be a straight man. However, he recognizes that the range of his actions are still limited: "I still feel my voice deepen when I enter a room. The level of flamboyance I exude in engineering spaces is still somewhat limited. But I get to express my queerness by talking about my boyfriend, making jokes that aren't heteronormative in engineering spaces (to balance the landscape) with heterosexual individuals, and wearing nail polish and sauntering around with rainbow stickers on my water bottle." It is important to note that not everyone wants or needs to move from coping into activism. As our experiences demonstrate, that might be an intimidating prospect, because affecting the people around us requires reflecting on and dealing with power dynamics. For us, however, moving from coping into activism was a way in which we could see ourselves fulfilling our roles as educators and researchers without losing our gay or engineering identities.

CONTRIBUTIONS TO ENGINEERING EDUCATION RESEARCH AND PRACTICE

The experiences and stories and our interpretation of them in this chapter are not universal experiences of gay, Latinx, engineering students. However, the narratives we have shared could be happening to your students, inside or outside the classroom. Any one of us could be a colleague or a learner in your classroom. Following our intention at the beginning of this chapter for engaging with our writing, we ask you, the reader, to deeply reflect on yourself as an educator about your students and/or colleagues in light of our stories. We could have been any of them.

For Engineering Education Research

In this chapter, we discussed the sense-making we employed as we dealt with the hostilities we experienced. The broader engineering education literature discusses persistence (Meyer & Fang, 2019) and grit (Direito et al., 2021) as qualities students leverage to advance through challenging engineering environments. Throughout writing this chapter, we felt the terms "dealing" and "coping" suitably represented those experiences. A broader range of words to describe the process of building resilience could prove useful in accurately communicating the experiences of individuals at the intersection of several minoritized identities in engineering. Further, broader terms could provide individuals with immediate acknowledgment and recognition of their situations. Being seen as dealing and coping en route to developing grit and persistence could lessen the perceived individual burden, perhaps allowing a greater number of minoritized individuals to decide to take on and complete engineering degrees.

Case and Hunter (2012) describe adaptive responding as "coping, resilience, and resistance" (p. 259), which directly maps to our findings about internal and external coping. Our findings exemplify adaptive responses in a minoritized individuals subgroup (Littman, 2022). We constructed our educator identities as a culmination of our experiences and desires to help other minoritized individuals succeed. Understanding our individual and intersecting identities as assets in engineering took too many needless steps. While some may argue that living through our experiences led to our resilience and agency, grit and persistence, we can't help but wonder: What if we did not spend time unlearning and healing from the bifurcation of our identities and building mechanisms to mitigate the "psychological consequences of oppression" (Case & Hunter, 2012, p. 259)? What if we immediately felt a sense of belonging (Osterman, 2000) by understanding our intersecting identities as assets (Surace et al., 2022; Vaughan & Rodriguez, 2014) and mapping our identities' funds of knowledge (Vélez-Ibáñez & Greenberg, 1992; Verdín et al., 2019; Wilson-Lopez & Acosta-Feliz, 2021) directly onto our engineering practice?

Writing this autoethnographic book chapter helped us develop stronger connections with one another, creating an academically and friendly supportive space. To understand our fellow coauthors more accurately, we had to understand each other's stories from their viewpoints. We learned about their viewpoint by listening, reflecting on the other's narratives from our viewpoints, asking clarifying questions, and relaying back to the person telling their narrative what we were hearing. Over the course of this project, we built bridges across boundaries of understanding, developing a shared language to describe our experiences. Our team's trilingual nature allowed us to find literature from different places and in different languages that best accompanied and supported our narratives. Whenever possible, we encourage engineering education researchers to make a space for their participants' intersecting identities and find supporting literature from different contexts and places that explicate ideas from the participant's viewpoint. Valuing participants in a study as translators (not just in terms of language but also in terms of thought and experience) with whom we collaborate throughout research can create more impactful, accurate, respectful, and equitable research.

Additionally, fostering a space where participants and researchers can unfurl intersecting identities that are not explicitly at the center of the research inquiry adds nuanced explanatory power to research projects. Throughout our process, we felt it was important to inquire about and document the sociohistorical context within which our experiences and identities formed. The countries we originate from have their own histories of racism, colorism, and colonialism. When talking with a group homogenously categorized by the United States, we encourage researchers to unpack race and skin color differences. Even though we are not considered white in the United States, we are considered white in Latin America, affecting our experiences becoming engineers. Lastly, the four of us self-identify as gay Latinx engineers, and our narratives in this chapter highlight the similarities and differences between similarly held identity categories as adopted differently by us. Having the space to further develop our own understanding of our identities alongside others throughout the research process provided more intimate access to the project so that the project felt less United States-centric and more inclusive of all coauthors.

If society and engineering culture could value an individual's *conocimiento* [knowledge] (Anzaldúa, 2015) of their identities from the beginning, individuals would not need to develop coping mechanisms under hostile conditions to navigate threats from other identities. Instead, they could openly build bridges (Anzaldúa, 1996; Keating, 2015) across identities in community, rather than having to do so in a sheltered closet. Perhaps they could seamlessly walk the borderlands

(Anzaldúa, 1996) of their intersecting identities without needing to prove their validity to themselves in light of other identities. We will work unwaveringly until that day.

For Engineering Education Practice

We contribute to engineering education practice broadly. Specifically, one of our main aims is to caution practitioners about letting identity labels lead to the idea that individuals who embody a label share the same experiences. This chapter shows that we are all classified under the same ethnic, racial, and non-visible (gay) umbrellas. Yet, we have shown how our experiences embodying these labels and identities differ. In writing this chapter, we intended to problematize using these labels and identities as all-encompassing umbrellas. While these labels provide broad information and insight into superficial aspects of a person's life, the labels should not be used to avoid getting to know a learner more deeply and considering them individually. We urge educators to consider the heterogeneity of individuals in their classes.

Our chapter describes moments when we coped and dealt with hostilities faced throughout our professional preparation. Unfortunately, most engineering educators are not formally taught how to tackle hostilities when they emerge and may not feel comfortable addressing them. Tackling the discomfort of shifting power dynamics in classrooms to foster more equitable and inclusive spaces is critical (Farrell et al., 2021). We echo Holly's (2021) exhortation: the time to feel uncomfortable has passed, and we need to use our tools to implement changes in engineering. Furthermore, considering hostilities beyond the scope of an engineering classroom supports social–technical dualism (Cech & Waidzunas, 2011; Faulkner, 2000), enabling dismissive attitudes toward social phenomena and further excluding minoritized learners. Farrell et al. (2021) offer strategies for implementing inclusive chemical engineering learning spaces for LGBTQ+ learners that include acknowledging the hidden curriculum (Villanueva et al., 2018) by laying out explicit curricular roadmaps. Exposing the hidden curriculum and cultivating growth mindsets (Theard et al., 2021) can work together to bring recognition and validity to knowledge not typically considered valuable in engineering spaces.

These practices help disrupt notions of fixed or innate engineering ability, lessening hostilities experienced by minoritized students. Holly (2021) questioned the epistemological and ontological foundations of engineers and found that building relationships between students and the community, showing political clarity in curriculum and pedagogy, and incorporating students' lived experiences as valuable in engineering were explicit ways to address sociopolitical disconnects in engineering—necessary to create equitable learning spaces for Black male youths. Wilson-Lopez and Acosta-Feliz (2021) worked to reframe the skills and competencies of Latinx youths as assets by recognizing their funds of knowledge as applicable epistemological resources to solve engineering problems. As engineering educators, we should critically reflect: *What can and should an engineering educator be aware of to identify hostilities? When identified, how will engineering educators address them?*

As Cristián showed us in the "institutional hostility" section, his middle school teacher tried to discuss the bullying around him in her classroom, but the results were not as she expected. Her well-intentioned intervention to address bullying could have led to positive learning in the class for some students; however, in this case, the action revictimized Cristián (Daigle & Hawk, 2022). He was bullied again and felt exposed at the cost of his safety, becoming the focus of the class's attention. It can be difficult to highlight societal issues or identities broadly without bringing attention to students' traumas or minority stresses. *How can engineering educators shift their classrooms to include the identities and experiences of minoritized students without retraumatizing them beyond what these students face on a daily basis?*

Formal institutions have the potential to create safe spaces for people of all identities. However, when safe spaces are not present, LGBTQ+ people and individuals of other minoritized identities may not feel like they have a supportive place or person to whom to reach out. As we saw in the "non-institutional hostility," "institutional hostility," and "I do for others" sections, Leo made careful decisions about to whom to come out, to feel comfortable in spaces he inhabited. He felt that his friends provided support, but feared that societal norms would influence those whom he told to react negatively. At all educational levels, spaces that support individuals who experience hostilities at the intersection of their minoritized identities should be clear and visible. For example, Butterfield et al. (2018) cite understanding terminology that's important to their LGBTQ+ students and setting the classroom tone by having students voluntarily include their pronouns, advocating for LGBTQ+ STEM groups that validate engineering students' intersecting identities, and supporting LGBTQ+ colleagues. With targeted, visible, and clear institutional resources, every human has the capacity to be resilient and thrive. Instead of spending time resisting societal and cultural forces to feel like we could peacefully and validly practice engineering, we could spend our time evolving our engineering and engineering education practice from a place of wholeness. *What could safe spaces look like for people of various minoritized intersecting identities at your institution?*

CONCLUSION

In this collaborative autoethnography, we shared our stories to demonstrate what it took us to become engineering educators at some of our intersecting identities: gay, engineer, Latinx, and educator. We used multiple sources of data—including radar charts and narratives—to explore and share the stories of our identities and the hostilities we endured to become engineers. These hostilities manifested within and outside institutional boundaries. We had to learn how to cope and deal, to survive in engineering. Coping manifested through actions that initially served only ourselves but then progressed to serve both ourselves and others around us through conscious decision making. Our processes of coping evolved as we built confidence and resilience to deal with hostilities, which to us are key to being educators and changing engineering. We educate the individuals that comprise engineering: these same individuals will embody, change, and define engineering culture. In this chapter, we highlight bringing change to engineering education by encouraging educators to intentionally and caringly embrace individuals' identities and their corresponding, diverse experiences—rather than expecting those individuals to cope and deal to find their place in engineering.

Acknowledgments

We want to thank the editors for providing the opportunity to contribute to this book and for their detailed feedback. We also want to thank Dra. Emma B. Simmonds for suggesting that we explore the differences between our identity labels. Lastly, special thanks to Jacqueline Rohde, Yangyang Zhu, and Nathan McBurnett, who helped to revise earlier versions of this chapter.

NOTE

1. We follow each authors' original choice of acronym. Each derivative from the LGBTQ+ acronym has implications for the individual published research, although many social identity labels apply to the same larger community.

REFERENCES

Anzaldúa, G. (Ed.). (1996). *Making face, making soul = Haciendo caras: Creative critical perspectives by feminists of color* (1st ed.). Aunt Lute Books.
Anzaldúa, G. (2015). *Light in the dark/luz en lo oscuro: Rewriting identity, spirituality, reality* (A. Keating, Ed.). Duke University Press. https://doi.org/10.1215/9780822375036
Baker, J. M. (2002). *How homophobia hurts children: Nurturing diversity at home, at school, and in the community*. Taylor & Francis Group.
Bandura, A. (1989). Human agency in social cognitive theory. *American Psychologist, 44*(9), 1175–1184. https://doi.org/10.1037/0003-066X.44.9.1175
Bethell, L. (2010). Brazil and "Latin America." *Journal of Latin American Studies, 42*(3), 457–485. https://doi.org/10.1017/S0022216X1000088X
Bilimoria, D., & Stewart, A. J. (2009). "Don't Ask, Don't Tell": The academic climate for lesbian, gay, bisexual, and transgender faculty in science and engineering. *NWSA Journal, 21*(2), 85–103.
Butterfield, A., McCormick, A., & Farrell, S. (2018). Building LGBTQ-inclusive chemical engineering classrooms and departments. *Chemical Engineering Education, 52*(2), 107–113.
Case, A. D., & Hunter, C. D. (2012). Counterspaces: A unit of analysis for understanding the role of settings in marginalized individuals' adaptive responses to oppression. *American Journal of Community Psychology, 50*(1–2), 257–270. https://doi.org/10.1007/s10464-012-9497-7
Cech, E. A., & Waidzunas, T. J. (2011). Navigating the heteronormativity of engineering: The experiences of lesbian, gay, and bisexual students. *Engineering Studies, 3*(1), 1–24. https://doi.org/10.1080/19378629.2010.545065
Chang, H., Ngunjiri, F., & Hernandez, K.-A. C. (2012). *Collaborative autoethnography* (Vol. 8). Taylor & Francis Group.

Cwir, D., Carr, P. B., Walton, G. M., & Spencer, S. J. (2011). Your heart makes my heart move: Cues of social connectedness cause shared emotions and physiological states. *Journal of Experimental Psychology, 47*(3), 661–664. https://doi.org/10.1016/j.jesp.2011.01.009

Daigle, L. E., & Hawk, S. R. (2022). Sexual orientation, revictimization, and polyvictimization. *Sexuality Research and Social Policy, 19*(1), 308–320. https://doi.org/10.1007/s13178-021-00543-4

Direito, I., Chance, S., & Malik, M. (2021). The study of grit in engineering education research: A systematic literature review. *European Journal of Engineering Education, 46*(2), 161–185. https://doi.org/10.1080/03043797.2019.1688256

Eisenberger, N. I., Lieberman, M. D., & Williams, K. D. (2003). Does rejection hurt? An fMRI study of social exclusion. *Science, 302*(5643), 290–292. https://doi.org/10.1126/science.1089134

Eliason, M. J., & Schope, R. (2007). Shifting sands or solid foundation? Lesbian, gay, bisexual, and transgender identity formation. In I. H. Meyer & M. E. Northridge (Eds.), *The Health of Sexual Minorities: Public Health Perspectives on Lesbian, Gay, Bisexual and Transgender Populations* (pp. 3–26). Springer. https://doi.org/10.1007/978-0-387-31334-4_1

Ellis, C., Adams, T. E., & Bochner, A. P. (2010). Autoethnography: An overview. *Forum Qualitative Sozialforschung Forum: Qualitative Social Research, 12*(1), 273–290. https://doi.org/10.17169/fqs-12.1.1589

Farrell, S., Godwin, A., & Riley, D. M. (2021). A sociocultural learning framework for inclusive pedagogy in engineering. *Chemical Engineering Education, 55*(4), 192–204. https://doi.org/10.18260/2-1-370.660-128660

Faulkner, W. (2000). Dualisms, hierarchies and gender in engineering. *Social Studies of Science, 30*(5), 759–792. https://doi.org/10.1177/030631200030005005

Foster, E., & Riley, D. (2019). From feminist hacker meet-ups to engineering educators: Implications of social movements in technology for change-making at the level of higher education. *2019 ASEE Annual Conference & Exposition Proceedings,* 32863. https://doi.org/10.18260/1-2--32863

Holly Jr, J. (2021). Equitable pre-college engineering education: Teaching with racism in mind. *Journal of Pre-College Engineering Education Research (J-PEER), 11*(1). https://doi.org/10.7771/2157-9288.1282

Holly Jr., J. S. (2018). *"Of the coming of James": A critical autoethnography on teaching engineering to Black boys as a Black man* [Doctoral dissertation, Purdue University Graduate School]. https://hammer.purdue.edu/articles/thesis/_Of_The_Coming_Of_James_A_Critical_Autoethnography_On_Teaching_Engineering_To_Black_Boys_As_A_Black_Man/7495913

Keating, A. (Ed.). (2015). Now let us shift . . . Conocimiento . . . Inner work, public acts. In G. Anzaldúa, *Light in the Dark/Luz en lo Oscuro: Rewriting Identity, Spirituality, Reality* (pp. 117–160). Duke University Press. https://doi.org/10.1215/9780822375036

Littman, D. M. (2022). Third places, social capital, and sense of community as mechanisms of adaptive responding for young people who experience social marginalization. *American Journal of Community Psychology, 69*(3–4), 436–450. https://doi.org/10.1002/ajcp.12531

Mejia, C. (2016). *Siendo mujer: An autoethnography on identity formation, intersectionality, y nuestras historias.* Our Lady of the Lake University.

Meyer, M., & Fang, N. (2019). A qualitative case study of persistence of engineering undergraduates. *International Journal of Engineering Education, 35*(1), 99–108.

Montagno, M. J., & Garrett-Walker, J. J. (2022). LGBTQ+ engagement in activism: An examination of internalized heterosexism and LGBTQ+ community connectedness. *Journal of Homosexuality, 69*(5), 911–924. https://doi.org/10.1080/00918369.2021.1898802

Osterman, K. F. (2000). Students' need for belonging in the school community. *Review of Educational Research*, 70(3), 323–367. https://doi.org/10.3102/00346543070003323

Paul, R. M. (2022). The changing landscape of mechanical engineering: Learning to embrace my ecofeminist identity within the elitism of engineering. In M. Bailey & L. Shackelford (Eds.), *Women in Mechanical Engineering* (pp. 135–147). Springer.

Rodríguez-Simmonds, H. E. (2022). Transcending the margins and boundaries as Latin-American engineer. In K. J. Cross, S. Farrel, & B. Hughes (Eds.), *Queering STEM Culture in US Higher Education* (pp. 36–56). Routledge. https://www.routledge.com/Queering-STEM-Culture-in-US-Higher-Education-Navigating-Experiences-of/Cross-Farrell-Hughes/p/book/9780367769895

Rodriguez-Simmonds, H., & Hira, A. (2021). A collaborative autoethnographic dialog exploring the soul of engineering education. *2021 ASEE Virtual Annual Conference Content Access Proceedings*, 1–17. https://doi.org/10.18260/1-2--36559

Rotter, J. B. (2017). *Social learning and clinical psychology*. Prentice Hall. (Original work published 1954)

Smith, D.-M. (2015). *Effective use of radar charts*. Model Systems Knowledge Translation Center. https://msktc.org/lib/docs/KT_Toolkit/Charts_and_Graphs/Charts_and_Graphics_Radar_508c.pdf

Surace, A., Kang, A., Kahler, C. W., & Operario, D. (2022). "I'm gay with an asterisk": How intersecting identities influence LGBT strengths. *Journal of Homosexuality*, 1–21. https://doi.org/10.1080/00918369.2022.2132579

Theard, M. A., Marr, M. C., & Harrison, R. (2021). The growth mindset for changing medical education culture. *EClinicalMedicine*, 37, 100972. https://doi.org/10.1016/j.eclinm.2021.100972

Tonso, K. L. (2014). Engineering identity. In A. Johri & B. M. Olds (Eds.), *Cambridge Handbook of Engineering Education Research* (pp. 267–282). Cambridge University Press; Cambridge Core. https://doi.org/10.1017/CBO9781139013451.019

Trinchero, R. (2014). Five research principles to overcome the dualism quantitative-qualitative. *Education Sciences & Society*, 5(1), 45–65.

Uribe Mallarino, C., & Ramirez Moreno, J. (2019). Clase media y movilidad social en Colombia. *Revista Colombiana de Sociología*, 42(2), 229–255. https://doi.org/10.15446/rcs.v42n2.50749

Vaughan, M. D., & Rodriguez, E. M. (2014). LGBT strengths: Incorporating positive psychology into theory, research, training, and practice. *Psychology of Sexual Orientation and Gender Diversity*, 1, 325–334. https://doi.org/10.1037/sgd0000053

Vélez-Ibáñez, C. G., & Greenberg, J. B. (1992). Formation and transformation of funds of knowledge among U.S.-Mexican households. *Anthropology & Education Quarterly*, 23(4), 313–335. https://doi.org/10.1525/aeq.1992.23.4.05x1582v

Verdín, D., Smith, J., & Lucena, J. (2019). Recognizing engineering students' funds of knowledge: Creating and validating survey measures. *2019 ASEE Annual Conference & Exposition Proceedings*, 33226. https://doi.org/10.18260/1-2--33226

Villanueva Alarcón, I., Mejia, J. A., Mejia, J., & Revelo, R. (2022). Latiné, Latinx, Latina, Latino, or Hispanic: Problematizing terms often used in engineering education. *Journal of Engineering Education*, jee.20486. https://doi.org/10.1002/jee.20486

Villanueva, I., Gelles, L., Di Stefano, M., Smith, B., Tull, R., Lord, S., Benson, L., Hunt, A., Riley, D., & Ryan, G. (2018). What does hidden curriculum in engineering look like and how can it be explored? *2018 ASEE Annual Conference & Exposition Proceedings*, 31234. https://doi.org/10.18260/1-2--31234

Wilson-Lopez, A., & Acosta-Feliz, J. (2021). Transnational Latinx youths' workplace funds of knowledge and implications for assets-based, equity-oriented engineering education. *Journal of Pre-College Engineering Education Research (J-PEER)*, 11(1). https://doi.org/10.7771/2157-9288.1289

9 · STUDYING LATINAS' EXPERIENCES IN ENGINEERING COURSES

Toward a Conceptual Framework

SELYNA PÉREZ BEVERLY AND
LISA R. LATTUCA

Women and people of color are especially needed in engineering fields to develop products and systems that serve the needs of all people and not just dominant or powerful populations. In 2018, women earned approximately half of all science and engineering bachelor's degrees, yet there are significant disparities across the various science and engineering disciplines (National Center for Science and Engineering Statistics [NCSES], 2022). According to the NCSES, compared to fields like psychology and biological sciences, which have the highest number of women who have graduated, engineering is among the lowest. Moreover, the field of engineering has only made minimal gains in recent years, with women comprising 18 percent of graduates in 2008 and only 22 percent in 2018. The 2018 data also show that people of color receive fewer bachelor's degrees in engineering than they do in science (NCSES, 2022).

Of the 12 percent of Latinx students who received degrees in engineering in 2018, 3 percent were Latinas. Although Latinas enroll in higher education in greater numbers than Latinos, they remain severely underrepresented in engineering compared to their Latino counterparts (Flores, 2011). Representation in engineering also declines in the workforce. According to NCSES data from 2017, women comprised 16 percent of science and engineering occupations but only 9 percent were Latinas. These baccalaureate completion and workforce data demonstrate that, as a field, engineering lacks significant representation—not only of women but of Latinas, and the trends are disconcerting.

In a scholarly literature review conducted on Latinx students in engineering, Revelo et al. (2017) describe how practices such as mentoring, family involvement, peer support, and positive faculty support contribute to the retention of Latinx students. Perceptions of a negative classroom climate and racialized and gendered interactions with peers and instructors may be particularly influential on Latinas' socioemotional outcomes in engineering (Banda, 2020; Garriott et al., 2019). These findings are congruent with studies across STEM fields, which indicate that unfavorable interactions with instructors and peers can negatively influence self-efficacy and sense of belonging for women of color (e.g., Lester et al., 2016; Ong et al., 2011; Rainey et al., 2018; Riegle-Crumb et al., 2020; Stewart et al., 2020; Winterer et al., 2020).

Both Banda (2020) and Garriott et al. (2019) provide further insight into how Latinas in engineering perceive the climate in engineering classrooms and how that perception affects their socioemotional well-being. Banda (2020), in a qualitative study on Latinas in engineering, found that perceptions of a "chilly" departmental climate were influenced not only by instructors' lack of care about student success but also by meritocratic grading policies, such as grading on a curve, that encouraged competition. Garriott et al. (2019), in their qualitative study, also found that Latinas in engineering encountered racialized, gendered microaggressions from men instructors and peers, which contributed to self-doubt and negatively affected their class participation.

While scholars have directly or indirectly identified engineering courses as important sites for further study, few frameworks for studying marginalized students' experiences in engineering focus on their perceptions of instructors, instruction, or peer experience in engineering courses and the corresponding effects (for an exception, see McGee and Martin's [2011] study of African American students' experience in engineering). This chapter proposes a conceptual framework to guide research designed to understand the effects of instruction on Latinas and their persistence in engineering. Such studies may provide greater understanding of the complexities Latinas may be encountering in engineering courses. At a time when the number of studies on Latinx students in engineering is growing, many scholars have discussed—but not systematically examined—Latinas' experiences in undergraduate engineering courses. These studies often include references to influential classroom experiences (e.g., Banda, 2020; Banda & Flowers, 2018; Garriott et al., 2019; Revelo & Baber, 2018; Rincón & Rodriguez, 2021; Rodriguez et al., 2020a; Rodriguez et al., 2022; Verdín & Godwin, 2018; Villa et al., 2020) and thus suggest the need for greater research attention to Latinas' experiences in their engineering courses. We draw on these studies whenever possible to acknowledge the evidence they provide and its contribution to the elements of the proposed framework.

We also base our framework on several other sources of information. First, it is informed by theory and research on the culture of engineering and similar STEM disciplines. We also utilize findings from empirical studies of students of color in STEM courses, but we supplement this broader literature with findings from the first author's mixed methods study of women's experiences of instruction in two online undergraduate chemical engineering courses at a Predominately White Institution. Selyna Pérez Beverly's (2022) study yields insights into how women's perceptions of instruction and interactions with peers and instructors in two focal classrooms were related to perceptions of classroom climate, sense of belonging, engineering self-efficacy, and desire to remain in the field—topics that have been of enduring concern among scholars studying the experiences of Latinx students. For the purposes of this chapter, we limit our discussion to engineering courses and classrooms as spaces in which students are learning and by which they have interactions with a primary instructor and engineering peers. This can include courses that are lecture-based, do or do not include lab experiences, or are primarily online. We also consider students' experience with a wide range of instructors, including both full-time and part-time faculty.

Working from a theoretically and empirically derived conceptual framework can help guide future studies on Latinas and the effects of instruction in engineering courses, suggesting concepts of interest and relationships to explore using a variety of research methodologies and methods. The following sections outline the conceptual framework's dimensions, as well as the methodologies that shaped its development. Analyzing Latinas' experiences in engineering courses—where both race and gender constructs exist—can illuminate underlying messages associated with engineering culture that may be contributing to Latinas' decisions to leave engineering. Studying the experiences of Latinas in engineering courses can not only demonstrate how oppression may manifest but also reveal ways that it can be altered.

RESEARCHER POSITIONALITY

The first author is a cisgender, non-STEM Latina, whose professional and research experiences include working within engineering departments and colleges. She has experienced sexism and witnessed the subjugation of women in these contexts, leading to her desire to engage in research on women in engineering and other STEM contexts. The second author—a cisgender white woman who studies curriculum, teaching, and learning in higher education—has collaborated with engineers on studies of undergraduate engineering education for twenty years, with the goal of improving students' educational experiences and learning. Both authors are cautiously optimistic that the growing emphasis on critical perspectives and the experiences of marginalized students in research and practice will yield the cultural changes long needed in engineering programs.

INTERSECTIONS OF RACE AND GENDER

Feminism and the recognition of the sexist and racist power structures that exist in engineering influenced the development of the conceptual framework presented in this chapter. Exploring the development of science culture can give credence to how engineering culture became situated in its current cultural values that "other" and exclude marginalized populations. Harding (1991) argues that the scientific methods used in science, focused on eliminating social influences and values, support the values that are part of the predominant white, Eurocentric, and masculine culture of science. Feminist standpoint theory, Black feminist standpoint theory, and feminist critiques of pedagogy can expose the mechanisms of dominance that operate in an environment such as an engineering classroom, creating power structures through dynamics between instructors, peers, and women.

Historically, science culture has centered the voices of white men while actively excluding women in the process (Roychoudhury et al., 1995). Taking a feminist standpoint is to oppose the positivistic approach that attempts to define women's low levels of participation in scientific and technical fields by placing blame on the women (e.g., lack of interest and spatial skills). A feminist standpoint argues that women are not the problem; rather, structures in place in social institutions prevent women from being considered equal to men.

Black feminist standpoint theory illuminates the interlocking issues of dominance that can emerge in social structures such as an engineering classroom. Collins (1986) describes economic, racial, and gender oppression as interlocking, thus indicating that by only focusing on one of these, we neglect oppression in the other areas. These interlocking forms of domination create a dualistic pattern that places categories at odds with one another (e.g., White–Black, men–women, reason–emotion), in which the categories "gain meaning only in relation to their difference from their oppositional counterparts" (Collins, 1986, p. 520). It is therefore important to consider how race and gender intersect to shape Latinas' perceptions of their experiences in engineering courses.

The dual systems of oppression that Latinas face is necessary to consider when engaging in research. Researchers on Latinas in engineering have studied how the intersections of racism and sexism contribute to the marginalization of Latinas in engineering, thus creating barriers to persistence (Banda, 2020; Banda & Flowers, 2018; Garriott et al., 2019; Rodriguez et al., 2022; Rodriguez et al., 2020a; Verdín & Godwin, 2018; Villa et al., 2020). Rodriguez et al. (2020a) and Villa et al. (2020) conducted phenomenological studies that found that Latinas in engineering resisted sexist, racist, and classist structures by focusing on positively impacting their communities and finding support through Latinx affinity groups, respectively. Verdín and Godwin (2018) also found that Latinas benefited from the peer group support of those with similar backgrounds, helping them to overcome challenges associated with the intersection of their race and gender. Because engineering has historically engaged in both gendered and racialized exclusionary practices, feminist standpoint theory is useful to understand how dominance in the engineering classroom might operate.

AN EMERGING CONCEPTUAL FRAMEWORK: WOMEN'S EXPERIENCES WITH ENGINEERING INSTRUCTION

The conceptual frameworks utilized in studies of engineering and STEM fields to examine the persistence and retention of students have typically focused on students' career interest, self-efficacy, and science/engineering identity. For example, scholars have used social-cognitive career theory to understand students' persistence in engineering (Lent et al., 2016), women's persistence in engineering majors (Lent & Hackett, 1987), and students' choice of engineering majors (Lent et al., 2008). Other studies have used Bandura's (1977) theory of self-efficacy, which posits that a person's belief in themselves influences their ability to complete a task, to understand engineering persistence in the major, career self-efficacy, and career behaviors beyond college. Still, other studies have centered the role of science/engineering identity formation. Godwin et al. (2016) developed the critical engineering agency framework to understand women's internalization of their STEM identities before entering the engineering major, while Carlone and Johnson (2007) utilized social practice theory to understand the science identities of women of color and their negotiation of the culture of science. Rodriguez et al. (2020b) specifically developed a conceptual framework to study Latina undergraduates' computing identities and influential experiences in computing contexts.

Each of these frameworks has supported studies that provide evidence of the challenges marginalized students encounter in STEM programs. The framework we propose takes a different approach. It centers analytical attention on how instructional experiences in engineering courses contribute to women's socio-emotional outcomes and therefore to their persistence. This approach places emphasis on teaching practices and pedagogical approaches that higher-education institutions have the power to change, by providing key professional development to instructors. Instructors' awareness of their effects on students may be instrumental in motivating them to adapt their instruction to be more inclusive. Thus, this framework centers the students' experience and perceptions of instruction to reveal how the classroom context can contribute to women's affective states, which have been linked to persistence in the major (e.g., Riegle-Crumb et al., 2020; Stewart et al., 2020, Wilson et al., 2015; Winterer et al., 2020).

Focusing on the role of instruction and classroom climate can provide insight into how instructors can create environments that can either enhance or diminish the educational experiences of Latinas in engineering classrooms. Of particular interest are the consequences of the intersection of race and gender that Latinas encounter in classroom spaces. Here, we briefly describe the conceptual framework that guided Beverly's (2022) study. A discussion of the supporting literature follows this description.

The proposed framework is depicted in figure 9.1. It posits that aspects of science and engineering culture—such as the values placed on meritocracy, objectivity,

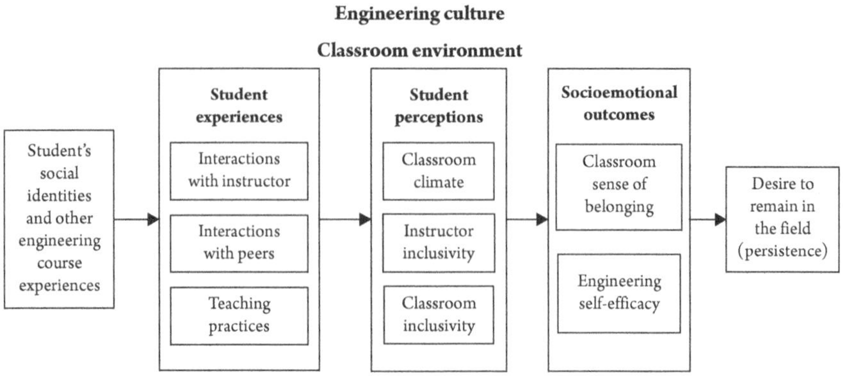

Engineering culture

Classroom environment

FIGURE 9.1. The Effects of Instruction on Women in the Engineering Classroom

and neutrality, as well as a history of patriarchy—infiltrate engineering class-rooms. These cultural values are manifested by instructors and men peers, and potentially by women students as well. Specifically, the framework assumes that engineering culture shapes instructors' pedagogy and their interactions with students, as well as students' interactions with their peers in the classroom. Accordingly, women's interactions with instructors and peers shape their perceptions of the classroom climate and affect their classroom sense of belonging, engineering self-efficacy, and desire to remain in engineering (a measure for persistence).

Both quantitative and qualitative findings from Beverly's (2022) study support several assumptions of this conceptual framework and provide additional insights into women's experiences and socioemotional outcomes. Of particular importance, gendered interactions in engineering classrooms communicate negative messages about women's abilities and skills in engineering and negatively affect women's academic self-confidence. In addition, women enter the classrooms with negative perceptions of themselves and their abilities.

Findings also support the assumption that instructors' pedagogical strategies affect women's perceptions of classroom climate. Specifically, findings indicate that inclusive teaching practices—such as providing positive reinforcement when students ask and answer questions as well as showing care by allowing students to provide feedback in the course and by having positive personal interactions with students in office hours—contribute to women's perceptions of a positive classroom climate. Based on qualitative evidence, a positive classroom climate, which women described as including a sense of comfort with the instructor and other students, contributed to women's desire to participate and engage in class.

Qualitative findings also suggest that other experiences in the classroom—such as group interactions into which women can self-select and positive connections made in the classroom—contributed to women's sense of belonging in the classroom. Women who reported engaging in groups where they felt safe, often due to group composition by gender and/or race, felt more comfortable and less intimidated. As a result of the pedagogical strategies in which the instructors engaged, women developed (a) a belief in their ability to learn the course content (self-efficacy), (b) a desire to take other engineering courses, and (c) a belief that they could finish a degree in and pursue a career in engineering.

In the following sections, we elaborate on the conceptual framework by reviewing the theoretical and empirical literature that supports its assumptions, incorporating literature specific to Latinas when possible. We organize this review according to the major components of the framework and its key constructs: namely, engineering culture; classroom climate, including instructor and peer influences; classroom sense of belonging; engineering self-efficacy; and desire to remain in engineering.

STEM Culture/Engineering Culture

Both STEM and engineering culture, rooted in masculinity and white supremacy, have unique ways of socializing individuals and work to exclude both women and people of color (Carter et al., 2019; Harding, 1991; Ladson-Billings, 2000). Engineering culture, then, can shape Latinas' experiences in higher-education institutions and their academic programs and may permeate and be observed in engineering courses where interactions occur between Latinas and their instructors and peers.

Engineering disciplinary culture shapes what occurs both inside and outside the engineering classroom, contributing to how women and people of color are perceived and treated. Critical scholars (e.g., Haraway, 1988; Harding, 1991; Ladson-Billings, 2000) posit that both white supremacy and masculinity are embedded in science. They argue that white men have had a significant role in equating scientific values with white, masculine attributes—thus placing women and people of color as outsiders, since they are viewed as lacking the attributes of white men that make them successful in STEM. According to Ladson-Billings (2000), these dominant paradigms within science can limit the voices of those with differing cultures and identities. Kishimoto (2018) contends that the hidden curriculum in higher education is shaped by Eurocentric values and male privilege and assimilates students to dominant epistemologies around objectivity and truth; in this way, it oppresses marginalized populations, such as students of color and women. As such, it is important to consider the interlocking nature of race and gender and how this may apply to the experiences of Latinas in engineering.

The historical, institutionalized practices of science contribute to the idea that science is universal and that a person's social identity does not influence scientific practice. Carter et al. (2019) discuss how meritocracy, which they argue values intelligence over other characteristics, has permeated higher education, creating and maintaining a social hierarchy. Ong et al. (2011) argue that the culture of STEM departments includes a structure of meritocracy that focuses on grades, classroom performance, and research while ignoring the intersections of racism and sexism in the science environment.

Professional majors such as engineering are particularly influenced by certain ideologies and values that contribute to its exclusionary nature. Farrell et al. (2021) explain that engineering culture consists of elitism through practices of epistemological dominance (devaluing other ways of knowing), majorism (placing higher value on STEM over the liberal arts), technical social dualism (the belief that diversity, equity, and inclusion [DEI] should not be part of engineering), and stereotype threat (negative stereotyping's psychological effect on identity groups). Kuzawa (2017) points specifically to engineering curricula as manifestations of particular values that shape the behaviors and attitudes of both students and instructors.

Women and people of color are also subjected to a socialization process within engineering culture that encourages conformity to social norms (see Johnson, 2019). Engineering also engages in dominant discourses that actively exclude others who do not conform to the culture. Research reveals how engineering culture contributes to the negative treatment of women through sexist behavior of both instructors and peers (see for example Tonso, 1996).

Engineering culture also has a considerable role in shaping course experiences for Latinas because of the focus on individualism instead of the collective community (López et al., 2019). The influences of Latinas' cultural values and ideals are important to consider when studying their experiences in engineering courses and their relation to the experiences of marginalized students such as Latinas.

Classroom Climate: Instructor and Peer Influences on Latinas in Engineering

Research suggests that engineering culture is likely a contributing factor to the academic and socioemotional challenges women can encounter while pursuing engineering degrees. Instructors and peers, for example, may believe that characteristics stereotypically possessed by men (e.g., logic and reason) are what make individuals successful in engineering and thus engage in practices and behaviors that are exclusionary of women, whom they perceive as not having these traits (e.g., Hottinger, 2016; Ridgeway & Correll, 2004).

Racial climate within engineering contexts has been discussed as a contributor to Latina students' experiences and their development (Banda & Flowers, 2018). In their qualitative study of eleven undergraduate Latinas in engineering, Banda and Flowers found that the "double bind" Latinas experience because of their race and gender contributes to their perceptions of hostile climates within engineering. They found that Latinas sought others who related to their race and/or gender because of their vast underrepresentation in engineering, experienced sexism within all areas of engineering contexts, and had diminished desire to enter the engineering workforce because of marginalizing experiences within engineering and internships. To complement studies like these that examine the experiences of Latinas in broad engineering contexts, researchers need to study other environments at a localized level that may be contributing to Latinas' experiences in higher education. Courses are an environment where negative interactions regularly occur with peers and faculty. Studying such interactions may reveal the extent to which such experiences, as well as positive interactions, contribute to Latina persistence in engineering. Given the frequency of students' interactions with instructors and peers in courses, focusing on this environmental context can provide important evidence that is largely missing from our current understanding of Latinas' experiences in engineering.

Instructor interactions. Instructors and their pedagogical strategies, as well as the way they interact with students, create a classroom climate that can have an impact on Latinas. According to Dewsbury (2020), creating inclusive classrooms can help mediate the challenges that underrepresented students, such as Latinas, encounter because of traditional exclusionary STEM methods of teaching. He contends that instructors' self-awareness (i.e., understanding their social position and the personal histories they bring to the classroom) and empathy (i.e., listening to and understanding the needs of students) combined with pedagogy (i.e., incorporating strategies to maximize "deep learning") contribute to the development of a positive classroom climate. Espinoza (2013) found a connection between Latinx students' perceptions of professors as unsupportive and inaccessible, which contributed to whether Latinx students felt validated, or worthy of the professor's attention, while Park et al. (2020) found that discrimination from instructors had a strong negative effect on STEM retention for Latinx students. These studies demonstrate that instructors can certainly have a role in shaping the experiences of Latinas and their persistence in engineering.

Latinas in engineering courses can also encounter both racial and gender microaggressions. Lee et al. (2020), in a large quantitative study, found that women of color in STEM experienced racial microaggressions from their instructors at a higher rate than men. Ong et al. (2011) discuss that women of color, such as Latinas, face intersecting systems of oppression due to their gender and race. Based on their review of the literature, they conclude that women of color's persistence is linked to the role of faculty in the classroom and their pedagogical approaches, which can affect women of color's participation. Negative treatment from instructors in STEM classrooms, such as singling out women in the classroom or ignoring them, can lead to isolation and fewer interactions between women and instructors (Lester et al., 2016; Murray et al., 1999).

Importantly, instructors can also influence student engagement and self-efficacy. Gasiewski et al. (2012), in a large mixed methods study, discovered that courses in which the instructor indicated openness to student questions and believed that part of their role was to help students succeed had more engaged students. Validation from instructors also appears to contribute to Latinas' self-efficacy in engineering. Women in Beverly's (2022) study discussed feeling validated by the instructors, which made them feel confident about themselves and their ability to do the work and be an engineer. Similarly, Espinoza (2013) found that validation from engineering professors positively influenced Latinx student self-efficacy.

Peer interactions. Besides faculty interactions, interactions with peers are also influential to the success of Latinas in engineering. Ong et al. (2011) found that relationships were important to women of color, which included not only faculty but also peers. The authors found that women of color found it difficult to form relationships with peers within their majors, so they developed relationships with peers outside of their majors that mirrored their racial/ethnic identity. Ong et al.'s study reveals that peers may have a significant role in improving or decreasing the engineering self-efficacy of Latinas. Also at the peer level, Lee et al. (2020) found that students indicated not feeling like they belonged because of racial microaggressions as a result of a lack of representation, such as being mistaken for being in the wrong class. Behaviors of white and Asian students toward Black and Latinx students made them feel that they did not want to work or be in a group with them and that they were viewed as unintelligent—a negative stereotype attributed to students of color in STEM. In Beverly's (2022) study in chemical engineering, Black women discussed the need to be in groups with others of a similar race and gender because of the ways they were negatively treated by those outside of their race and gender.

Espinoza (2013) found that peers were important to the success of Latinos in engineering. The students reported that they often sought help from classmates before going to the instructor. Similarly, Gasiewski et al. (2012) found that positive, collaborative interaction with peers positively predicted students' levels of engagement in STEM courses. Research in STEM classrooms also reveals that, despite women outnumbering men in a classroom, women continue to encounter negative treatment (see Grunspan et al., 2016). Negative peer interactions also appear to affect women. Murray et al. (1999) found that women were interrupted more often than men when speaking, had their ideas dismissed more often, took on "secretary" roles in lab teams, and encountered sexist remarks and jokes more than men. These behaviors were also evident in Beverly's (2022) study, in which women consistently spoke about men dominating groups and subjugating them by relegating them to diminished group roles. Therefore, it is important to also consider the potential influences of peers on Latinas in the classroom, as they too can have a significant effect on their experiences.

Classroom Sense of Belonging

Students' positive perceptions of the classroom climate, as informed by their experiences with instructors and peers, can lead to women's development of a classroom sense of belonging. Sense of belonging, or the connectedness a student has to their community, can contribute to positive outcomes, such as persistence for students in the STEM classroom (Wilson et al., 2015). Much of the literature on sense of belonging in college has focused on the campus level: for example, examining how campus racial climate shapes the sense of belonging of students of color (e.g., Strayhorn, 2012). Less has been written about how other localized environments affect sense of belonging in women of color in STEM disciplines, such as engineering. Exploring the experiences of Latinas in engineering is particularly important because empirical studies suggest that they struggle with a sense of belonging in engineering environments, and studies are beginning to identify positive influences on this socioemotional outcome.

For example, in their phenomenological study of five Latinas, Rodriguez et al. (2022) found that Latinas in engineering found a sense of belonging in engineering through Latinx professional engineering organizations. These organizations contributed to a sense of community and support, which contributed to Latina engineering identity development. Similarly, Villa et al. (2020) found that Latinas in engineering sought out affinity groups and spaces outside of the classroom, which helped them cultivate a sense of belonging that helped mediate the negative messages they received regarding their participation and contributions in engineering. Yet, in a phenomenological study of the social and academic experiences of seventeen undergraduate Latinas in STEM, Rodriguez and Blaney (2020) found that Latinas felt isolated from their respective STEM communities because of negative treatment they encountered based on their gender and race/ethnicity. Understanding how academic experiences can affect Latinas' sense of belonging— both positively and negatively—can contribute to needed cultural change in engineering programs.

According to Rainey et al. (2018), who interviewed 201 college seniors, for women of color, negative interpersonal relationships with individuals within their STEM departments (including instructors) negatively affected their sense of belonging. In another large-scale study, Johnson (2012) similarly found that, in a sample of 1,722 women in STEM, being a woman of color had a negative relationship with sense of belonging. However, she also found a relationship between academic self-confidence and sense of belonging; academic and socially supportive environments were the strongest contributors to the sense of belonging of women of color in STEM. Harben and Bix (2019) also found that interventions that facilitated interaction with peers and the instructor increased student sense of belonging in a first-year packaging[1] course. In addition, the self-selected small groups that were formed for the course may also have contributed to a sense of belonging in the classroom—a finding that is consistent with what Beverly (2022) found in her study of engineering courses.

Studies that examine the influence of instruction and instructors can help us better understand how these affect Latina sense of belonging in engineering. Sax et al. (2018) used a cross-institutional data set of 1,355 students in introductory computing courses to study the experiences of women and underrepresented minorities in computing. They evaluated students' classroom experiences, examining the role of instructors' pedagogy, inclusivity, and communication with students' sense of belonging. They found a relationship between sense of belonging and college environments and experiences, particularly how faculty members contributed to belonging. Analyses show a positive relationship between inclusive pedagogy and students' feelings of support from the department. Overall, however, women entered computing with lower levels of sense of belonging in their introductory computing course, and their sense of belonging decreased by the end of the course.

Women and people of color's experiences interacting with STEM instructors can contribute to self-efficacy and sense of belonging in different ways, which can offer perspective for understanding the link between engineering instructors and Latina persistence (e.g., Lawson et al., 2018; Robnett, 2016). Lawson et al. (2018) found a negative relationship between women's sexist experiences in male-dominated fields (which included STEM) and social belonging. They also found that advising time served as a buffer between high encounters of sexism and social belonging.

Two studies in engineering conducted by Verdín offer additional findings on women's sense of belonging in engineering. In the first (a multi-institutional quantitative study), Verdín (2021a) found that minoritized women in engineering who received outside recognition from instructors and peers had increases in their sense of belonging in the major and the classroom. In her second study, Verdín (2021b) conducted a longitudinal inquiry on the experience of one Latina in an engineering classroom. The Latina's sense of belonging was negatively influenced by instructors who actively promoted values of engineering culture, such as a fixed mindset that rejected the idea that all students can become capable engineers. Counter-spaces made up of supportive peers were especially important for the Latina's development of a sense of belonging. As Latinas navigate classrooms, their interactions with instructors and to an extent peers may have a significant impact on their sense of belonging, and further study is needed to examine the relationships suggested by the empirical literature to date.

Self-Efficacy

Women in engineering appear to have disparities in self-efficacy (confidence in oneself to complete a task) compared to men. In a longitudinal and multi-institutional study, Marra et al. (2013) found that women in engineering had lower levels of self-efficacy compared to their men peers because they felt a lack of inclusion in the engineering community. Similarly, in a survey of 363 first-year engineering students at a large state university, Jones et al. (2010) found that women had lower levels of self-efficacy and expectancy for success in engineering than men; both were predictive of engineering GPAs. The authors of both studies note that low self-efficacy can negatively affect persistence.

Literature on women of color in engineering appears to indicate a relationship between self-efficacy and sense of belonging in the broader engineering context and suggests that experiences in engineering courses may be sources of influence on self-efficacy for Latinas (Ong et al., 2020). Ong et al. discuss how institutional inequities influence the experiences, participation, and advancement of women of color in engineering as well as their persistence. They argue that inequities related to race and gender cause women of color in engineering to lack a sense of belonging and have low self-efficacy.

Studies of instruction can provide needed insight into course components that promote or hinder women's sense of belonging and self-efficacy. Beverly (2022) found a positive, significant, bidirectional relationship between engineering self-efficacy and classroom sense of belonging for both men and women in engineering; this underscores the need for further study on how engineering classrooms that foster a sense of belonging might also contribute to women of color's engineering self-efficacy. In addition, in a survey of women engineers, Verdín (2021a) found that women who had stronger beliefs in their competence and performance, which she argues are congruent to self-efficacy, also had a higher sense of belonging in the classroom and major.

Desire to Remain in STEM Fields

Underrepresentation of Latinas in degree completion, despite having the academic ability to succeed, suggests that environmental factors are influencing the attrition of engineering majors. A qualitative analysis of Latinx students' experiences in the Society of Hispanic Professional Engineers by Revelo and Baber (2018) found that collective membership through professional Latinx organizations appears to be important for Latinx students to develop familial ties with other Latinx students, which can aid in resisting stereotypes and persisting in engineering. Focusing on the course experiences could reveal whether and how Latinas' desire to remain in engineering might be enhanced or diminished by peer interactions in classrooms. Using data from a longitudinal survey, Espinosa (2011) analyzed women of color's reasons for persisting in STEM and found that their interactions with peers outside the classroom that were focused on course content were beneficial to their persistence. Espinosa concluded that, for women of color, academic relationships may be equally as important as social ones. Focusing on course experiences specifically, Beverly (2022) found that relationships with peers cultivated in courses appeared to contribute to Latinas' persistence in engineering.

The role of family in shaping Latinas' interest and persistence in engineering may be beneficial when considering how to make engineering courses more inclusive. Talley and Ortiz (2017) conducted a mixed methods study at a Hispanic-Serving Institution (HSI) with a sample composed mostly of African American[2] and Latina STEM majors. The focus of the study was to understand the interest and motivations of the women in the study toward pursuing a STEM career. The authors found that African American and Latina women cited family expectations as a prime motivator to pursue a STEM career. As previously discussed, Latinas also find relationships with peers and instructors instrumental to their success in courses. Cultivating classrooms where peer and instructor relationships are fostered in engineering could be important to not only enhance the experiences of Latinas in the classroom but also contribute to their desire to remain in the field of engineering.

IMPLICATIONS FOR LATINAS IN ENGINEERING PROGRAMS AND SCHOOLS

Clear tensions exist between engineering culture and Latinx culture. Specifically for Latinas, engineering values that place importance on the individual conflict with the importance of social connections with instructors and peers. Latinas appear to need meaningful relationships with instructors and peers to overcome hostile engineering environments and to persist. As a result of hostile engineering environments, marginalized students are stripped away from their identities while engineering culture champions a colorblind approach. This colorblind approach can ignore the needs of marginalized students, especially those who enter spaces with intersecting -isms, such as Latinas. The conceptual framework discussed in this chapter provides a guide for researchers who want to further understand course dynamics and the corresponding effects on Latinas. Both instructors and peers appear to have negative effects on women in courses, affecting their sense of belonging, engineering self-efficacy, and desire to remain in the field, which provides a potential template for exploring these effects on Latinas.

Beverly's (2022) study was conducted at a highly selective Predominately White Institution. Research is needed to determine whether the proposed framework holds across four-year institutions, including those that have recently been designated as HSIs and serve large numbers of Latinx students. Engineering culture may persist even within HSIs due to the socialization of engineering faculty within a strongly embedded engineering culture. Particularly at institutions that have sought HSI designation as their enrollments of Latinx populations have increased, institutional commitments to serve Latinx students may be lacking or in nascent stages (Garcia et al., 2019). In their study, Garriott et al. (2019) provide evidence that students at HSIs are still susceptible to toxic aspects of engineering culture.

Research suggests that men peers and faculty can have a significant effect on women's internalization of negative self-appraisals of their engineering abilities, but instructors who understand the importance of inclusion and equity can design courses and teach in ways that may mitigate or preclude women's negative self-perceptions. It is not clear whether engineering instructors recognize the influence they have in courses and classrooms, but it may be beneficial to present information like that which is presented in this chapter to engineering instructors. Making instructors aware of the power they have in shaping student experiences and their further pursuit of engineering may be instrumental in efforts to change the landscape of engineering education. Instructors' pedagogical strategies shape classroom climate, which in turn shapes whether women feel like they belong in the classroom. Emerging research suggests that positive perceptions of instructors and the teaching practices in which they engage appear to influence the way women feel about their abilities to be successful in engineering coursework and perhaps even feel more efficacious about their career pursuits in engineering.

Studies that identify how engineering instructors can contribute to equitable learning experiences for Latinas are needed to highlight the tangible ways instructors can create inclusive classroom environments. Inclusive strategies that provide comfort and care to students, such as validating student questions and answers in the classroom and asking students to provide feedback on the trajectory of the course, may help Latinas feel that they are part of a welcoming and supportive community. Because Latinas value the familial nature of a community, making spaces (such as classrooms) collegial may provide a better environment, which can help support Latina persistence in engineering.

The conceptual framework in this chapter can inform future studies that seek to generate insights into how instructors and what they do in the classroom influence Latinas and their eventual success in engineering—both academically and professionally. By placing the responsibility on those who create engineering classroom spaces, we hope to make changes that do not take a deficit approach but rather a strengths-based approach. The classroom environment can be changed by the active players that create those spaces, particularly instructors and the cultural messages they receive from their departments and colleges. Although changing culture takes time, instructors can have a significant role in taking the first steps toward making environments more inclusive for marginalized students such as Latinas. More research that applies this framework can help determine the impacts of engineering classroom experiences on Latinas across subdisciplines and potentially at different institutional contexts. Such studies are critical if educators are to identify and create classroom spaces that are more inclusive.

Authors' Note

Selyna Pérez Beverly ORCID: 0000-0003-0504-3918
Lisa Lattuca ORCID: 0000-0003-1547-5992
We have no known conflicts of interest to disclose. Correspondence concerning this article should be addressed to Selyna Pérez Beverly, Eastern Michigan University, 310 Porter Building, Ypsilanti, MI 48197
Email: sbeverl2@emich.edu

NOTES

1. A packaging degree is a STEM degree that involves studying physics, chemistry, mathematics, materials science, and business to prepare students to work in consumer goods industries.
2. Although I have chosen to use the term "Black" in this chapter to describe those of African-American descent, the author in this article used the term "African-American," which I am using as I describe the study.

REFERENCES

Banda, R. M. (2020). From the inside looking out: Latinas intersectionality and their engineering departments. *International Journal of Qualitative Studies in Education, 33*(8), 824–839. https://doi.org/10.1080/09518398.2020.1735565

Banda, R. M., & Flowers, A. M. (2018). Critical qualitative research as a means to advocate for Latinas in STEM. *International Journal of Qualitative Studies in Education, 31*(8), 769–783. https://doi.org/10.1080/09518398.2018.1479046

Bandura, A. (1977). Self-efficacy: toward a unifying theory of behavioral change. *Psychological Review, 84*(2), 191. https://doi.org/10.1037/0033-295X.84.2.191

Beverly, S. P. (2022). *Raising their voices: Exploring women's experiences with instruction in engineering courses.* [Doctoral Dissertation, University of Michigan]. Deep Blue Documents.

Carlone, H. B., & Johnson, A. (2007). Understanding the science experiences of successful women of color: Science identity as an analytic lens. *Journal of Research in Science Teaching: The Official Journal of the National Association for Research in Science Teaching, 44*(8), 1187–1218. https://doi.org/10.1002/tea.20237

Carter, D. F., Dueñas, J. E. R., & Mendoza, R. (2019). Critical examination of the role of STEM in propagating and maintaining race and gender disparities. In M. B. Paulsen, & L. W. Perna (Eds.), *Higher Education: Handbook of Theory and Research* (34th ed., pp. 39–97). Springer.

Collins, P. H. (1986). Learning from the outsider within: The sociological significance of Black feminist thought. *Social Problems, 33*(6), s14-s32. https://doi.org/10.2307/800672

Dewsbury, B. M. (2020). Deep teaching in a college STEM classroom. *Cultural Studies of Science Education, 15*(1), 169–191. https://doi.org/10.1007/s11422-018-9891-z

Garriott, P. O., Navarro, R. L., Flores, L. Y., Lee, H.-S., Carrero Pinedo, A., Slivensky, D., Muñoz, M., Atilano, R., Lin, C.-L., Gonzalez, R., Luna, L., & Lee, B. H. (2019). Surviving and thriving: Voices of Latina/o engineering students at a Hispanic serving institution. *Journal of Counseling Psychology, 66*(4), 437–448. https://doi.org/10.1037/cou0000351

Grunspan, D. Z., Eddy, S. L., Brownell, S. E., Wiggins, B. L., Crowe, A. J., & Goodreau, S. M. (2016). Males under-estimate academic performance of their female peers in undergraduate biology classrooms. *PloS ONE, 11*(2), 1–16. https://doi.org/0.1371/journal.pone.0148405

Espinosa, L. (2011). Pipelines and pathways: Women of color in undergraduate STEM majors and the college experiences that contribute to persistence. *Harvard Educational Review, 81*(2), 209–241. https://doi.org/10.17763/haer.81.2.92315ww157656k3u

Espinoza, A. (2013). The college experiences of first-generation college Latino students in engineering. *Journal of Latino/Latin American Studies, 5*(2), 71–84. https://doi.org/10.18085/llas.5.2.p38569tj26k6w972

Farrell, S., Godwin, A., & Riley, D. M. (2021). A sociocultural learning framework for inclusive pedagogy in engineering. *Chemical Engineering Education, 55*(4), 192–204. https://doi.org/10.18260/2-1-370.660-128660

Flores, G. M. (2011). Latino/as in the hard sciences: Increasing Latina/o participation in science, technology, engineering and math (STEM) related fields. *Latino Studies, 9*(2), 327–335. https://doi.org/10.1057/lst.2011.36

Garcia, G. A., Núñez, A., & Sansone, V. A. (2019). Toward a multidimensional conceptual framework for understanding "servingness" in Hispanic-Serving Institutions: A synthesis of the research. *Review of Educational Research, 89*(5), 745–784 https://doi.org/10.3102/0034654319864591

Gasiewski, J. A., Eagan, M. K., Garcia, G. A., Hurtado, S., & Chang, M. J. (2012). From gatekeeping to engagement: A multicontextual, mixed method study of student academic engagement in introductory STEM courses. *Research in Higher Education, 53*(2), 229–261. https://doi.org/10.1007/s11162-011-9247-y

Godwin, A., Potvin, G., Hazari, Z., & Lock, R. (2016). Identity, critical agency, and engineering: An affective model for predicting engineering as a career choice. *Journal of Engineering Education, 105*(2), 312–340. https://doi.org/10.1002/jee.20118

Haraway, D. (1988). Situated knowledges: The science question in feminism and the privilege of partial perspective. *Feminist Studies, 14*(3), 575–599. https://doi.org/10.2307/3178066

Harben, A., & Bix, L. (2019). Student sense of belonging in a large, introductory STEM course. *NACTA Journal, 64*, 288–297. https://www.jstor.org/stable/27157804

Harding, S. (1991) *Whose science? Whose knowledge? Thinking from women's lives.* Cornell University Press. http://www.jstor.org/stable/10.7591/j.ctt1hhfnmg

Hottinger, S. N. (2016). *Inventing the mathematician: Gender, race, and our cultural understanding of mathematics.* State University of New York Press.

Johnson, A. M. (2019). 'I can turn it on when I need to': Pre-college integration, culture, and peer academic engagement among Black and Latino/a engineering students. *Sociology of Education, 92*(1), 1–20. https://doi.org/10.1177/0038040718817064

Johnson, D. R. (2012). Campus racial climate perceptions and overall sense of belonging among racially diverse women in STEM majors. *Journal of College Student Development, 53*(2), 336–346 https://doi.org/10.1353/csd.2012.0028

Jones, D., Paretti, M., Hein, S., & Knott, T. (2010). An analysis of motivation constructs with first-year engineering students: Relationships among expectancies, values, achievement, and career plans. *Journal of Engineering Education, 99*(4), 319–336. https://doi.org/10.1002/j.2168-9830.2010.tb01066.x

Kishimoto, K. (2018). Anti-racist pedagogy: From faculty's self-reflection to organizing within and beyond the classroom. *Race, Ethnicity and Education, 21*(4), 540–554. https://doi.org/10.1080/13613324.2016.1248824

Kuzawa, D. (2017). More than recruitment and outreach: Diversity and inclusion in engineering education curricula and classrooms. *American Society for Engineering Education Annual Conference Proceedings.* American Society for Engineering Education. http://se.asee.org/proceedings/2017%20Zone%20II%20Proceedings/papers/proceedings/3/162.pdf

Ladson-Billings, G. (2000). Racialized discourses and ethnic epistemologies. In N. Denzin & Y. Lincoln (Eds.), *Handbook of qualitative research* (2nd ed., pp. 257–277). Sage Publications.

Lawson, K. M., Kooiman, L. Y., & Kuchta, O. (2018). Professors' behaviors and attributes that promote U.S. women's success in male-dominated academic majors: Results from a mixed methods study. *Sex Roles, 78*(7–8), 542–560. https://doi.org/10.1007/s11199-017-0809-0

Lee, M. J., Collins, J. D., Harwood, S. A., Mendenhall, R., & Huntt, M. B. (2020). "If you aren't White, Asian or Indian, you aren't an engineer": racial microaggressions in STEM education. *International Journal of STEM Education, 7*(1), 1–16. https://doi.org/10.1186/s40594-020-00241-4

Lent, R. W., & Hackett, G. (1987). Career self-efficacy: Empirical status and future directions. *Journal of vocational Behavior, 30*(3), 347–382. https://doi.org/10.1016/0001-8791(87)90010-8

Lent, R. W., Miller, M. J., Smith, P. E., Watford, B. A., Lim, R. H., & Hui, K. (2016). Social cognitive predictors of academic persistence and performance in engineering: Applicability across gender and race/ethnicity. *Journal of Vocational Behavior, 94*, 79–88. https://doi.org/10.1016/j.jvb.2016.02.012

Lent, R. W., Sheu, H., Singley, D., Schmidt, J. A., Schmidt, L. C., & Gloster, C. S. (2008). Longitudinal relations of self-efficacy to outcome expectations, interests, and major choice goals in engineering students. *Journal of Vocational Behavior, 73*, 328–335. https://doi.org/10.1016/j.jvb.2008.07.005

Lester, J., Yamanaka, A., & Struthers, B. (2016). Gender microaggressions and learning: The role of physical space in teaching pedagogy and communication. *Community College Journal of Research and Practice, 40*(11), 909–926. https://doi.org/10.1080/10668926.2015.1133333

López, E. J., Basile, V., Landa-Posas, M., Ortega, K., & Ramirez, A. (2019). Latinx students' sense of familismo in undergraduate science and engineering. *The Review of Higher Education, 43*(1), 85–111. https://doi.org/10.1353/rhe.2019.0091

Marra, R. M., Rodgers, K. A., Shen, D., & Bogue, B. (2013). Women engineering students and self-efficacy: A multi-year, multi-institution study of women engineering student self-efficacy. *Journal of Engineering Education, 98*(1), 27–38. https://doi.org/10.1002/j.2168-9830.2009.tb01003.x

McGee, E. O., & Martin, D. B. (2011). "You would not believe what I have to go through to prove my intellectual value!" Stereotype management among academically successful Black mathematics and engineering students. *American Educational Research Journal, 48*(6), 1347–1389. http://www.jstor.org/stable/41306389

Murray, S. L., Meinholdt, C., & Bergmann, L. S. (1999). Addressing gender issues in the engineering classroom. *Feminist Teacher, 12*(3), 169–183. https://www.jstor.org/stable/40545825

National Center for Science and Engineering Statistics [NCSES]. (2022). *Women, minorities, and persons with disabilities in science and engineering: 2022* (NSF 19–304). https://www.nsf.gov/statistics/wmpd.

Ong, M., Wright, C., Espinosa, L., & Orfield, G. (2011). Inside the double bind: A synthesis of empirical research on undergraduate and graduate women of color in science, technology, engineering, and mathematics. *Harvard Educational Review, 81*(2), 172–209. https://doi.org/10.17763/haer.81.2.t022245n7x4752v2

Ong, M., Jaumot-Pascual, N., & Ko, L. T. (2020). Research literature on women of color in undergraduate engineering education: A systematic thematic synthesis. *Journal of Engineering Education, 109*(3), 581–615. https://doi.org/10.1002/jee.20345

Park, J. J., Kim, Y. K., Salazar, C., & Hayes, S. (2020). Student–faculty interaction and discrimination from faculty in STEM: The link with retention. *Research in Higher Education, 61*(3), 330–356. https://doi.org/10.1007/s11162-019-09564-w

Pawley, A. L., Schimpf, C., & Nelson, L. (2016). Gender in engineering education research: A content analysis of research in JEE, 1998–2012. *Journal of Engineering Education, 105*(3), 508–528. https://doi.org/10.1002/jee.20128

Rainey, K., Dancy, M., Mickelson, R., Stearns, E., & Moller, S. (2018). Race and gender differences in how sense of belonging influences decisions to major in STEM. *International Journal of STEM Education, 5*(1). https://doi.org/10.1186/s40594-018-0115-6

Revelo, R. A., & Baber, L. D. (2018). Engineering resistors: Engineering Latina/o students and emerging resistant capital. *Journal of Hispanic Higher Education, 17*(3), 249–269. https://doi.org/10.1177/1538192717719132

Revelo, R. A., Mejia, J. A., & Villanueva, I. (2017, June). *Who are we? Beyond monolithic perspectives of Latinxs in engineering* [Conference session]. 2017 ASEE Annual Conference & Exposition, Columbus, Ohio.

Riegle-Crumb, C., Peng, M., & Russo-Tait, T. (2020). Committed to STEM? Examining factors that predict occupational commitment among Asian and White female students completing STEM US postsecondary programs. *Sex Roles, 82*(1), 102–116. https://doi.org/10.1007/s11199-019-01038-8

Ridgeway, C. L., & Correll, S. J. (2004). Unpacking the gender system: A theoretical perspective on gender beliefs and social relations. *Gender & Society, 18*(4), 510–531. https://doi.org/10.1177/0891243204265269

Rincón, B. E., & Rodriguez, S. (2021). Latinx students charting their own STEM pathways: How community cultural wealth informs their STEM identities. *Journal of Hispanic Higher Education, 20*(2), 149–163. http://dx.doi.org/10.1177/1538192720968276

Robnett, R. D. (2016). Gender bias in STEM fields: Variation in prevalence and links to STEM self-concept. *Psychology of Women Quarterly, 40*(1), 65–79. https://doi.org/10.1177/036168 4315596162

Rodriguez, S. L., & Blaney, J. M. (2020). "We're the unicorns in STEM": Understanding how academic and social experiences influence sense of belonging for Latina undergraduate students. *Journal of Diversity in Higher Education, 14*(3), 441–455. https://doi.org/10.1037 /dhe0000176

Rodriguez, S. L., Bukoski, B. E., Cunningham, K. J., & Jones, A. (2020a). Critiquing oppression and desiring social justice: How undergraduate Latina students in STEM engage in acts of resistance. *Journal of Women and Gender in Higher Education, 13*(3), 251–267. https://doi.org /10.1080/26379112.2020.1838297

Rodriguez, S. L., Doran, E. E., Sissel, M., & Estes, N. (2022). Becoming *la ingeniera*: Examining the engineering identity development of undergraduate Latina students. *Journal of Latinos and Education, 21*(2), 181–200. https://doi.org/10.1080/15348431.2019.1648269

Rodriguez, S. L., Lu, C., & Ramirez, D. (2020b). Creating a conceptual framework for computing identity development for Latina undergraduate students. In E. Gonzalez, F. Fernandez, & M. Wilson (Eds.), *An asset-based approach to advancing Latina students in STEM* (pp. 25–39). Routledge.

Rodriguez, S., Pilcher, A., & Garcia-Tellez, N. (2021). The influence of *familismo* on Latina student STEM identity development. *Journal of Latinos and Education, 20*(2), 177–189. https:// doi.org/10.1080/15348431.2019.1588734

Roychoudhury, A., Tippins, D. J., & Nichols, S. E. (1995). Gender-inclusive science teaching: A feminist-constructivist approach. *Journal of Research in Science Teaching, 32*(9), 897–924. https://doi.org/10.1002/tea.3660320904

Sax, L. J., Blaney, J. M., Lehman, K. J., Rodriguez, S. L., George, K. L., & Zavala, C. (2018). Sense of belonging in computing: The role of introductory courses for women and underrepresented minority students. *Social Sciences, 7*(8), 122. https://doi.org/10.3390 /socsci7080122

Stewart, J., Henderson, R., Michaluk, L., Deshler, J., Fuller, E., & Rambo-Hernandez, K. (2020). Using the social cognitive theory framework to chart gender differences in the developmental trajectory of STEM self-efficacy in science and engineering students. *Journal of Science Education and Technology, 29*(6), 758–773. https://doi.org/10.1007/s10956-020-09853-5

Strayhorn, T. L. (2012). *College students' sense of belonging: A key to educational success for all students.* Routledge. https://doi.org/10.4324/9781315297293

Talley, K. G., & Ortiz, A. M. (2017). Women's interest development and motivations to persist as college students in STEM: a mixed methods analysis of views and voices from a Hispanic-Serving Institution. *International Journal of STEM Education, 4*(1), 1–24. https://doi.org /10.1186/s40594-017-0059-2

Tonso, K. L. (1996). Student learning and gender. *Journal of Engineering Education, 85*(2), 143–150. https://doi.org/10.1002/j.2168-9830.1996.tb00223.x

Verdín, D. (2021a). The power of interest: Minoritized women's interest in engineering fosters persistence beliefs beyond belongingness and engineering identity. *International Journal of STEM Education, 8*(1), 1–19. https://doi.org/10.1186/s40594-021-00292-1

Verdín, D. (2021a, July). *Negotiating belongingness: A longitudinal narrative inquiry of a Latina first-generation college student's experience in the engineering culture* [Conference paper]. American Society for Engineering Education, Virtual Annual Conference. https://peer.asee.org /37524

Verdín, D., & Godwin, A. (2018). Exploring Latina first-generation college students' multiple identities, self-efficacy, and institutional integration to inform achievement in engineering.

Journal of Women and Minorities in Science and Engineering, 24(3). https://doi.org/10.1615 /JWomenMinorScienEng.2018018667

Villa, E. Q., Esquinca, A., Hampton, E., & Guerra, H. M. (2020). "Is engineering for me?": Examining Latinas' narratives of resilience and agency to confront enduring struggles and challenges in undergraduate engineering studies. *Peace and Conflict: Journal of Peace Psychology,* 26(4), 403. https://doi.org/10.1037/pac0000427

Wilson, D., Jones, D., Bocell, F., Crawford, J., Kim, M. J., Veilleux, N., Floyd-Smith, T., Bates, R., & Plett, M. (2015). Belonging and academic engagement among undergraduate STEM students: A multi-institutional study. *Research in Higher Education,* 56(7), 750–776. https://doi .org/10.1007/s11162-015-9367-x

Winterer, E. R., Froyd, J. E., Borrego, M., Martin, J. P., & Foster, M. (2020). Factors influencing the academic success of Latinx students matriculating at 2-year and transferring to 4-year US institutions—implications for STEM majors: A systematic review of the literature. *International Journal of STEM Education,* 7(1), 1–23. https://doi.org/10.1186/s40594-020-00215-6

10 · THE ROLE OF DUAL CREDITS IN SOCIALIZING LATINX STUDENTS FOR ENGINEERING FIELDS

TARYN OZUNA ALLEN, CHRISTINE HALL, AND COURTNEY MATTHEWS

In recent years, policymakers and postsecondary institutions have focused on college degree completion. Students with a college degree are more likely to increase their earning potential, have better job opportunities, and gain specialized knowledge in a particular field (Association of Public & Land-Grant Universities [APLU], n.d.; Carnevale et al., 2011). Degree completion also leads to better health outcomes, higher civic engagement, and safer communities (Frase & Bauldry, 2022; Trostel, 2017). While Texas' completion agenda (Texas Higher Education Coordinating Board, 2023) is focused on increasing postsecondary credentials, persistence and completion challenges remain.

Dual credit—sometimes referred to as "dual enrollment"—is one strategy to promote college readiness, retention, and degree completion (Lee et al., 2022; Struhl & Vargas, 2012; Tobolowsky & Allen, 2016). Dual credit is defined as high school courses that simultaneously count for college credit if a passing grade is earned (Thomas et al., 2013). Dual credit programs have grown across the country, and in Texas (the location of this study), dual credit increased from 42,000 high school students in 2000 to 151,000 high school students in 2017 (Troutman et al., 2018). Previous research suggests that students gain benefits by earning dual credit (Giani et al., 2014), such as learning college-level expectations (Duncheon, 2020) and saving money on college tuition (Hoffman, 2005). In addition to earning college credit, dual credit programs allow students to learn the hidden curriculum of being a college student. By learning the "how to" of being a college student (Bailey et al., 2002; Kanny, 2014; Tobolowsky & Allen, 2016), dual credit earners are better prepared to transition to college and earn a bachelor's degree (An, 2013; Edwards et al., 2011).

Thus, exposing Latinx students to rigorous curricula via dual credit programs may better prepare them for college-level rigor and socialize students to expectations in STEM programs such as engineering. Indeed, research shows that early participation in STEM classes can bolster interest and persistence in STEM fields (Bottia et al., 2017; McClure et al., 2017), yet no prior research has examined how dual credit programs socialize Latinx students to be successful in engineering fields. Therefore, the purpose of this qualitative study was to understand how dual credit participation shaped and informed Latinx students' engineering interests and socialized them to college and program norms. The research questions guiding this study were as follows: (1) How does earning dual credit influence Latinx students' interest in engineering fields? and (2) What messages, experiences, and individuals motivated Latinx students with dual credits to pursue engineering programs?

LITERATURE REVIEW

The literature review begins with an overview of dual credit programs and the benefits and challenges associated with these programs. We then focus on the literature related to motivations for STEM and specifically engineering fields.

Inequity in College Enrollment and Completion

Equity gaps persist in degree completion rates for historically minoritized populations (i.e., students of color, low-income students, and first-generation college students; Association of American Colleges and Universities [AAC&U], 2018; Bragg et al., 2005); *Excelencia* in Education, 2020; Nichols & Schack, 2018). When compared to their White peers, African American and Latinx students are less likely to earn a four-year postsecondary degree (*Excelencia* in Education, 2020; Nichols & Schack, 2018). Even as the Latinx population has increased in degree completion from 19 percent in 2010 to 24 percent in 2020, an equity gap remains (*Excelencia* in Education, 2020). Given the projected growth of the Latinx population in particular, the potential impact on the US economy warrants special attention to the resources available for Latinx students.

This is particularly evident in states like Texas, where the Latinx population represents 40.2 percent of the total population (United States Census Bureau, 2022) and is expected to represent the majority of Texans by 2035 (Davis, 2016). In the Texas public school system, approximately 53 percent of children are Hispanic (Texas Education Agency, 2021), which should translate to increased representation in the higher education system in upcoming years. Yet, these demographic trends are not translating to postsecondary access and completion, particularly in high-demand STEM fields, such as engineering (Gates et al., 2011; NSF, 2017). Participation in dual credit provides early exposure to college-level expectations, norms, and behaviors which can help promote college going and success in demanding engineering majors (Bailey et al., 2002; Kanny, 2014, 2015; Tobolowsky & Allen, 2016).

Dual Credit

Dual credit programs are designed for high school students to earn college credits and learn college-level rigor (Hofmann, 2012). However, these programs are highly variable in their implementation. For example, courses may be taught online, on a high school campus, or on a community college campus (Tobolowsky & Allen, 2016). Community college professors or high school teachers have the freedom to instruct their classes according to their teaching philosophy (Tobolowsky & Allen, 2016).

More US secondary schools, particularly in Texas, are adopting dual credit programs and marketing them as a cost-effective way to earn college credit and pursue a degree (Mansell & Justice, 2014). Edwards, Hughes, and Weisberg (2011) further highlight that students who participate in dual enrollment may have additional access when exploring their aspirations for future educational and employment opportunities. Additionally, students completing dual credit programs have the ability to improve enrollment and student success rates when considering the decline of degrees awarded at two- and four-year colleges (Holten & Pierson, 2016; Mansell & Justice, 2014).

Since more students are entering postsecondary education with college credits, the time spent on degree completion could decrease and credit transfers may increase, impacting college enrollment and student success indicators. Tobolowsky and Allen (2016) also discovered that students preferred dual credit classes compared to advanced-placement (AP) programs "because [students] could earn the credit without taking an exam" (p. 48). Although dual credit courses provide many benefits to students in secondary education—that is, little to no cost and access to college credit—there are some caveats that may create challenges.

A primary concern of dual enrollment initiatives includes the level of rigor and college-level teachings offered by secondary school educators. Teachers' qualifications may be questioned since dual credit courses are typically instructed at high schools and therefore void of interactions with other college faculty (Tobolowsky & Allen, 2016). Since specific teacher criterion and certification requirements to instruct dual credit courses were not established, the educational content quality varies from school to school and from classroom to classroom (Edwards et al., 2011).

Some dual credit programs do not include on-campus exposure for high school students (Phelps & Chan, 2016). Alongside the difficulties of accessing a college campus, dual credit programs are typically offered to upperclassmen (juniors and seniors). High school students' disinterest in dual credit courses sometimes stems from common attitudes and actions—dismissive or easy-going—seen toward the latter part of their education, potentially affecting enrollment and accessibility to college courses (Mansell & Justice, 2014). The benefits and challenges related to dual enrollment programs provide key information when supporting students and fostering academic aspirations and motivations.

Pursuing Engineering

How and why students choose to enter STEM majors and careers is an area of growing research. Students who receive encouragement in pursuing their interests tend to interpret STEM career possibilities based on their academic preparation and whether they liked their high school math and science courses (Bonous-Hammarth, 2000; Bottia et al., 2018; Cole & Espinoza, 2008; Jahn & Myers, 2015). In a similar manner, students who performed well in certain subjects leaned toward careers in areas where they had certain talent or ability and conversely avoided careers in areas where they did not feel competent (Jahn & Myers, 2015). Seeing how STEM concepts can be used in real-world applications and having the opportunity to interact with STEM professionals may increase students' interest in these subjects (Jahn & Myers, 2015; Kitchen et al., 2018).

Despite their interest in a STEM field, particularly engineering, research reveals that undergraduate students may encounter academic and personal challenges. For example, they may question their abilities to be successful, struggle with their perceptions of their identity as an engineer, and become overwhelmed and experience self-doubt (Cruz & Kellam, 2018; Matusovich et al., 2010; Meyer & Marx, 2014). In addition to these obstacles, Latinx students, more specifically, may experience inadequate academic advising and insufficient financial aid (Long III et al., 2018). They may also encounter microaggressions or racial discrimination from faculty (Camacho & Lord, 2011; Park et al., 2022).

Peer relationships, mentorship, and academic support can help counter these challenges (Cole & Espinoza, 2008; Hernandez & Lopez, 2004; Hernandez et al., 2016; Main et al., 2018). Fostering a culture in engineering organizations that demonstrates an appreciation for Latinx students' ethnic identity will provide opportunities to build community—and create networks for mentorship (Alonso Revelo, 2015; Banda & Flowers, 2016; Rodriguez et al., 2022). Families are also integral to Latinx engineering students' success, as they provide support and counter insecurities or doubts (Rodriguez et al., 2022; Villa et al., 2016; Wilson-Lopez et al., 2016).

THEORETICAL FRAMEWORK

Utilizing the theoretical framework of vocational anticipatory socialization (VAS) can shed light on how students start to learn about the world of work through various kinds of vocational messaging (Levine & Hoffner, 2006; Myers et al., 2011). VAS suggests that STEM classes, socializing messages, and direct experiences with STEM-related material/curricula shape what students know about similar careers and, therefore, influence their interests in those fields (Jahn & Myers, 2015). Part of the VAS process includes students gaining occupational information and comparing this information with their idea of self (Jablin, 2001).

A student's sense of self can be further shaped through parental and familial interactions, since parents and family members play a significant part of vocational messaging (Jahn & Myers, 2015). As such, they have an impact as the student navigates the decision to attend college and decides which majors will lead to meaningful and worthwhile careers. This messaging can include advice on seeking self-sufficiency as well as pursuing careers in fields where the student has interest and talent (Jahn & Myers, 2014). For example, if a student has an interest and natural ability in math, their family may encourage them to pursue an engineering degree to ultimately become an engineer. However, while the family may provide part of the vocational messaging foundation, they may not be aware of the specifics when pursuing STEM careers (Jahn & Myers, 2014; Myers et al., 2011). One benefit of a dual enrollment program is access to those with first-hand knowledge of engineering career pursuits. Therefore, learning engineering content while in high school could enhance students' level of motivation.

METHODOLOGY

This study is one part of a broader three-year, longitudinal, qualitative study funded by the Greater Texas Foundation. This larger study sought to understand the influence of dual credit on the college transitions and experiences of Latinx engineering students. The purpose of the present study was to understand the individuals and experiences that motivated Latinx students to enroll in engineering programs and how early participation in STEM courses socialized students to college and program norms.

The findings for this research project are primarily based on one-on-one, semi-structured interviews with eleven Latinx dual credit-earning undergraduate engineering students. Our research questions included the following: (1) How does dual credit inform or foster Latinx students' interests in engineering fields? and (2) What messages, experiences, and individuals motivated Latinx students with dual credits to pursue engineering programs?

Site Selection

Our study was conducted at North Texas Research University (NTRU, a pseudonym), a somewhat selective (88 percent) institution located in an urban area. NTRU has demonstrated support for transfer students, and it is ranked among the top ten institutions that enroll the most transfer students. We employed purposeful sampling, which is a procedure that allows researchers to select a site that can "purposefully inform an understanding of the research problem and central phenomenon in the study" (Creswell & Poth, 2017, p. 326). NTRU was purposefully selected because of its engineering program and its substantial representation of Latinx students. NTRU offers numerous undergraduate and graduate engineering programs (e.g., civil engineering, computer science and engineering, industrial engineering, and materials science engineering). In addition to excellent academic programs and state-of-the-art research facilities, engineering students have the opportunity to participate in engineering-related student organizations or conduct research with faculty members affiliated with engineering centers and institutes.

Further, as a recognized Hispanic-Serving Institution (HSI), NTRU enrolls a substantial number of Latinx students. Recent data reveal that Latinx students account for approximately 30 percent ($n = 1{,}018$) of all engineering students at NTRU. These demographic trends, coupled with NTRU's stellar engineering programs, qualify it as a unique context to understand the motivations of Latinx engineering students.

Participant Recruitment and Selection

We also employed purposeful sampling (Creswell & Poth, 2017) to identify and select the participants for this study. Participants were recruited in fall 2015 and spring 2017. A campus administrator provided the contact information for students who self-identified as Latinx and were currently enrolled in an engineering program. These students were contacted via email and invited to participate in the study. All eligible and interested students were invited to participate in fall 2015, but this process yielded few women. As a result, we specifically sought additional women in spring 2017 to promote gender parity in the study. Eleven students in total agreed to participate. Participants received a $20 Walmart gift card as compensation. Table 10.1 includes additional demographic and background information about each participant.

Data Collection

Participants in this project were interviewed annually for three years. The findings from the present study were based primarily on their first- and second-year inter-

TABLE 10.1   Participants' Pseudonyms, Backgrounds, and Dual Credits Earned

Participant's pseudonym	Gender	First-generation college student	Academic major	Hours of dual credit earned	Enrolled immediately after high school
Vanessa	Female	No	Industrial engineering	21 hours	Yes
Adrian	Male	Yes	Civil engineering	6 hours	Yes
Debbie	Female	Yes	Electrical engineering	12 hours	Yes
Erin	Female	Yes	Computer science	60 hours	Yes
Dominic	Male	No	Civil engineering	24 hours	Yes
Evan	Male	Yes	Electrical engineering	28 hours	Yes
Isaac	Male	Yes	Aerospace engineering	9 hours	Yes
Jaime	Male	Yes	Civil engineering	12 hours	Yes
Mark	Male	Yes	Mechanical engineering	77 hours	Yes
Orlando	Male	Yes	Mechanical engineering	69 hours	Yes

views. Through one-on-one, semi-structured interviews, we asked students to describe their dual credit experiences while in high school. We also inquired into the participants' initial motivations for enrolling in an engineering program and asked them to reflect on how dual credit participation informed their lives as engineering students. During these interviews, students were asked questions such as (1) Why did you decide to pursue engineering? (2) What do you like about engineering? and (3) How have your experiences in dual credit influenced your time at NTRU? All the student interviews occurred at an on-campus location, and they were digitally recorded for transcription. On average, the interviews lasted approximately forty-five minutes.

Data Analysis

Once the audio recordings were transcribed verbatim, we began our analysis. First, we listened to the audio recordings and reviewed the transcripts to ensure accuracy. We began by conducting an open-coding process to identify emerging categories of data (Creswell & Poth, 2017). We then conducted axial coding to begin establishing a relationship between emerging themes (Creswell & Poth, 2017). After conducting open coding and axial coding independently, we met as a research team to discuss our emerging findings. We discussed these findings until consensus was reached. As a group, we conducted selective coding (Creswell & Poth, 2017) to create our comprehensive narrative of Latinx students' motivations for enrolling in engineering programs and the role of dual credit in pursuing engineering fields. The theoretical framework and research questions guided our analysis for the study.

Positionality

It is important to identify and describe how the researchers can directly influence the project. This research team was led by a Latina woman with support from a White woman and a Latina woman. All of the researchers were affiliated with the NTRU campus in various capacities, such as student, faculty member, and administrator. Although we have our individual research interests (e.g., college access and persistence, transfer students, and community development and change in urban areas), our scholarly and professional interests align, as we are focused on K–16 transitions for traditionally underrepresented students. To counter the potential for bias with our project, we debriefed during bi-weekly meetings to discuss our process. We also debriefed with external peers, as described in the next section.

Trustworthiness

To promote trustworthy findings, we first conducted our individual analysis of the data. We then triangulated our findings across research team members (Lichtman, 2006). In addition, we conducted member checks (Creswell & Poth, 2017) for participants to review their transcripts and provide edits or omissions to their responses. Finally, we debriefed with other higher-education researchers who have experience conducting studies on Latinx students in engineering programs. These peer debriefings helped counter our biases and provided additional clarification on our findings (Creswell & Poth, 2017).

FINDINGS

We found that dual credit did not directly influence the students' interest in and ultimate pursuit of an engineering—perhaps in part because the Latinx students interviewed had limited opportunity in high school to engage in engineering and other STEM coursework; rather, they completed dual credit courses in other fields. However, dual credit socialized students to college-level expectations and allowed them to focus on core classes, so they could more quickly begin engineering-specific coursework and advance in their major.

Pursuit of Dual Credit

Students in this study pursued dual credit opportunities in order to complete less-desirable classes, transfer college credit that would afford some flexibility in course scheduling, and help their family financially in the long term. The majority of students in this study completed dual credit coursework outside STEM disciplines, such as English and history, to fulfill general education course requirements. Several mentioned that dual credit coursework in "the hard sciences" were not offered. Additionally, in this study, dual credit did not facilitate a specific interest in pursuing engineering degrees, but it did allow students to focus on engineering coursework once they were attending college.

For example, Evelyn wanted to take advantage of earning transferrable college credit through a dual credit program so that she could immediately start taking classes in college that she truly enjoyed. Here, her goal was to be able to "focus [more] on the computer/engineering side. And [she] like[s] it a lot." In Evelyn's case, dual credit created a timely pathway to engineering classes. By getting her core requirements out of the way while in high school, she could focus on the course content that most interested her once she arrived at college.

Diana was also motivated to get the classes in which she was least interested out of the way. She was ready to focus on her passion for engineering. Her motivation "was the free college" experience as well as "getting it done." Dual credit saved her not only "a lot of money" but also "a lot of time." She was ahead of everyone else in terms of course credits and remarked, "I'm glad I did dual credit." Like Evelyn, Diana was able to immerse herself in her major classes more quickly, since the general education requirements were already met. Additionally, the dual credit decision was an economical one for Diana, in terms of both the price of college and opportunity costs.

Oscar discussed the flexibility allowed in his college schedule due to transferring in a high number of dual credits. Namely, he was able to focus his attention on fewer courses "at the university" in any given semester, without falling behind his classmates. "Dual credit. Um, well, I think I found myself at an advantage I would say over some of my, like, classmates because they would I don't have to take those basics, you know. Um, those basic classes so when other people were taking, like, five classes a semester, I'd be taking like three or four, and we would still be at the same level." Oscar may not be graduating early, unlike many other dual credit students in non-STEM majors. However, he gave his full attention to his engineering courses since he completed many of his core college requirements while still in high school.

On the other hand, José was inspired by his mother's work ethic and pursued dual credits as an opportunity to honor her efforts. He wanted to do his part to contribute, and he viewed dual credit as a small way to help his family in the long term. "Um, as I grew up, I always saw my mom work hard. Work hard to give us a better life until this day. She works two jobs, and so she always taught me that, and, well, working hard for something like she worked for us, she worked for her kids, like, can really put you further in life. And so, I always thought that's who, so that's why I looked into AP or honors and dual credit." José also discussed the importance of trying to save time and money so as not to be an additional burden to his family as he pursued his dream of attending college and majoring in civil engineering. He believed that hard work was an investment into his and his family's future.

Love of Tinkering and Solving Real-World Problems

Participants shared the value of hands-on experiences in fostering curiosity and learning about the field of engineering. For example, Miguel liked to tinker at home, which led to his decision to major in mechanical engineering. "When I was 12, I opened up a welder. Since it was broken, nobody used it, so I opened it up. I had two transformers. So basically, I just took two transformers out of two microwaves, made a welder out of that, and I saw it work. I'm like, 'Yeah, it works.' So, this is my thing. I know this is my thing." This hands-on experience allowed Miguel to discover the inner workings of a piece of equipment, and the success of his experiment solidified for him that engineering was the path he wanted to pursue. He recognized that it was the enjoyment received from tinkering that led him to hone in on his abilities by choosing an engineering major.

Another student, José, also liked to solve problems through a creative lens. He wanted to make a lasting impact on his community and felt that civil engineering would allow him to do so. He saw engineering as a broad major and career path that would allow him to tap into his creative side, as well as help others solve problems—"especially when it comes to big things." Architecture was an initial career interest for José, "but then [he] started seeing that, well, [he] like[d] to solve problems, and [he] looked at the civil engineering [field]." Through his chosen major and ultimate career choice, José wanted to use his talents in solving real-world, tangible problems to better the lives of individuals and to help guide companies in making the world a better place.

Diego also wanted to make an impact on the world. He therefore started to make the connection between how science and engineering coursework would allow him to do so. Diego considered himself to be strong in math. He had an early revelation that math and science could be used for real-word applications, especially since he did not "want to just teach math all [his] life, so [he'd] rather use it for something that could change the world." He recognized that, while he enjoyed math, math education was not a field that interested him. This helped him choose the path toward a civil engineering degree.

Career and Employment Opportunities

Participants expressed that they wanted to pursue higher education with the goal of gainful employment after graduation. That is, while they enjoyed the process of learning, they were looking forward to putting their education into practice. In thinking about long-term career goals, Diana, a first-generation college student, expressed the following, "I just don't like the idea of, like, working in a field and doing manual labor." Knowing family and friends who were employed in manual labor positions was a motivating factor for Diana to want more, in terms of career possibilities. Regarding her family's perception of the engineering field, Diana further shared that they felt "you can get hired everywhere." While Diana was already interested in engineering as a course of college study, she and her family were keenly aware of the many opportunities that would be available to her on graduation. As a first-generation college student, a degree in electrical engineering had a direct connection to employment after graduation.

Another student, Edward, mentioned not wanting to follow in his father's footsteps when it came to career choices. As to why he chose electrical engineering as a major, he shared, "So I don't have to be a painter when I grow up. I'm, that's why I'm doing this." Like Diana, Edward wanted to expand his career options. He chose a major that utilized his talents and spoke to his interests and passions while always keeping his career trajectory in mind.

IMPLICATIONS FOR RESEARCH, POLICY, AND PRACTICE

The present study focused on the role of dual credit programs in high schools, which led to exploring students' motivations to enroll in engineering programs. We found that Latinx students learned about the college experience and expectations through their dual credit programs. Dual credit classes also affirmed students' interests and allowed them to align their curiosities with a specific field of study.

Students agreed that family interactions directly or indirectly influenced or conveyed a certain message concerning their pursuit of an engineering degree. In alignment with previous research (Hernandez et al., 2016; Rodriguez et al., 2022), familial support and parental involvement were critical to their engineering pursuits. A student's natural or learned ability in a STEM-related area may be further cultivated through familial support and vocational messaging, and the validation to explore tinkering led to our second finding. As a result, K–12 educational leaders may consider sharing engineer career information that is accessible and free or affordable: they could do so by using asynchronous workshops or media, through which parents and guardians can learn about the field of engineering and how they can support their students' interests at home. These resources may be saved to an online platform that is available at the parent or guardian's convenience.

The hands-on experience that participants encountered outside of the classroom greatly influenced their desire to major in engineering, which also resulted in wanting to solve real-world issues. For instance, the participants highlighted their interests in deconstructing and reconstructing household items or wanting to better their local and global environments. Similar to prior research (Jahn & Myers, 2015), participants highlighted the importance of practical experiences for engaging with local issues and addressing mechanical problems. School leaders may support these out-of-class activities with school-wide engineering demonstrations, which can highlight the direct and positive implications of pursuing an engineering career. Educators may also consider inviting guest speakers or starting a mentorship program for aspiring engineers to learn more about the practical engineering problems.

On completing their degree, job and career stability were highly prioritized when considering the offers engineering graduates may receive. Enrolling in and completing dual credit courses are known to increase degree obtainment, employment opportunities, income, and health care (Tobolowosky & Allen, 2016). The exposure to college-level coursework in high school and the transition as an incoming college student concurrently establish marketable skills for job employment and student achievement. Likewise, participants in this study majored in engineering to avoid the kinds of manual-labor jobs their family members held, and to expand their employment options. Although the participants understood the immediate and longstanding impacts of earning dual credit, the courses were not motivating factors for their engineering major, since their programs did not offer saturated STEM-related content. In an effort to develop an early interest in engineering, community college instructors and high school teachers may consider how to infuse engineering topics, assignments, or field trips in their dual credit courses. Field trips or visits to engineering firms or federal contractors can support students' engineering interest and broaden their understanding of the engineering field.

Finally, unlike previous research (Bonous-Hammarth, 2000; Bottia et al., 2018; Cole & Espinoza, 2008; Jahn & Myers, 2015), we found that enrolling in dual credit courses through their high school did not increase interest in their respective engineering concentrations. This finding may be attributed to Latinx students' limited access to high school STEM classes (Tyson et al., 2007) and STEM high schools (Rogers-Chapman, 2013). Participants described dual credit as a method to (a) improve class schedule flexibility by completing general requirements prior to entering college and (b) supplement the cost of higher education. Dual enrollment courses assisted students in pursuing a four-year college degree and allowed them to focus on requirements pertaining to their major and specific concentration earlier in their program; however, they did not specifically ignite an interest in engineering.

LIMITATIONS AND FUTURE RESEARCH

Although this study contributes to the understanding of Latinx students' decisions to pursue engineering, it is limited in certain facets. First, we did not explore differences between men's and women's motivations and experiences or the intersectionality of gender with major/career decision-making. Therefore, our findings may not reflect the nuances of motivations and experiences among students of different genders. A second limitation was that this study explored why Latinx students pursued engineering from an overall perspective without considering the possible differences among the many engineering disciplines offered at NTRU. Future research could include such a distinction to explore any differences or similarities in motivations. Finally, most students in this study did not take STEM dual credit coursework. Future studies could focus on students who explored engineering in a more structured, intentional format while in high school.

CONCLUSION

As the literature suggests, there are many factors that influence Latinx students' decision to pursue engineering. In this study, vocational messaging from parents and family members impacted students' exploration of engineering from an early age. Tinkering and solving problems progressed the students' interests in the field, thus guiding their enjoyment of discovery and fine-tuning their abilities. A student's competencies and values related to engineering tasks were also instrumental in their final major and career decision. Finally, intentional, achievement-related choices—such as participating in dual credit programs—helped families financially as well as prepared students for what to expect in a college academic setting.

Author Note

Taryn Ozuna Allen: t.o.allen@tcu.edu
Christine Hall: christine.hall@tcu.edu
Courtney Matthews: courtney.matthews-289@tccd.edu
Correspondence concerning this article should be addressed to Taryn Ozuna Allen, PhD, Texas Christian University, TCU Box 297900, Fort Worth, TX 76019–057.

REFERENCES

An, B. P. (2013). The influence of dual enrollment on academic performance and college readiness: Differences by socioeconomic status. *Research in Higher Education, 54*(4), 407–432. https://doi.org/10.1007/s11162-012-9278-z

Association of Public & Land-Grant Universities (APLU). (n.d.). *How does a college improve graduates' employment and earnings potential?* APLU. https://www.aplu.org/our-work/4 -policy-and-advocacy/publicuvalues/employment-earnings/

Bailey, T. R., Hughes, K. L., & Karp, M. M. (2002). *What role can dual enrollment programs play in easing the transition between high school and postsecondary education?* U.S. Department of Education.

Banda, R. M., & Flowers III, A. M. (2016). Birds of a feather do not always flock together: A critical analysis of Latina engineers and their involvement in student organizations. *Journal of Hispanic Higher Education, 16*(4), 359–374. https://doi.org/10.1177/1538192716 66296

Bonous-Hammarth, M. (2000). Pathways to success: Affirming opportunities for science, mathematics, and engineering majors. *Journal of Negro Education, 69*(1–2), 92–111.

Bottia, M., Mickelson, R., Giersch, J., Stearns, E., & Moller, S. (2018). The role of high school racial composition and opportunities to learn in students' STEM college participation. *Journal of Research in Science Teaching, 55*(3), 446–476.

Bottia, M., Stearns, E., Mickelson, R., & Moller, S. (2017). Boosting the numbers of STEM majors? The role of high schools with a STEM program. *Science Education Policy, 102,* 85–107.

Bragg, D. D., Kim, E., & Rubin, M. B. (2005, November 17–19). *Academic pathways to college: Policies and practices of the fifty states to reach underserved students* [Conference paper]. Association for the Study of Higher Education annual meeting, Philadelphia, PA.

Camacho, M. M., & Lord, S. M. (2011). "Microaggressions" in engineering education: Climate for Asian, Latina, and White women. *Proceedings from Frontiers in Education Conference.* https://doi.org.10.1109/FIE.2011.6142970

Carnevale, A. P., Rose, S. J., & Sheah, B. (2011). *The college payoff.* The Georgetown University Center on Education and the Workforce.

Cole, D., & Espinoza, A. (2008). Examining the academic success of Latino students in science, technology, engineering, and mathematics (STEM) majors. *Journal of College Student Development, 49*(4), 285–300.

Creswell, J. W., & Poth, C. N. (2017). *Qualitative inquiry and research design: Choosing among five approaches* (4th ed.). Sage.

Cruz, J., & Kellam, N. (2018). Beginning an engineer's journey: A narrative examination of how, when, and why students choose the engineering major. *Journal of Engineering Education, 107*(4), 556–582.

Davis, T. (2016, June 23). Census data shows rising young Latino workforce, whether Texas is ready or not. *The Dallas Morning News.* https://www.dallasnews.com/business/personal -finance/2016/06/23/census-data-shows-rising-young-latino-workforce-whether-texas-is -ready-or-not/

Duncheon, J. C. (2020). "We are exposed to that college environment": Exploring the socialization of early college high school students. *Community College Review, 48*(2), 173–194.

Eccles, J. S. (2005). Subjective task value and the Eccles et al. model of achievement-related choices. In A. J. Elliot, & C. S. Dweck (Eds.), *Handbook of competence and motivation* (pp. 105–121). The Guilford Press.

Edwards, L., Hughes, K. L., & Weisberg, A. (2011). *Different approaches to dual enrollment: Understanding program features and their implications.* Community College Research Center.

Excelencia in Education. (2020). *Latino college completion: United States.* https://www .edexcelencia.org/research/latino-college-completion

Frase, R. T., & Bauldry, S. (2022). The expansion of higher education and the education-health gradient in the United States. *Social Currents, 9*(1), 70–86. https://doi.org/10.1177/232949652 11021645

Gates, A. Q., Hug, S., Thiry, H., Aló, R., Beheshti, M., Fernandez, J., . . . Adjouadi, M. (2011). The computing alliance of Hispanic-serving institutions: Supporting Hispanics at critical transition points. *ACM Transactions on Computing Education (TOCE), 11*(3), 1–21.

Giani, M., Alexander, C., & Reyes, P. (2014). Exploring the variation in the impact of dual-credit coursework on postsecondary outcomes: A quasi-experimental analysis of Texas students. *The High School Journal, 97*(4), 200–218.

Hernandez, D., Rana, S., Alemndar, M., Rao, A., & Usselman, M. (2016). Latino parents' educational values and STEM beliefs. *Journal for Multicultural Education, 10*(3), 354–367.

Hernandez, J. C., & Lopez, M.A. (2004). Leaking pipeline: Issues impacting Latina/o college student retention. *Journal of College Student Retention, 6*(1), 37–60.

Hoffman N. (2005). *Add and Subtract: Dual Enrollment as a State Strategy to Increase Postsecondary Success for Underrepresented Students.* Jobs for the Future.

Hofmann, E. (2012). Why dual enrollment? *New Directions for Higher Education, 2012*(158), 1–8.

Holten, B., & Pierson, A. (2016). *Getting ahead with dual credit: Dual-credit participation, outcomes, and opportunities in Idaho.* Education Northwest, Regional Educational Laboratory Northwest.

Institute of Education Sciences (IES). (2019). *Dual enrollment: Participation and characteristics.* U.S. Department of Education. https://nces.ed.gov/pubs2019/2019176.pdf

Jablin, F. M. (2001). Organizational entry, assimilation, and disengagement/exit. In F. M. Jablin & L. L. Putnam (Eds.), *The new handbook of organizational communication* (pp. 732–818). Sage.

Jahn, J. L. S., & Myers, K. K. (2014). Vocational anticipatory socialization of adolescents: Messages, sources, and frameworks that influence interest in STEM careers. *Journal of Applied Communication Research, 42*(1), 85–106.

Jahn, J. L. S., & Myers, K. K. (2015). "When will I use this?" How math and science classes communicate impressions of STEM careers: Implications for vocational anticipatory socialization. *Communication Studies, 66*(2), 218–237.

Johnson, C. C. (2012). Implementation of STEM education policy: Challenges, progress, and lessons learned. *School Science and Mathematics, 112*(1), 45–55.

Joint Economic Committee. (2012). *STEM education: Preparing for the jobs of the future.* United States Congress.

Kanny, M. A. (2014). Dual enrollment participation from the student perspective. *New Directions for Community Colleges, 2014*(167), 59–70.

Kanny, M. A. (2015). Dual enrollment participation from the student perspective. *New Directions for Community Colleges, 2015*(169), 59–70.

Kitchen, J. A., Sonnert, G., & Sadler, P. M. (2018). The impact of college- and university-run high school summer programs on students' end of high school STEM career aspirations. *Science Education, 102*(3), 529–547.

Lee, J., Hernandez, F., Ro, H. K., & Suh, H. (2022). Does dual enrollment influence high school graduation, college enrollment, choice, and persistence? *Research in Higher Education, 63,* 825–848.

Levine, K. J., & Hoffner, C. A. (2006). Adolescents' conceptions of work: What is learned from different sources during anticipatory socialization? *Journal of Adolescent Research, 21*(6), 647–669.

Lichtman, M. (2006). *Qualitative research in education: A user's guide.* Sage.

Long III, L., Henderson, T. S., & Williams, M. S. (2018). Institutional barriers to Black and Latino male collegians' success in engineering and related STEM fields. *Proceedings of the American Society for Engineering Education.* https://doi.org.10.18260/1-2--30673

Main, J. B., Griffith, A. L., Xu, X., & Dukes, A. M. (2018). Choosing an engineering major: A conceptual model of student pathways into engineering. *Journal of Engineering Education, 111,* 40–64.

Mansell, N., & Justice, J. (2014). Learning from the past: Dual credit. *Administrative Issues Journal: Connecting Education, Practice, and Research, 4*(1), 1–10.

Matusovich, H. M., Streveler, R. A., & Miller, R. L. (2010). Why do students choose engineering? A qualitative, longitudinal investigation of students' motivational values. *Journal of Engineering Education, 99*(4), 289–303.

McClure, E., Guernsey, L., Clements, D., Bales, S., & Nichols, J. (2017). How to integrate STEM into early childhood education. *Science and Children, 55*(2), 8–10.

Meyer, M., & Marx, S. (2014). Engineering dropouts: A qualitative examination of why students leave engineering. *Journal of Engineering Education, 103*(4), 525–548.

Myers, K. K., Jahn, J. L. S., Gailliard, B. M., & Stoltzfus, K. (2011). Vocational anticipatory socialization (VAS): A communicative model of adolescents' interests in STEM. *Management Communication Quarterly, 25*(1), 87–120. https://doi.org/10.1177/0893318910377068

National Science Foundation, National Center for Science and Engineering Statistics. (2017). *Women, minorities, and persons with disabilities in science and engineering: 2017.* Special Report NSF 17–310.

Nichols, A. H., & Shak, J. O. (2018). *Degree attainment for Black adults: National and state trends.* The Education Trust.

Park, J. J., Kim, Y. K., Salazar, C., & Eagan, M. K. (2022). Racial discrimination and student-faculty-interaction in STEM: Probing the mechanisms influencing inequality. *Journal of Diversity in Higher Education, 15*(2), 218–229.

Pew Research Center. (2014). *The rising cost of not going to college.* https://www.pewresearch.org/social-trends/2014/02/11/the-rising-cost-of-not-going-to-college/

Phelps, L. A., & Chan, H. (2016). Optimizing technical education pathways: Does dual-credit course completion predict students' college and labor market success? *Journal of Career and Technical Education, 31*(1), 61–84.

Rodriguez, S. L., Doran, E. E., Sissel, M., & Estes, R. (2022). Becoming *la ingeniera*: Examining the engineering identity development of undergraduate Latinas. *Journal of Latinos and Education. 21*(2), 181–200.

Rogers-Chapman, M.F. (2013). Accessing STEM-focused education: Factors that contribute to the opportunity to attend schools across the United States. *Education and Urban Society, 46*(6), 716–737.

Struhl, B., & Vargas, J. (2012). *Taking college courses in high school: A strategy guide for college readiness—The college outcomes of dual enrollment in Texas.* Jobs for the Future.

Texas Education Agency. (2021). *Enrollment in Texas public schools 2020–21.* Texas Education Agency. https://tea.texas.gov/sites/default/files/enroll-2020-21.pdf

Texas Higher Education Coordinating Board. (2017). *Phase II: Ensuring effective scaling of dual-credit education programs in Texas.* https://reportcenter.highered.texas.gov/reports/data/dual-credit-education-programs-in-texas-phase-ii/

Thomas, N., Marken, S., Gray, L., Laure, L., & Ralph, J. (2013, February). *Dual credit and exam-based courses in U.S. public high schools: 2010–2011: First Look* (NCES 2013–001). U.S. Department of Education.

Tobolowsky, B., & Allen, T. O. (2016). On the fast track: Understanding the opportunities and challenges of dual enrollment. *Association for the Study of Higher Education Higher Education Report Series, 42*(3). Jossey-Bass.

Trostel, P. (2017, January 29). Beyond the college earnings premium. Way beyond. *The Chronicle of Higher Education.* https://www.chronicle.com/article/beyond-the-college-earnings-premium-way-beyond/

Troutman, D. R., Hendrez-Soto, A., Creusere, M., & Mayer, E. (2018). *Dual credit and success in college.* The University of Texas System. https://www.utsystem.edu/documents/docs/ut-system-reports/2018/dual-credit-and-success-college

Tyson, W., Lee, R., Borman, K. M., & Hanson, M.A. (2007). Science, technology, engineering, and mathematics (STEM) pathways: High school science and math coursework and post-secondary degree attainment. *Journal of Education for Students Placed at Risk, 12*(3), 243–270. https://doi.org/10.1080/10824660701601266

United States Census Bureau. (2022). *QuickFacts: Texas.* https://www.census.gov/quickfacts /fact/table/TX/RHI725221

Villa, E. Q., Wandermurmen, L., Hampton, E. M., & Esquinca, A. (2016). Engineering education through the Latina lens. *Journal of Education and Learning, 5*(4), 113–129.

Wilson-Lopez, A., Mejia, J. A., Hasbún, I. M. and Kasun, G. S. (2016), Latina/o Adolescents' Funds of Knowledge Related to Engineering. *Journal of Engineering Education, 105*(2), 278–311. https://doi.org/10.1002/jee.20117

11 · WHAT ABOUT DISABLED LATINO/A/XS IN ENGINEERING?

LISETTE E. TORRES AND KRYSTAL PERALEZ

The Latino/a/x population in the United States is spreading throughout the country from historic population centers. As of 2020, it has grown to 62.1 million people (Passel et al., 2022). Of those, up to 11.8 million Latino/a/xs (or one out of every six people; Centers for Disease Control and Prevention [CDC], 2020) identify with some type of disability—the two most frequent disability types being mobility/ambulatory impairments (about 16 percent) and cognitive impairments (about 18 percent) among working Latino/a/xs with disabilities (Costa & Rosenblum, 2022). Approximately 43% of employed disabled Latino/a/xs, in fact, have two or more disabilities (Costa & Rosenblum, 2022). The Latino/a/x population is also the nation's youngest racial/ethnic group, with a median age of twenty-eight (Lopez et al., 2018), and it is getting younger, with newborns driving recent population growth (Krogstad & Noe-Bustamante, 2021). Based on these statistics, it is reasonable to assume that the number of Latino/a/x undergraduates, including those with disabilities, will increase. Disability is already on the rise in Latino/a/x communities, as the overall disability rate and the prevalence of disability among working-age (sixteen to sixty-four years) Latino/a/xs is higher than for other racialized groups (World Institute on Disability, 2006). Unfortunately, there is scant research on students with disabilities at the undergraduate and graduate levels generally (Wells & Kommers, 2020), let alone on the intersection of Latinidad and disability in engineering.

Data on Latino/a/x undergraduates with disabilities in STEM are important because these students are an untapped source of ingenuity and resilience and may be needed to maintain and expand the STEM workforce in the United States. A recent analysis by the Bureau of Labor Statistics projects continued significant growth in the coming years for STEM occupations in the United States (Ice et al., 2021), including an increase of nearly 140,000 new jobs expected for engineers from 2016 to 2026 (Torpey, 2018). Despite this occupational growth, according to the Pew Research Center, "Hispanic workers remain underrepresented in STEM jobs compared with their share of the U.S. workforce" (Kennedy et al., 2021, para. 2). They also continue to remain a small percentage of engineering students (Garcia-Felix, 2019), although the percentage of Latino/a/xs who have earned engineering bachelor's degrees has increased a bit, to approximately 13 percent as of 2018 (National Center for Science and Engineering Statistics [NCSES], 2021). Based on a 2017 to 2019 analysis, Hispanics constitute 9 percent of the engineering and architecture workforce, while white-bodied workers are overrepresented at 71 percent (Fry et al., 2021). Any further change in the composition of the engineering workforce depends on how the American education system addresses the matriculation and retention of diverse student populations.

The challenges of retaining racial and ethnic diversity in engineering education and training pathways and the increased need to address the enrollment and effective support of Latino/a/x students with disabilities pursuing higher education—particularly along pathways to engineering—require in-depth examination. Analysis may help engineering educators and professionals better target and eliminate barriers affecting students' access to and success in the field. Thus, this chapter presents a scoping review (Grant & Booth, 2009) of extant literature related to disabled Latino/a/xs in engineering. It is an initial assessment of the size and scope of peer-reviewed literature, reports, dissertations, and conference proceedings that exist about the lived experiences of Latino/a/x students in engineering; individuals with disabilities majoring in engineering; and Latino/a/xs with disabilities. We discuss the patterns of research on these topics, offer recommendations, and articulate the critical need for change in addressing these topics to understand and improve the experience of Latino/a/xs with disabilities pursuing engineering degrees and careers.

THEORETICAL FRAMEWORK AND POSITIONALITY

LatDisCrit (Padilla 2021, 2022) is an integration of the theoretical frameworks of LatCrit theory (Bernal, 2002; Dávila & de Bradley, 2010; Solórzano & Bernal, 2001; Valdés, 1999, 2000; Yosso, 2006) and DisCrit (Annamma et al., 2013; Annamma et al., 2016) within the broader Critical Race Theory (CRT) lineage. Both LatCrit and DisCrit interrogate the interplay of race/ethnicity, diasporic cultures, historical social and political factors, and disability. LatDisCrit, in particular, honors and complicates the multifacetedness of Latino/a/x identities, acknowledging the "waves and wakes of converging diasporas" (Alexis Padilla, personal communication, 2022) and the historic and contemporary manifestations of coloniality. It highlights how trans-Latino/a/x identities are construed as "other" by dominant groups, as well as how these identities are spread beyond the boundaries of nation-states. Consequently, trans-Latino/a/x identities are simultaneously tenuously connected to the hegemonic forces of the Global North and Global South, a bridge between worlds that can be ruptured and can promote the possibility of new imaginaries and mobilization. LatDisCrit also aims to enact rightful presence (Calabrese Barton & Tan, 2020), rejecting inclusion as an extension of rights that seeks to perpetuate guest–host relationships, rather than disrupting and altering the status quo. Lastly, LatDisCrit privileges alternative modes of intersectional subaltern knowing and doing (Padilla, 2022).

We use LatDisCrit in this chapter as a way to think about the complexity of Latino/a/x identities and, consequently, the unique ways of problem solving that arise from those trans-Latinidades (Padilla, 2022) that STEM disciplines—like engineering—are ignoring and marginalizing. We want to know whether disabled Latino/a/x students exist in the engineering education literature and, if so, are they viewed as having a rightful presence in undergraduate engineering class-rooms? Calabrese and Tan (2019) state that students are denied rightful presence when they are made invisible by being "positioned as outsiders because of who they are and the cultural assets they bring to learning" (p. 4). Based on our review of the literature, we would argue that rightful presence does not exist for disabled Latino/a/xs in engineering, specifically, but also within STEM generally.

Both authors identify as Latinas in the field of higher education, specifically STEM education, student affairs, and workforce development. We are both dedicated to greater access, equity, and social justice for Latino/a/x populations in the United States. Lisette identifies as a disabled Boricua motherscholar–activist. As a trained scientist who now conducts STEM education research, she focuses on addressing racialized gender justice and disability in science and higher education. She subscribes to CRT, intersectionality, women of color epistemologies, and other critical frameworks that seek to reveal and dismantle systems of power, privilege, and oppression. Lisette is also a cofounder and former executive board member of the National Coalition for Latinxs with Disabilities (CNLD), an organization whose mission is to "affirm, celebrate, and collectively uplift Latinxs with disabilities through community building, advocacy, protection of rights, resources, and education."[1]

Krystal identifies as a Mexican American practitioner–scholar. Her career's focus on student development and transition, as well as her interest in education equity and access, led to roles managing STEM research and education outreach, overseeing school-to-career programming for economic and workforce development organizations in San Antonio, Texas, and managing precollege and academic support programs at the University of Texas at San Antonio and the University of Texas at Austin. Krystal also serves as adjunct faculty in the Education Leadership and Policy Studies Department at the University of Texas at San Antonio. CRT, intersectionality, and Chicana and postcolonial feminism inform her examination of and response to systemic and institutional inequities that affect young adults of color.

Despite conducting an exhaustive search, we found that the engineering education and STEM education literature on Latino/a/x engineering undergraduates

(a) often have Latino/a/x students grouped with other engineering undergraduates of color and (b) have yet to look at the intersection of race/ethnicity and disability as it relates to the lived experiences of disabled Latino/a/x undergraduate engineering students. In our initial attempt to understand how race/ethnicity and disability influence disabled Latino/a/x undergraduate engineering students, we provide an overview of the literature related to disabled Latino/a/xs within engineering, trying to thread together isolated engineering education research findings that pertain to disability and Latino/a/xs. We hope to highlight the research gaps that need to be addressed and argue that an intersectional approach to engineering and STEM education research, in general, must include disability.

Sociopolitical Context

As described earlier, Latino/a/x populations are growing in the United States, and the rate of disability within Latino/a/x communities is increasing. More recently, the onset of the COVID-19 pandemic has created a mass disabling event that has disproportionately negatively affected Latino/a/xs and collectively traumatized entire communities of color (Kira et al., 2021). A lack of access to effective health care, increased virus exposure, and financial losses and hardships have especially hit students who are undocumented or have undocumented parents (Enriquez et al., 2023). Financial instability, isolation, stress from the pandemic, and symptoms of what is often called long COVID have created new Latino/a/x members of the disability community, many of whom are college-aged and may not acknowledge or identify as being disabled. This is important to consider when looking at recruitment, retention, and sense of belonging within engineering education. For example, Latino/a/x engineering students have recently conveyed stories "focused on the need to juggle academics with less than accommodating instructors, jobs (as support for school, family, or both), and new responsibilities as caretakers" (Sealey et al., 2021, p. 14). Thus, engineering education researchers must use an intersectional lens if they want to center and understand Latino/a/x lived experiences.

Additionally, hostile sentiment toward Latino/a/x communities has affected students' experiences. Institutional and systemic barriers, as well as negative interpersonal interactions, can impede student learning and progress (Banda & Flowers, 2017). In a recent study on the microaggressions experienced by Latino/a/x engineering students, students conveyed that they believed that "microassaults [e.g., racial slurs and jokes] were the result of increased racism across the United States due to the Trump Administration's immigration policies and derogatory comments" (Smith et al., 2022, p. 15) and that those situations decreased their academic self-efficacy, academic performance, and mental health. Greater attention to the social, cultural, and emotional contexts of the student experience is imperative; STEM departments have a responsibility to address these aspects of learning to best ensure that students are retained and complete their degrees.

Latino/a/xs in Engineering

Historically, engineering has consisted mainly of "white, middle class, typically functioning men" (Spingola, 2018, p. 2). Over time, social and cultural movements have changed these demographics. Currently, 9 percent of engineers are Latino/a/x, while 13 percent of engineering degrees are earned by Latino/a/x students (Fry et al., 2021; NCSES 2021). Student populations will continue to change. Specifically, the population of college-aged Latino/a/xs will increase in the coming decades; however, enrollment of Latino/a/xs in postsecondary education is not likely to match that growth (Cole & Espinoza, 2008). This stagnation has the potential to limit any continued growth of Latino/a/x students enrolling in STEM and, specifically, engineering education and training pathways.

Factors contributing to this challenge include student recruitment and enrollment practices; cultural and organizational characteristics used to systematize the student experience and create a more uniform body of future engineers; the level and type of social and academic capital incoming students possess; and the recent COVID-19 pandemic and increase in xenophobia (Camacho & Lord, 2013; Cole & Espinoza, 2008; Nazempour et al., 2022; Park et al., 2020; Ramirez, 2021). Engineering curricula and overall education and training pathways were originally designed with a rigidity "to ideologically 'keep men in line'" (Camacho & Lord, 2013, p. 35), and, eventually, this rigor became permanent and highly valued. Engineering programs continue to require some of the most difficult prerequisites and place students in a gauntlet of coursework that can force a premature decision regarding their major at the start of their academic career (Camacho & Lord, 2013; Schreffler et al., 2019; Cole & Espinoza, 2008). Students lacking the academic and social capital to navigate this aspect of engineering culture tend to suffer the most from it.

More recently, the COVID-19 pandemic has changed the delivery of curricula in engineering fields, presenting new challenges to academic departments and students (Park et al., 2020; Nazempour et al., 2022). The unpredictability of the pandemic and lack of institutions' preparedness for teaching on virtual platforms, in combination with psychological stressors, may have affected student health, learning, and degree completion. In addition, higher-achieving students may have struggled with the transition to remote teaching, indicating a further need to explore student outcomes from this period (Nazempour et al., 2022). Coupling these challenges with the challenges already faced by Latino/a/x students, achieving an engineering degree can be even more difficult.

Cultural capital can also significantly affect a student's experience. Latino/a/x students commonly enter engineering programs and college life with less preparation for some of the developmental challenges they will encounter. In many cases, these students will face a hostile campus climate, forms of gender discrimination and ableism, and assumptions about their ability to succeed in an engineering program (Camacho & Lord, 2013; Cardoso et al., 2013; Cole & Espinoza, 2008). Despite barriers, Latino/a/x students, when supported, can and do perform as well as their white peers. However, there is a paucity of research detailing the factors supporting Latino/a/x student success, especially those with Latino/a/x identities that intersect with other identities (Camacho & Lord, 2013; Spingola, 2018). Without a deeper understanding of these factors, improving the student experience and expanding the population of Latino/a/x engineers will be difficult.

Racist and xenophobic ideologies are more significantly affecting the student experience, particularly for Latino/a/x students (Banda & Flowers, 2017; Kocaturk & Bozdag, 2020; Ramirez, 2021). Institutional and systemic barriers can prevent access to higher education, and they can discourage students from accessing much needed support with navigating the rigor and challenges of college life (Smith et al., 2022; Ramirez, 2021). Additionally, students may experience social challenges from peers and faculty, discouraging their engagement in coursework. STEM and, specifically, engineering education require a closer examination of the marginalization of students to eliminate bias and improve enrollment and the overall experience of minoritized students (Banda & Flowers, 2017).

Though research on this topic is limited, especially in the context of students with disabilities, some studies explore aspects of the Latino/a/x STEM or engineering student experience. Recent research emphasizes the need for students to build cultural capital within higher education; calls for an updated curriculum that is inclusive of students from diverse academic backgrounds (such as those transferring from community colleges); and prioritizes building self-efficacy in students to promote success (Banda & Flowers, 2017; Hawley et al., 2013; Cardoso et al., 2013; Cole & Espinoza, 2008; Smith et al., 2022; Spingola, 2018). These emphases improve our understanding of the student experience but do not fully elucidate the experiences and needs of those Latino/a/x students with intersecting identities.

Some studies focus on gender, exploring the experience of Latinas in engineering. They suggest that Latinas tend to be well-prepared for engineering studies, though they face gender discrimination, cope with mental illness, and withstand (sometimes self-imposed) pressure for perfection (Banda & Flowers, 2017; Camacho & Lord, 2013; Cole & Espinoza, 2008). Even more sparse are studies examining the intersection of Latinidad and disability in engineering. Spingola (2018) clarifies that "there is a blatant need for more research including the perspectives of disabled individuals in both engineering education and engineering practice" (p. 7). They further elucidate that the literature reveals that individuals with disabilities are viewed within engineering as objects to be fixed, rather than participants and contributors with a unique and valid perspective. These problematic views of disability in the academic space will be further discussed in the next section.

An effective way to address the declining enrollment in STEM and, specifically, engineering disciplines is to focus on the recruitment and retention of women, students of color, and students with disabilities. These populations can have the greatest impact on the growth of STEM—particularly engineering—fields (Banda & Flowers, 2017; Hawley et al., 2013; Cardoso et al., 2013; Cole & Espinoza 2008; Smith et al., 2022). Additionally, the engineering profession and academic programs should consider sourcing these students from less "traditional" education pathways, including community college (Camacho & Lord, 2013). Though these are helpful solutions to the larger problem of the recruitment and retention of Latino/a/xs in STEM and engineering, a significant concern is the lack of well-studied and well-addressed explorations of the issue, with subsequent recommended practices and solutions (Camacho & Lord, 2013; Spingola, 2018). Further qualitative and quantitative research is necessary to identify specific factors influencing the enrollment activities of students and schools; the barriers to and supports for academic, social, and emotional growth and development; and how engineering, academically and professionally, may adjust to remove those barriers and expand on supports.

Disability in Engineering

According to National Center for Education Statistics 2019 data, approximately 21 percent of college students identify as individuals with disabilities (NCES, 2023). A majority of these individuals disclose some type of non-apparent disability, such as a learning disability (31 percent), attention-deficit/hyperactivity disorder (ADHD; 18 percent), and psychiatric disability (15 percent; Raue & Lewis, 2011). We also know that approximately 21 percent of undergraduates with disabilities identify as Latino/a/x (NCES, 2023). However, these statistics are most likely underestimates, as only 24 percent of undergraduates who have a documented individualized education plan (IEP) or 504 plan[2] in high school decide to register with their university's disability services office prior to the start of classes and only 60 percent of registered students at four-year colleges receive accommodations (Newman et al., 2011). It is possible that true estimates of the number of disabled undergraduates are two to three times higher than reported (DiversAbility at Yale, 2022). Accurate data are hard to come by because many disabled college students do not want to disclose their disability, and obtained estimates vary depending on the "survey methodology, how disability is defined, and how the questions are phrased" (Booksh, 2018, p. 625). Yet, data analysis by Friedensen et al. (2021) suggests that there are multiple small bottlenecks by which disabled students can be prevented from pursuing and staying in STEM, including in engineering. The first bottleneck occurs at the transition between high school and college.

Despite having a similar desire to pursue science and engineering majors and careers as students without disabilities, disabled undergraduate students tend to have high dropout rates between the transition from high school to college and from undergraduate to graduate education (Booksh, 2018). First-year students with disabilities are not properly prepared for the challenge of navigating a different accommodation process in college as compared to high school (Scott, 2019). During their secondary education journey, students who qualify as disabled according to the Individuals with Disabilities Education Act (IDEA) and need special education accommodations are ideally involved in the design of their IEP with their parents, teachers, and school administrators. Alternatively, if they have a disability but do not qualify for an IEP, the student will work with their parents, teachers, and principal to develop a 504 plan. Neither the IDEA nor 504 plan processes occur at the postsecondary level. This does not take into account that racial disparities exist in the accommodations process prior to college enrollment, where students of color are disproportionately excluded from protection under Section 504 (Erevelles, 2016). The bottom line is that disability accommodation services are largely meant to ensure Americans with Disabilities Act (ADA) compliance. Inadequate or missing accommodations and difficulty navigating college, especially during their first year, can leave disabled students "in an early academic hole that is difficult to escape" (Booksh, 2018, p. 627), negatively influencing their ability to pursue an engineering degree. A lack of accommodations can make degree attainment within engineering incredibly difficult, particularly since disabled students in engineering often face more challenges academically and have more difficulty getting help from engineering faculty and advisors (Lezotte et al., 2020).

Students who choose to disclose must navigate bureaucratic red tape that is often uncomfortable and marginalizing (Smith et al., 2021). They must seek and provide medical documentation to prove that they have a disabling impairment that affects their learning (Evans et al., 2017). This process can be costly, as students may have to spend money on doctor visits to receive a diagnosis confirmation. Once they obtain this information, they must fill out other documentation and then go to their university's disability resources office (if there is one) to meet with a staff member about their accommodation needs. The staff member then notifies the student's faculty, but the onus is still on the student to have a one-on-one discussion with the faculty, which can be incredibly uncomfortable. In a survey of 155 students from two universities on disclosure and the requesting and use of accommodations, disabled students reported that they often could not privately disclose to faculty members, which increased their discomfort with the disclosure process (Smith et al., 2021). Disabled students with non-apparent disabilities related to mental health expressed significantly greater discomfort and more negative peer interactions than their peers with apparent disabilities or non-apparent learning disabilities (Smith et al., 2021). Unfortunately, this discomfort is something they must endure, as the disclosure process and accommodation requests occur every semester, leading to access fatigue (Konrad, 2021). Further research is needed on how engineering students with disabilities, particularly those with intersectional identities, negotiate disclosure, accommodations, and self-advocacy.

Disabled undergraduates are often deterred from pursuing engineering degrees by the dearth of disabled role models in engineering, assumptions by faculty about students' abilities, insufficient career counseling, physical barriers, and the lack of STEM coursework and preparation in high school (Dunn et al., 2012). Even if a disabled undergraduate receives accommodations, has successfully completed their first year of college, and has declared an engineering major, they still face stigma, mistrust, stereotypes, and low expectations related to their disability from their faculty and peers within engineering. Friedensen et al. (2021) identify this as the second bottleneck for disabled students. For example, Rao and Gartin (2003) found evidence that engineering and law faculty members were significantly less willing to provide accommodations compared to other faculty. Jensen et al. (2004) report similar findings, highlighting how faculty were reluctant to provide accommodations to students with learning disabilities because they were not convinced of the diagnoses and were preoccupied with fairness related to extended time on exams and assignments. This faculty mistrust is common for individuals with non-apparent disabilities, such as learning disabilities, depression, and anxiety. They may experience a different kind of stigma and are more often questioned about their impairments and access needs as compared to those with more obvious impairments, such as those who identify as blind or d/Deaf and hard of hearing or are wheelchair users (Hawley et al. 2013; Booksh, 2018; Zongrone et al., 2021).

As part of a larger study on professional identity formation in undergraduate civil engineering students with disabilities, Zongrone et al. (2021) interviewed four engineering students and found that participants with non-apparent disabilities utilized "subjective severity" to determine the "extent to which their disability influenced their experiences as undergraduate engineering students and their ability to adhere to expectations and norms of higher education" (p. 11). They essentially spent time worrying and comparing themselves to their peers and a fictional notion of normality; this could lead them to perceive a disconnect between their identity and STEM (or engineering, specifically)—the final bottleneck identified by Friedensen et al. (2021). Disabled students also had to practice forms of self-advocacy to receive accommodations. The decision to disclose and the types of self-advocacy practices used are highly influenced by race (Julia Karpicz, her dissertation work, personal communication, March 2022) and other marginalized, intersectional identities (Scott, 2019). Lastly, Zongrone et al. (2021) found that participants often described their disability as being in conflict with their professional identities as engineers.

Along with interpersonal challenges and barriers to success in engineering, there has been limited work on examining the accessibility of physical lab and classroom spaces for disabled students. For example, Jeannis (2018) conducted a comprehensive review of literature between 1991 and 2015, noting that there were only twenty-two peer-reviewed articles that looked at barriers and supports for students with physical disabilities in labs and much of the evidence was anecdotal. The author also designed and implemented a survey to determine the barriers and facilitators in lab settings for physically disabled students. They found that 75 percent of survey respondents reported that entrances to science and engineering buildings were accessible, but major barriers inside the buildings included unclear or narrow aisles and pathways (46.8 percent), cramped workspaces

(31.9 percent), and items being too close together or too high (19 percent). Similarly, over half of respondents stated that they could not independently use lab tools and equipment because the tools and equipment could not be modified for their needs to complete the required lab tasks. Beyond the use of the physical lab space, disabled students reported that, when they did engage in lab activities, their participation was often limited by their peers to more passive roles, such as note-taking, writing papers, writing software, dishwashing, or observation (Jeannis, 2018; Jeannis et al., 2020).

The specific challenges and supports for disabled engineering students are still largely unknown. What we do know is that two-thirds of students with disabilities, at least at the secondary level, are more likely to be men (Schaeffer, 2020; Trainor et al., 2019). We also know that students with disabilities tend to be older than the traditional undergraduate student and are more likely to be financially independent (Wells & Kommers, 2020). They are also more likely to attend a two-year institution or a four-year college part-time rather than full-time, which makes it less likely that they will be exposed to research opportunities (Booksh, 2018). Women undergraduate students with disabilities are significantly less likely than their male counterparts to enroll in a STEM major, while disabled students from low-income backgrounds are more likely to select STEM majors (Lee, 2014). We do not know whether these selection processes are similar for engineering disciplines.

Due to the systemic challenges presented earlier, students with disabilities are less likely to graduate with a STEM major than their peers (Wells & Kommers, 2020). In a National Center for College Students with Disabilities (NCCSD) study using data from the 2018 Multi-Institutional Study of Leadership survey, disabled undergraduate students had a significantly lower sense of belonging and experienced significantly more discrimination on their campuses compared to their able-bodied peers (Soria, 2021). Despite these challenges, according to a recent and thorough systematic review of college students with disabilities, those who are academically successful tend to have the following personal traits: self-determination, self-advocacy, self-awareness, self-discipline, self-esteem, and executive functioning (Moriña & Biagiotti, 2021). Yet, frequent negative personal experiences along with ableist policies and practices can wear down and discourage disabled students from pursuing STEM careers (Friedensen et al., 2021), like engineering.

A recent literature review related to systemic and personal barriers to success for disabled engineering students highlights that "relatively little emphasis has been placed on students with disabilities, especially in STEM disciplines, engineering in particular" (Weatherton et al., 2017, p. 8). The authors also recommend further research on the intersection of disability and other social identities and their influence on the persistence of engineering students (Weatherton et al., 2017). Other studies have also underscored the dearth of literature related to people with disabilities in STEM (Lee, 2014; Wells & Kommers, 2020). In particular, Wells and Kommers (2020) call for more research along the K–20 pathway, including collecting data on the labor market and across disability types.

Disabled Latino/a/x Engineers

Our initial examination of the literature resulted in no peer-reviewed articles or other literature on the experiences of disabled Latino/a/x engineers at any education level (i.e., undergraduate, graduate, professional). This lacuna is important to address because of the increase in individuals identifying as having a disability in higher education and the responsibility of institutions to ensure equity and access to these individuals by providing support and mitigating barriers to academic success. Specifically, a large portion of English language learners (ELLs) with disabilities identify as Latino/a/x (71 percent), and almost 85 percent of ELLs with disabilities live in households that are considered below the poverty line (Trainor et al., 2019). As mentioned in previous sections, these are students who may lack the social and academic capital of their peers to navigate their academic programs in engineering.

Citing data from SRI International, Povenmire-Kirk et al. (2010) state that disabled Latino/a/x youth are more likely to attend vocational or trade schools as compared to their Black and white peers. However, if they do receive the opportunity to attend a two- or four-year institution, they are the most likely to complete a certificate or degree. Based on individual and focus group interview data, the authors found that disabled Latino/a/x high school youth and their families wanted to be involved in college summer programs but were barred from participating if they had low grades or multiple behavioral referrals. Latino/a/x families also remarked that—often due to a lack of culturally appropriate practices and language interpreters—they did not receive needed guidance on how to effectively navigate and participate in their child's special education services (Povenmire-Kirk et al., 2010) or information about education options for their child after high school (Trainor et al., 2019). In addition, parents of disabled ELLs, 71 percent of whom identified as Latino/a/x, were more likely than parents of other students with disabilities to report that their child's school had mostly come up with postgraduation goals during their transition planning meeting, and one-third or fewer of disabled ELLs received any supports beyond a review of the results of their college entrance exams (Trainor et al., 2019). Thus, systemic barriers at the secondary level, in effect, eliminate any opportunity for disabled Latino/a/xs to learn about college and potential careers and programs in engineering.

The only article that clearly dealt with undergraduate students of color with disabilities in STEM, including engineering, was a qualitative study by da Silva Cardoso et al. (2016). They interviewed six participants who were part of the MIND Alliance program, a recruitment and retention program hosted by Hunter College (New York) and Southern University (Louisiana). The authors identified four "domains" that spoke to the experiences of disabled college students of color: interpersonal factors (including family, peers, faculty, and staff), accommodations (specifically, faculty or staff response and peer attitudes toward accommodations), individual factors (such as self-motivation, difficulty forming relationships, help-seeking, participation in college organizations, and confidence), and, not surprisingly, the MIND Alliance program. Despite family and peers being supportive much of the time, participants shared several negative comments that they received from faculty and staff related to their access needs. Since some of them had non-apparent disabilities, they often made the decision to not draw attention to themselves and their accommodations. Unfortunately, students in the study were not explicitly asked about their intersectional experiences; they were generally asked about supports and barriers to their persistence in STEM (not within a specific subdiscipline or major). Participants' racial identities were not shared, and there was variability in disability type. It is evident that more research is needed about the lived experiences of disabled Latino/a/xs in engineering.

IMPLICATIONS AND CONCLUSIONS

We find that research on Latino/a/xs with disabilities in engineering programs and professions is limited. While research most often addresses the topic of Latino/a/xs in engineering or engineering professionals and students with disabilities separately, it rarely acknowledges the intersection of multiple identities and their combined impact. Therefore, we call for the collection and analysis of disaggregated data by race/ethnicity, gender, and disability for students within engineering. We also want greater attention paid to the varied cultural backgrounds that comprise the Latino/a/x community. The disabled Latino/a/x population is made up of trans-Latinidades and pan-disability identities (Padilla, 2022). We have different migration and immigration experiences influenced by colonialism, racism, and ableism. We speak different languages, come from different racialized lineages, practice different religions, and have different relationships with disability. It is time for engineering education and higher-education researchers to acknowledge that the Latino/a/x community is not a monolith.

Based on the limited number of publications we were able to locate in the literature, we also call for more critical qualitative research on students with disabilities—especially disabled Latino/a/xs—in engineering. We need to question the power dynamics that affect the educational journeys of disabled students with multiple marginalized identities and prevent them from becoming engineers. We need to engage interdisciplinary and transdisciplinary methodological approaches, such as photo elicitation, photovoice, and participatory action research, as well as critical theoretical frameworks such as CRT, intersectionality, DisCrit, multiracial feminist theory, and LatDisCrit. Currently, studies on disabled students tend to use positivistic and quantitative approaches or qualitative research based on a small number of individuals. If we truly want all students, including disabled Latino/a/xs, to have a rightful presence in engineering, then we need to look at how they are studied and portrayed; push for more detailed accounts; and effectively disseminate the information among engineering and higher education professionals. As Edlyn Vallejo Peña (2014) underscores in her article on the marginalization of published scholarship on students with disabilities in higher education journals, "When scholars, researchers, and editors of top tier journals do not engage in or include scholarship on students with disabilities, even if unintentionally, they communicate that understanding these needs and interests is less important than other issues in higher education. In other words, the indifference of the scholarly community to inquire into the growing needs of this student population in privileged scholarly venues has effectively deemed the research in this area as marginally important" (p. 38). Marginalizing this research, whether or not it is intentional, marginalizes the disabled Latino/a/x population, creating a repeating cycle of ignoring critical, relevant issues in higher education and, specifically, engineering and STEM fields.

Creating a body of research that includes disaggregated data along with qualitative studies of participants from an intersectionality framework (Crenshaw, 1989) will not only shift research foci to be more inclusive and practical but will inform necessary changes to improve the education and professional experience of Latino/a/xs with disabilities—and ultimately address the urgent workforce needs of the twenty-first century (Spingola, 2018). This information is vital in this sociopolitical climate, particularly given the increasing number of disabled Latino/a/xs due to COVID-19, immigration stress, xenophobia, and racial battle fatigue. Our hope is that this work is a beginning to a greater understanding of the lack of research on students with intersecting identities and a step forward to establishing a baseline understanding of engineering students and professionals with intersecting identities. Continued, deeper exploration of this topic can prompt a culture shift and the development of resources for Latino/a/xs with disabilities as they continue to encounter barriers to fulfilling their academic and professional goals.

NOTES

1. National Coalition for Latinxs with Disabilities website - http://www.latinxdisabilitycoalition.com/history-of-cnld.html
2. An individualized education plan, or IEP, is a legal document that outlines the instruction, supports, and services for a child as part of special education in pre-K–12 public education in the United States. IEPs are a right under the Individuals with Disabilities Education Act (IDEA). A 504 plan is a plan for how public schools can support the learning of disabled students within "regular" classrooms, and it is guaranteed under Section 504 of the Rehabilitation Act, which prohibits the discrimination of people with disabilities in programs or activities that receive federal funding.

REFERENCES

Annamma, S. A., Connor, D. J., & Ferri, B. A. (2013). Dis/ability critical race studies (DisCrit): Theorizing at the intersections of race and dis/ability. *Race, Ethnicity and Education, 16*(1), 1–31.

Annamma, S. A., Connor, D. J., & Ferri, B. A. (2016). Introduction: A truncated genealogy of DisCrit. In D. J. Connor, B. A. Ferri, S. A. Annamma (Eds.), *DisCrit: Disability studies and critical race theory in education* (pp. 1–9). Teachers College Press.

Banda, R., & Flowers, A. (2017). Critical qualitative research as a means to advocate for Latinas in STEM. *International Journal of Qualitative Studies in Education, 31*(8), 769–783. https://doi.org/10.1080/09518398.2018.1479046

Bernal, D. D. (2002). Critical race theory, LatCrit theory, and critical race-gendered epistemologies: Recognizing students of color as holders and creators of knowledge. *Qualitative Inquiry, 8*(1), 105–126. https://doi.org/10.1177/107780040200800107

Booksh, K. S. (2018). Academic pipeline for scientists with disabilities. *MRS Bulletin, 43,* 625–631. https://doi.org/10.1557/mrs.2018.194

Calabrese Barton, A., & Tan, E. (2019). Designing for rightful presence in STEM: The role of making present practices. *Journal of the Learning Sciences,* 1–43. https://doi.org/10.1080/10508406.2019.1591411

Calabrese Barton, A., & Tan, E. (2020). Beyond equity as inclusion: A framework of "rightful presence" for guiding justice-oriented studies in teaching and learning. *Educational Researcher, 49*(6), 433–440. https://doi.org/10.3102/0013189X20927363

Camacho, M. M., & Lord, S. M. (2013). *The borderlands of education: Latinas in engineering.* Lexington Books.

Cardoso, E. da Silva, Dutta, A., Chiu, C., Johnson, E. T., Kundu, M., & Chan, F. (2013). Social-cognitive predictors of STEM career interests and goal persistence in college students with disabilities from racial and ethnic minority backgrounds. *Rehabilitation Research, Policy, and Education, 27*(4), 271–284. DOI: 10.1891/2168-6653.27.4.271

Cardoso, E. da Silva, Phillips, B. N., Thompson, K., Ruiz, D., Tansey, T. N., & Chan, F. (2016). Experiences of minority college students with disabilities in STEM. *Journal of Postsecondary Education and Disability, 29,* 375–388.

Centers for Disease Control and Prevention (CDC). (2020). *Adults with disabilities: Ethnicity and race.* https://www.cdc.gov/ncbddd/disabilityandhealth/materials/infographic-disabilities-ethnicity-race.html

Centers for Disease Control and Prevention (CDC), National Center on Birth Defects and Developmental Disabilities (NCBDDD), Division of Human Development and Disability. Disability and Health Data System (DHDS) Data [online]. Accessed May 6, 2022, from https://dhds.cdc.gov

Challen, K., Lee, A. C. K., Booth, A., Gardoi, P., Woods, H. B., & Goodacre, S. W. (2012). Where is the evidence for emergency planning: A scoping review. *BMC Public Health, 12*(542), 1–7. https://doi.org/10.1186/1471-2458-12-542

Cole, D., & Espinoza, A. (2008). Examining the academic success of Latino students in science technology engineering and mathematics (STEM) majors. *Journal of College Student Development, 49*(4), 285–300. https://doi.org/10.1353/csd.0.0018

Costa, J., & Rosenblum, D. (2022, October 12). *Hispanic workers with disabilities in the labor force* [Blog]. Accessed January 13, 2024, from https://blog.dol.gov/2022/10/12/hispanic-workers-with-disabilities-in-the-labor-force

Crenshaw, K. (1989). Demarginalizing the intersection of race and sex: A Black feminist critique of antidiscrimination doctrine, feminist theory and antiracist politics. *University of Chicago Legal Forum, 1989*(1), 139–167.

Dávila, E. R., & de Bradley, A. A. (2010). Examining education for Latinas/os in Chicago: A CRT/LatCrit approach. *Educational Foundations, 24*(1), 39–58.

DiversAbility at Yale. (2022). Disability facts. Accessed April 3, 2022, from https://day.yale.edu/disability-facts

Dunn, C., Rabren, K. S., Taylor, S. L., & Dotson, C. K. (2012). Assisting students with high-incidence disabilities to pursue careers in science, technology, engineering, and mathematics. *Intervention in School and Clinic, 48*(1), 47–54. https://doi.org/10.1177/10534512 12443151

Enriquez, L. E., Morales, A. E., Rodriguez, V. E., Chavarria, K., & Ro, A. (2023). Mental health and COVID-19 pandemic stressors among Latina/o/x college students with varying self and parental immigration status. *Journal of Racial and Ethnic Health Disparities, 10*, 282–295. https://doi.org/10.1007/s40615-021-01218-x

Erevelles, N. (2016). The right to exclude: Locating Section 504 in the disproportionality debate. In J. Allan, A. J. Artiles (Eds.), *World yearbook of education 2017: Assessment inequalities* (pp. 120–136). Routledge.

Evans, N. J., Broido, E. M., Brown, K. R., & Wilke, A. K (Eds.). (2017). *Disability in higher education: A social justice approach.* Jossey-Bass.

Friedensen, R., Lauterbach, A., Kimball, E., & Mwangi, C. G. (2021). Students with high-incidence disabilities in STEM: Barriers encountered in postsecondary learning environments. *Journal of Postsecondary Education and Disability, 34*(1), 77–90.

Fry, R., Kennedy, B., & Funk, C. (2021, March 30). *Racial and ethnic group representation in STEM occupations.* Pew Research Center. https://www.pewresearch.org/science/2021/04/01/diversity-in-stem-appendix/

Garcia-Felix, L. V. (2019). Latinos not engaging in science, technology, engineering and mathematics (STEM) careers. *Journal of Academic Perspectives, 2019*(4), 1–21. https://www.journalofacademicperspectives.com/back-issues/volume-2019/volume-2019-no-4/

Grant, M. J., & Booth, A. (2009). A typology of reviews: An analysis of 14 review types and associated methodologies. *Health Information and Libraries Journal, 26*(2), 91–108. https://doi.org/10.1111/j.1471-1842.2009.00848.x

Hawley, C. E., Cardoso, E., & McMahon, B. T. (2013). Adolescent to adulthood in STEM education and career development: The experience of students at the intersection of underrepresented minority status and disability. *Journal of Vocational Rehabilitation, 39*, 193–204. https://doi.org/10.3233/JVR-130655

Ice, L., Rieley, M. J., & Rinde, S. (2021, February). Employment projections in a pandemic environment. *Monthly Labor Review,* U.S. Bureau of Labor Statistics https://doi.org/10.21916/mlr.2021.3

Jeannis, H. (2018). *Barriers and facilitators in training students with physical disabilities in science and engineering laboratories* [Doctoral dissertation, University of Pittsburgh]. ProQuest Dissertations Publishing. http://d-scholarship.pitt.edu/33864/1/HJ_ETD_Final_Apr2018.pdf

Jeannis, H., Goldberg, M., Seelman, K., Schmeler, M., & Cooper, R.A. (2020). Barriers and facilitators to students with physical disabilities' participation in academic laboratory spaces. *Disability and Rehabilitation. Assistive Technology, 15*(2), 225–237. https://doi.org/10.1080/17483107.2018.1559889

Jensen, J. M., McCrary, N., Krampe, K., & Cooper, J. (2004). Trying to do the right thing: Faculty attitudes toward accommodating students with learning disabilities. *Journal of Postsecondary Education and Disability, 17*(2), 81–90

Kennedy, B., Fry, R., & Funk, C. (2021, April 14). *6 facts about America's STEM workforce and those training for it.* Pew Research Center. https://www.pewresearch.org/fact-tank/2021/04/14/6-facts-about-americas-stem-workforce-and-those-training-for-it/

Kimball, E., Wells, R. S., Lauterbach, A., Manly, C. A., & Ostiguy, B. (2016). Students with disabilities in higher education: A review of the literature and an agenda for future research. In M. Paulsen (Ed.), *Higher education: Handbook of theory and research* (Vol. 31, pp. 91–156). Springer.

Kira, I. A., Shuwiekh, H. A., Alhuwailah, A., Ashby, J. S., Sous Fahmy Sous, M., Baali, S. B. A., & Jamil, H. J. (2021). The effects of COVID-19 and collective identity trauma (intersectional discrimination) on social status and well-being. *Traumatology, 27*(1), 29–39. https://doi.org /10.1037/trm0000289

Konrad, A. M. (2021). Access fatigue: The rhetorical work of disability in everyday life. *College English, 83*(3), 179–199.

Kotacurk, M. & Bozdag, F. (2020). Xenophobia among university students: Its relationship with five factor model and dark triad personality traits. *International Journal of Educational Methodology, 6*(3), 545–554. https://doi.org/10.12973/ijem.6.3.545

Krogstad, J. M., & Noe-Bustamante, L. (2021, September 9). *Key facts about U.S. Latinos for National Hispanic Heritage Month.* Pew Research Center. https://www.pewresearch.org /fact-tank/2021/09/09/key-facts-about-u-s-latinos-for-national-hispanic-heritage-month/

Lee, A. (2014). Students with disabilities choosing science, technology, engineering, and math (STEM) majors in postsecondary institutions. *Journal of Postsecondary Education and Disability, 27*(3), 261–272.

Lezotte, S., Hartman, H., Farrell, S., & Forin, T. (2020). Disability and engineering: A case of "othering"? *American Society for Engineering Education (ASEE) Conference Proceedings.* ASEE Annual Conference & Exposition, Virtual.

Lopez, M. H., Krogstad, J. M., & Flores, A. (2018, September 13). *Key facts about young Latinos, one of the nation's fastest-growing populations.* Pew Research Center. https://www.pewresearch .org/fact-tank/2018/09/13/key-facts-about-young-latinos/

Moriña, A., & Biagiotti, G. (2021). Academic success factors in university students with disabilities: A systematic review. *European Journal of Special Needs Education,* 1–18. https://doi.org /10.1080/08856257.2021.1940007

Munn, Z., Peters, M. D. J., Stern, C., Tufanaru, C., McArthur, A., & Aromataris, E. (2018). Systematic review or scoping review? Guidance for authors when choosing between a systematic or scoping review approach. *BMC Medical Research Methodology, 18*(143), 1–7. https://doi .org/10.1186/s12874-018-0611-x

National Center for Education Statistics (NCES). (2023). Fast facts: Students with disabilities. Accessed January 13, 2024, from https://nces.ed.gov/fastfacts/display.asp?id=60

National Center for Science and Engineering Statistics (NCSES). (2021). *Women, Minorities, and Persons with Disabilities in Science and Engineering: 2021.* Special Report NSF 21–321. National Science Foundation. https://ncses.nsf.gov/wmpd

Nazempour, R., Darabi, H., & Nelson, P. (2022). Impacts on students' academic performance due to emergency transition to remote teaching during the COVID-19 pandemic: A financial engineering course case study. *Education Sciences, 12*(202), 1–14. https://doi.org /10.3390/educsci12030202

Newmann, L., Wagner, M., Knokey, A.-M., Marder, C., Nagle, K., Shaver, D., & Wei, X. (2011). *The post-high school outcomes of young adults with disabilities up to 8 years after high school: A report from the national longitudinal transition study-2* (NLTS2) NCSER 2011–3005, National Center for Education Statistics, US Department of Education.

Ong, M., Jaumot-Pascual, N., & Ko, L. T. (2020). Research literature on women of color in undergraduate engineering education: A systematic thematic synthesis. *Journal of Engineering Education, 109*(3), 1–35. https://doi.org/10.1002/jee.20345

Padilla, A. (2021). LatDisCrit: Exploring Latinx global south DisCrit reverberations as spaces toward emancipatory learning and radical solidarity. In S. A. Annamma, B. A. Ferri, D. J. Connor (Eds.), *DisCrit expanded: Inquiries, reverberations & ruptures.* Teachers College Press.

Padilla, A. (2022). *Disability, intersectional agency and Latinx identity: Theorizing LatDisCrit counterstories.* Routledge.

Park, J. J., Park, M., Jackson, K., & Vanhoy, G. (2020). Remote engineering education under COVID-19 pandemic environment. *International Journal of Multidisciplinary Perspectives in Higher Education, 5*(1), 160–166.

Passel, J. S., Lopez, M. H., & Cohn, D. (2022, February 3). *U.S. Hispanic population continued its geographic spread in the 2010s.* Pew Research Center. https://www.pewresearch.org/fact-tank/2022/02/03/u-s-hispanic-population-continued-its-geographic-spread-in-the-2010s/

Peña, E. V. (2014). Marginalization of published scholarship on students with disabilities in higher education journals. *Journal of College Student Development, 55*(1), 30–40. https://doi.org/10.1353/csd.2014.0006

Povenmire-Kirk, T. C., Lindstrom, L., & Bullis, M. (2010). De escuela a la vida adulta/From school to adult life: Transition needs for Latino youth with disabilities and their families. *Career Development for Exceptional Individuals, 33*(1), 41–51. https://doi.org/10.1177/0885728809359004

Ramirez, B. (2021). Racist nativism in the college access experiences of undocumented Latinx students. *Journal of College Access, 6*(2), 65–79.

Rao, S., & Gartin, B. C. (2003). Attitudes of university faculty toward accommodations to students with disabilities. *Journal for Vocational Special Needs Education, 25*(3), 47–54.

Raue, K., & Lewis, L. (2011). *Students with Disabilities at Degree-granting Postsecondary Institutions* (NCES 2011–018). U.S. Department of Education, National Center for Education Statistics. U.S. Government Printing Office.

Schaeffer, K. (2020, April 23). *As schools shift to online learning amid pandemic, here's what we know about disabled students in the U.S.* Pew Research Center. https://www.pewresearch.org/fact-tank/2020/04/23/as-schools-shift-to-online-learning-amid-pandemic-heres-what-we-know-about-disabled-students-in-the-u-s/

Schreffler, J., Vasquez III, E., Chini, J., & James, W. (2019). Universal design for learning in STEM education for students with disabilities: A systematic literature review. *International Journal for STEM Education, 6*(8), 1–10. https://doi.org/10.1186/s40594-019-0161-8

Scott, S. (2019). Access and participation in higher education: Perspectives of college students with disabilities. *NCCSD Research Brief, 2*(2). National Center for College Students with Disabilities, Association on Higher Education and Disability.

Sealey, Z., Lewis, R. S., & Fletcher, Z. (2021). *What I wish my instructor knew: Navigating COVID-19 as an underrepresented student—Evidence based research* [Conference paper]. American Society for Engineering Education (ASEE) Annual Conference & Exposition, Virtual Conference. https://peer.asee.org/38044

Smith, K., Pleacovschi, C., Feinstein, S., & Luster-Teasley, S. (2022). Ethnicity, race, and gender in engineering education: The nuanced experiences of male and female Latinx engineering undergraduates targeted by microaggressions. *Psychological Reports, 126*(5), 2345–2382. https://doi.org/10.1177/00332941221075766

Smith, S. A., Woodhead, E., & Chin-Newman, C. (2021). Disclosing accommodation needs: Exploring experiences of higher education students with disabilities. *International Journal of Inclusive Education, 25*(12), 1358–1374. https://doi.org/10.1080/13603116.2019.1610087

Solórzano, D., & Bernal, D. D. (2001). Examining transformational resistance through a critical race and LatCrit theory framework: Chicana and Chicano students in an urban context. *Urban Education, 36*(3), 308–342. https://doi.org/10.1177/0042085901363002

Soria, K. M. (2021). *Supporting undergraduate students with disabilities: A focus on campus climate and sense of belonging.* National Center for College Students with Disabilities.

Spingola, E. (2018, June 24–27). *Literature review on disability participation in the engineering field* [Conference paper]. American Society for Engineering Education Annual Conference & Exposition, Salt Lake City, UT.

Torpey, E. (2018, February). *Engineers: Employment, pay, and outlook.* U.S. Bureau of Labor Statistics. https://www.bls.gov/careeroutlook/2018/article/engineers.htm

Trainor, A. A., Newman, L., Garcia, E., Woodley, H. H., Traxler, R. E., & Deschene, D. N. (2019). Postsecondary education-focused transition planning experiences of English learners with disabilities. *Career Development and Transition for Exceptional Individuals, 42*(1), 43–55. doi: 10.1177/2165143418811830

Valdes, F. (1999). Afterword, theorizing "OutCrit" theories: Coalitional method and comparative jurisprudential experience—RaceCrits, QueerCrits and LatCrits. *University of Miami Law Review, 53,* 1265–1306.

Valdes, F. (2000). Race, ethnicity, and Hispanismo in a triangular perspective: The "essential Latina/o" and LatCrit theory. *UCLA Law Review, 48,* 1–41.

Yosso, T. J. (2006). *Critical race counterstories along the Chicana/Chicano educational pipeline.* Routledge.

Weatherton, Y. P., Mayes, R. D., & Villanueva-Perez, C. (2017, June 24–28). *Barriers to persistence for engineering students with disabilities* [Conference paper]. American Society for Engineering Education Annual Conference & Exposition, Columbus, Ohio. doi:10.18260/1-2-27650

Wells, R., & Kommers, S. (2020). Graduate and professional education for students with disabilities: Examining access to STEM, legal, and health fields in the United States. *International Journal of Disability, Development and Education.* https://doi.org/10.1080/1034912X.2020.1726299

World Institute on Disability. (2006). *Latinos with disabilities in the United States: Understanding and addressing barriers to employment.* World Institute on Disability.

Zongrone, C., McCall, C. J., Paretti, M. C., Shew, A., Simmons, D. R., & McNair, L. D. (2021, January 25–28). *"I'm looking at you, you're a perfectly good person . . .": Describing non-apparent disability in engineering* [Conference paper]. Collaborative Network for Engineering and Computing Diversity (CoNECD) Conference. https://peer.asee.org/36059

12 · LATINA RESILIENCE IN ENGINEERING

Strategies of Success in a Hispanic-Serving Institution

ELSA GONZÁLEZ AND EMMA CLAUDIA PÉREZ

Creating and reviewing scientific knowledge requires unique and varied perspectives, yet the United States continues to lack racial/ethnic and gender diversity among STEM stakeholders. According to the National Center for Science and Engineering Statistics (NCSES; 2021), women and certain racialized groups (Latine, Black, and Indigenous) continue to be underrepresented in STEM education and the STEM workforce.[1] This has been exacerbated by the COVID-19 pandemic (National Science Board, 2022). Increasing heterogeneity in these contexts is essential for the advancement of STEM, for meeting its labor demands, and in the name of social equity and justice. To do this, the barriers faced by women and underrepresented groups require examination in tandem with the success strategies such populations employ.

The Latine population has significantly increased in the last couple of decades. The number of Hispanic-Serving Institutions (HSIs) in the United States has more than doubled (*Excelencia* in Education, 2020), with approximately 3.6 million Latine students enrolled in postsecondary institutions in 2019 (Hispanic Association of Colleges and Universities, 2021). Yet, Latine students neither enroll (Lumpkin et al., 2021) nor graduate (Tajalli & Ortiz, 2018) in equitable numbers. Though there have been gains in representation, women also continue to be underrepresented in the science and engineering workforce (Martinez & Christnacht, 2021). The underrepresentation of women is particularly conspicuous in the fields of computer science and engineering. In 2021, women comprised only 26 percent of computer/mathematical scientists and 16 percent of engineers of the college-educated labor force (NCSES, 2023).

Among Latine students, Latinas enroll in greater numbers than Latinos, but fewer Latinas graduate with engineering degrees than Latinos (*Excelencia* in Education, 2015). Women of color's unique experiences, which are tied to their intersectional identities, likely influence this pattern. Among Hispanic/Latina women, 3,200 received bachelor's degrees in engineering in 2018, representing approximately 3 percent of engineering degrees earned that year or an additional 1 percent increase in ten years (NCSES, 2021).

A synthesis of decades of research conducted by Ong and colleagues (2011) show that minoritized women—Latinas, Black, and Indigenous—earn disproportionately fewer STEM degrees, a pattern partly driven by the social climate in STEM fields. Within engineering, Latinas and other women of color tend to enter college with high self-efficacy and, over time, may learn to navigate the stereotypically masculine culture, sometimes at the cost of emotional and cognitive resources (Ong et al., 2020). This is concerning, as Latinas are a growing female minoritized group.

However, Latinas in STEM do complete their degrees through their resilience and success strategies. In this chapter, the researchers address Latina students in engineering: participants who are part of an ongoing study about Latina undergraduate STEM majors. The following research questions guide this synthesis: (1) What role does resilience, as influenced by culture, background, and family and community support, play in the pursuit and completion of an engineering college degree among Latina students in an HSI in Texas? (2) How do experiences in an HSI influence the resilience of Latina students in engineering fields, and what strategies of success drive persistence?

LITERATURE REVIEW

Latine Students in Higher Education and STEM

Previous research about Latine and other marginalized students' educational pathways have demonstrated a deficit-based perspective, such as a lack of high-quality schooling prior to college or a greater likelihood of dropping out of college (Darling & Smith, 2007; Schneider et al., 2006). In addition, students of color face racial and ethnic stigmas and encounter messages about "natural ability" in STEM fields, academic fields in which they do not typically see others like themselves (National Academies of Sciences, Engineering, and Medicine, 2016). However, Latine students in the United States generally bring culturally relevant assets that contribute to their experience in higher education, including family support, an ability to navigate family life and college, and bilingualism (Rendón et al., 2014; Rendón et al., 2019). Coping strategies—an example of resilience—are used to deal with challenges, including positive reframing (optimistic views), utilizing low expectations from others as motivation, and seeking support (Cavazos et al., 2010). Latine undergraduate students majoring in STEM similarly demonstrate resilience through community cultural wealth (CCW; further discussed in the conceptual framework section of this chapter). For example, they credit their family as a motivation for pursuing their degree and may even establish relationships with faculty (Rincón et al., 2020). Cultural assets additionally inform their STEM identities, such as grounding their identity in community instead of individualism (Rincón & Rodriguez, 2021).

When applying an intersectional lens, unique challenges and assets are elucidated through their different identities. Latinas majoring in STEM have expressed unsatisfactory experiences in STEM, including cultural incompatibilities and discouragement related to pursuing their undergraduate STEM degrees (Gonzalez et al., 2020). Family is also typically an integral part of the Latina student journey, as they provide essential support (which may be complicated by the "good-daughter persona"; Rodriguez, Pilcher, et al., 2019). Finally, like other women of color, Latinas report having difficulty being taken seriously in engineering, computer science, and physical sciences (Contreras Aguirre et al., 2020): fields known to lack minoritized women. This can sour the STEM experience, as recognition from others can influence how women of color view themselves as scientists (Carlone & Johnson, 2007). After all, women of color in STEM are less likely to feel that they belong—particularly in the physical sciences, where they are most underrepresented (Rainey et al., 2018).

Latinas in Engineering and Related Fields. The underrepresentation of women is high in engineering (NCSES, 2023), exacerbating the underrepresentation of women of color in these fields. Ong and colleagues (2020) found that academic preparation in STEM areas before college often preceded interest in engineering among women of color.

Interestingly, women of color from low socioeconomic backgrounds have shown academic resilience by pursuing higher education despite their background and resistance from significant others (Morales, 2008). In addition to an interest in the field, women of color pursuing engineering degrees have cited "giving back" in some way to their communities, whereby Latinas particularly focused on giving back to their families (Trenor et al., 2008). Latina students in undergraduate engineering programs have expressed how family influenced their engineering persistence (Gonzalez et al., 2020) and identity (Rodriguez, Cunningham et al., 2019) through inspiration and conquering challenges. Identity as a STEM person seems to be particularly important in fields like engineering as a navigational strategy (Ong et al., 2020) and in terms of Latina persistence (Gonzalez et al., 2021).

First-generation Latinas majoring in engineering have used various forms of CCW by engaging communal goals, as well as familial and personal aspirations (Coronella, 2021). Latinas have revealed *testimonios* or counter-stories about racialized experiences as a source of this CCW, aspiring to be engineers even in the face of overt aggressions and microaggressions or using resistance as capital. A powerful example of resilience comes from a single case study of a nontraditional engineering Latina student in a community college who "learned how to divide at 25" years old, as she was highly underprepared in high school (Verdín, 2021, p. 147). Nevertheless, she persevered, and her case represents how agentic acts inform identity or seeing how oneself is (re)shaped by social participation and practice—assets that can be viewed as resilience.

Multiple forms of support are also crucial to resilience. In addition to family support, Latino/as in undergraduate engineering programs have reported social support, such as benefiting from student or professional organizations (e.g., the Society of Hispanic Professional Engineers) and academic support through tutoring, college advisors, and federal programs (Garriot et al., 2019).

Conceptual Framework

"Resilience" is a term often used to describe someone's strength in character or perseverance after a stressful event. This overlaps with how scholars define resilience, as the adjustment to adversity or trauma (Luthar & Cicchetti, 2000) or adaptive and positive functioning with the passage of time (Southwick et al., 2014). Resilience is thus a broad construct that spans multiple disciplines. It is additionally influenced by several internal and external factors that affect this adaptive response, including attitude (an internal factor) and viable community and family support (external factors; Greene et al., 2004). Regarding sociocultural factors that affect resilience to a traumatic experience, factors such as gender, race/ethnicity, and socioeconomic status play important roles (Bonanno et al., 2007). As previously discussed, Latinas in engineering show resilience through resistance, family, and/or other forms of support.

This chapter is based on an emerging model of Latina/o/e resilience (Gonzalez, 2020). The model centers resilience as a critical, communal response, focusing on how Latinas draw on their cultural background as a source of individual and communal strength. This resilience model continues and extends the work by Yosso (2005) on CCW, which challenges previous, traditional, deficit-based perspectives by introducing an asset-based approach. Yosso (2005) describes CCW as an "array of knowledge, skills, abilities, and contacts possessed and utilized by Communities of Color" through six forms of capital: linguistic, aspirational, familial, navigational, resistant, and social (p. 77). CCW is built on a foundation of critical race theory scholarship (Patton, 2015): a theory built upon critical legal studies and radical feminism (Delgado & Stefancic, 2017) which challenges the presumed neutrality and objectivity of research and education and thus exposes systems that silence or misinterpret communities of color (Ladson-Billings & Tate, 1995; Yosso, 2005).

The Latina engineering student's cultural response influences and is influenced by different contexts: home, college, school, community, and workplace. Consequently, these were utilized in the present study to determine how Latinas use their resilience in challenging contexts within STEM fields, with a focus on engineering. How Latinas use their resilience may inform resultant success strategies.

This chapter provides data on Latina experiences and persistence (success) in undergraduate engineering programs. It addresses a need for an asset-based approach to explore how Latinas employ their resilience. Recommendations, particularly in terms of success strategies are discussed at the end of the chapter.

METHODS

Qualitative analyses were performed on participant data (Gonzalez & Forister, 2020; Lincoln & Guba, 1985) to understand the nature of resilience among participants. Qualitative methods allow researchers to interpret "multiple realities," value the human experience (Lincoln & Guba, 1985, p. 296), understand the complexity of the issue, and build overarching concepts that arise from the data (Creswell, 2012; Saldaña, 2021). Inductive and deductive coding were utilized as categories, and themes were both emerging in nature and created based on a set of research questions and a conceptual framework (Bingham & Witkowsky, 2021; Saldaña, 2021).

Participants

Participants were Latina undergraduate students in engineering at a diverse, public, and research-intensive university in Texas. The institution is an HSI, which is an institution with 25 percent or more undergraduate Hispanic/Latine students (*Excelencia* in Education, 2020). At the time of the research, this institution had a student enrollment of 33.3 percent Hispanic, 21.7 percent White, 11.9 percent African American, 22.1 percent Asian, 7.7 percent international, and 3.4 percent other/unknown. All participants belonged to the engineering college of the HSI. Participants were sought out by (a) contacting the authors' faculty networks/colleagues in engineering about students they knew and (b) contacting student organizations registered at the HSI in which this study was conducted. Voluntary participation in our study was requested via email. An electronic interest form was distributed for recruitment of potential participants. Eligible participants identified as Latina and were over eighteen years old. Approval from the university's institutional review board was obtained.

Because the research team was looking for "persisting" students in engineering, undergraduate juniors and seniors were interviewed. Of the seven Latina engineering students that filled out the interest form (all eligible), five (71 percent) responded and completed an interview. Table 12.1 summarizes the characteristics of our student respondents. At the time of the interview, students were majoring in different engineering fields. Though a small sample, the five participants allowed for an in-depth investigation of the experiences and strategies utilized by Latina undergraduates majoring in engineering.

Data Collection and Analysis

Following trustworthiness qualitative techniques, semi-structured deep interviews were recorded for audio transcription; member checking was accomplished

TABLE 12.1   Participant Characteristics

Pseudonym	Undergraduate level	Major	First-generation student	Family member is engineer	Attributes persistence to family	Active in STEM student organization
Erica	Senior	Biomedical engineering	Yes	No	Yes	Yes
Lily	Senior	Mechanical engineering	No	No specific mention	Yes	Yes
Claudia	Junior	Mechanical engineering	Yes	No specific mention	Yes	No
Valeria	Junior	Chemical engineering	Yes	No	Yes	Yes
Mari Jose	Senior	Civil engineering	No	Extended family	No	Yes

NOTE. "Yes" under "First-generation student" indicates that parents do not have college degrees. Data from the right two columns are further discussed in the findings section of this chapter.

during interviews by repeating, summarizing, and/or confirming participants' thoughts, and a reflexive journal was regularly maintained to examine biases and assumptions (Gonzalez & Forister, 2020). Interviews were open-ended, audio/ video recorded, virtual or in-person, and lasted between fifty and eighty minutes. The Dedoose "cross-platform app for analyzing qualitative and mixed methods research" (Dedoose 9.0.46, 2021) was used to analyze transcribed interviews. Interviews were transcribed verbatim and then interpreted, organized, and coded utilizing the overarching contexts (home, college, school, community, and work-place) in the emerging Latine model (Gonzalez, 2020) as starter categories. Content analysis and a constant comparative method were utilized (Gonzalez & Forister, 2020; Lincoln & Guba, 1985). The data were unitized (Bhattacharya, 2017), and the first rounds of coding analysis applied strategies that fall under what Saldaña (2021) describes as structural and descriptive coding. Quotes chosen to represent the participants' voices are shown precisely as they were spoken.

Researchers Positionality

The authors are two cisgender, heterosexual women who identify as Latina. One identifies as a Mexican Peruvian American and first-generation college student with a STEM educational background (terminal degree in the life sciences). The other is a Mexican-born faculty member in higher education whose research during the last ten years has focused on the access, retention, graduation, and success of Latina students in STEM fields. We recognize that some participants' experiences and identities differ from that of the researchers, including second-generation status, field of study, and family background. We engaged in this study due to our passion and desire to impact the educational success of Latinas, as well as the success strategies that we have learned and practiced.

FINDINGS

Themes and subthemes that emerged from the data analysis are discussed in terms of their relationships with the research questions. These include student organizations and supportive peer networks in the college context; faculty and staff within the college context; family influence through encouragement and motivation; self-efficacy, overcoming challenges and persevering; engineering provides a path to a good career but is challenging; and gendered and racialized experiences.

Student Organizations and Supportive Peer Networks in the College Context

This overarching theme describes the different communities and interactions within this HSI and thus pertains to both research questions. All participants talked in detail about academic peers and friends. Most were involved in student organizations geared toward STEM/engineering and Latine students. A few participants discussed their role and the community provided by these organizations to great length. For example, Lily—a mechanical engineering student—credited her student peers in a Latine engineering student organization with playing a significant role in her academic success and supporting her throughout her engineering program: "I think that has definitely played a big part is [Latine engineering organization] and coming to college is where I found my community and my support system. Without that support system that I found here [at the university], I probably would not be where I am right now."

Peer networks were often a primary source of academic support and thus consisted of regular study networks. For the majority of participants, these were the same peers or friends as those involved in student organizations. Erica, a biomedical engineer involved in a health sciences organization, provided several clear examples of how this may look in action: "They also have, you know, study—study events, kinda like—what was it? I think during the pandemic one of our now officers had helped run a workshop where we're basically doing an all-nighter study session where everyone is just on Zoom. And if you're studying the same subjects, you could obviously be in different break rooms and study together, which is really helpful, and it's really fun." Erica joined this organization in her sophomore year and was active through her senior year of college, becoming an officer along the way. She thus gained a great deal of social capital through this student organization's academic and social support, as well as the leadership role she occupied, all of which allowed her to develop strength and confidence within a community of peers: "I've really enjoyed being part of this organization because, one, I get to be exposed to different leadership opportunities and just being the leader in general, helping run things—helping organize different events and contacting people to become sponsors or helping host events with us. It's been really great; it's been really fun." Even Claudia—the one participant who was not heavily involved in an organization—noted that she sought such a community in college: "I want to join a group that—I don't know. I want to see what they do. I kind of want to be a part of a big family."

Communal activities and group membership nourished resilience among Latinas, allowing them to gain confidence and social support in a safe community of student peers and friends, particularly when close connections formed. These student networks additionally provided the participants with a space to enhance their academic performance.

Faculty and Staff within the College Context

In addition to peer support, all participants noted having received support from faculty and staff, pointing to specific faculty as great teachers. They described certain professors as "fair and understanding" or as having a genuine "passion" for transmitting knowledge. "Helpful" staff advisors were also mentioned, particularly when seeking guidance regarding the course curriculum. In other words, faculty and staff provided support indirectly through their passion for teaching or directly by giving instructions on choosing semester courses. Additionally, Valeria reported receiving mentorship from STEM faculty before landing on chemical engineering as her pathway to a career, explaining, "I really looked in lots of places to try figure it out, but I also think I was super fortunate to have people that cared enough to tell me these things."

All but one participant reported having also experienced discouragement or doubt from department faculty or advisors. Students discussed how there was a disconnect between faculty and students and/or a need for faculty to be more helpful when students sought help outside of lecture time. Interestingly, some participants said nearly the same thing about the professors teaching their courses: Half were great, and half were not. Claudia said: "They're divided into, like, two groups. There's one group that's, like, really passionate about what they do and so when they teach, it, like, shows through and it makes learning with them better, because you, like, understand more; and there's the other group that's kind of just, like, more notes-driven, and it's kind of, like, Do you want to be [laughter] here? Do you want to be teaching me type-of-thing?" Claudia expressed her frustration with learning when professors seemed "unmotivated." Mari Jose, a mechanical engineering student, described a charged discussion in class due to the comments made by the professor, whereby the professor described an area in town as "hovels" or something resembling a pile of rocks: "'Have you ever been poor ma'am? Do you even know what it's like to not live in a two-story like mansion or whatever?' I don't know where she lives, but, like, there's been little things like that. So, in this course, it's supposed to be us learning this code of ethics." Mari Jose described her frustration with the lack of understanding and empathy from professors, pointing out how they may be highly disconnected from the population and environment surrounding the institution—an important issue, given the many first-generation students that attend the institution. Mari Jose perhaps also tapped into her resistant capital when she pointed out the irony of such comments, as they occurred in a class dedicated to learning a code of ethics.

A couple of participants reported that their department expressed more direct discouragement or doubt regarding their abilities to pursue engineering, whether it was someone pointing out that the participant was "incorrect" about her plans for an engineering career or an advisor setting a "tone" that "you might not make it and that's on you." However, as evidenced by their current status (upperclassmen engineering majors), the participants had persisted despite these messages, likely tapping into resistant capital. Mari Jose expressed her resistance to such messaging in frustration by saying, "Well, I'm going to be a [expletive] engineer."

While some participants credited other students with providing mentorship, it is notable that only one mentioned an engineering (or any STEM) faculty member when asked about a mentor or role model during their undergraduate years. Experiences (or the lack thereof) involving department leaders within the HSI may influence Latina resilience, whether in a positive or negative way, as these participants noted having received both support and discouragement from departmental leaders.

Family Influence through Encouragement and Motivation

Not surprisingly, family—particularly parents—were a salient source of support while participants completed their engineering degrees. They were often a source of both social support and motivation for students. Claudia, a mechanical engineering major, expressed that family is her "drive" in addition to her own ambition. When asked about what her family thought about her attending the HSI to pursue engineering, she responded: "They're really proud, I guess. Yes, they've always been supportive, like it doesn't matter—they've always told me, like, "It doesn't matter what you want to pursue, like, as long as you're happy." So, that's something I'm really grateful, because not a lot of parents are like that." Claudia expressed gratitude for the support she received from her parents, a sign of familial capital. Additionally, she demonstrated how family, particularly parents, can hold a special role in the Latina college experience. In fact, most participants in this study cited family as one of the top reasons for their persistence in their engineering degree (table 12.1).

In addition to social or emotional support, the subtheme "parents provide financial support" is also relevant here. Most participants expressed that their parents provided indirect financial support, typically by allowing them to focus on academics by not having to take on a job while attending the university. This was expressed by Claudia, a first-generation student (table 12.1), and Mari Jose. Erica, another first-generation college student, was able to obtain scholarships through her father's military veteran status.

Additionally, another subtheme appeared within the context of family influence: "family influence in engineering direction." Interestingly, most participants expressed that their family had some level of influence on their decision to pursue engineering. One participant pointed out that she had a family member who was an engineer or had recently studied engineering (table 12.1). Two participants stated that they did not have family in engineering but expressed that their parents had an influence in some way. Lily, for example, mentioned that her college-educated mother helped her get connected to organizations or professional networks during her first year of college. And Claudia described how her home represented an early context that allowed creative thinking:

> I think it's influenced the reason why I like engineering. My dad does construction, so, like, from a young age, like, he'd be doing, like, work in the backyard, and then he'd cut up, like—when he cut up with whatever, there's, like, leftover pieces or whatever. It didn't even, like, occur to me until I was, like, in college, and I was like, "Oh, my god." Like, I literally used to play with, like, those little leftover blocks, and I'd make, like, my own little, like, racetracks type of stuff. I was like, "Oh, my god. Maybe I have been influenced from my young age. Maybe there's a connection there that I don't know.

Lily and Claudia were utilizing familial capital as they tapped into resources provided by their parents, though in very different ways. Lily was directly aided by her mom's involvement in college (a reflection of her mother's social capital), while Claudia discussed developing curiosity and creativity indirectly through her father's work in construction.

Two participants also shared that their parents desired their daughters to stay at or near home while attending university. When asked about what her family thought about her pursuing her biomedical engineering degree, Erica said that they were happy about it in general and that her university was close to home, something her parents were particularly happy about. Mari Jose, an older student pursuing her second STEM degree in civil engineering, returned home to pursue her engineering degree. Her mom welcomed her back with open arms. Mari Jose had moved away from home for several years when she completed her first bachelor's degree (in life science) and got a job shortly after. Her mom was content with having her daughter at home again while Mari Jose pursued her second degree, regardless of the fact that Mari Jose would not be working while living with her: "She was very supportive. Honestly, to ask her, like, 'What do you think if I—like, quit my job and enroll in the school, be a bum.' She's like, 'Sounds great, sweetie.' [laughter] I'm like, 'Okay, cool.' That's unexpected but cool. She was happy to not be in an empty nest again [laughter]." Mari Jose and Erica demonstrated how family was likely an early source for the Latinas in our study to build resilience, as they provided a supportive environment: particularly one that allowed them to pursue their ambitions, albeit sometimes with a few restrictions (such as staying home).

Self-Efficacy, Overcoming Challenges, and Persevering

Though resilience goes beyond persisting in light of trauma, one of the simplest ways to define it is showing perseverance in the face of adversity. The Latinas interviewed for this study had persisted and were nearing graduation, despite having dealt with relationships ending, car accidents, and/or the COVID-19 pandemic. Valeria, a chemical engineering major, explained how, despite the long journey of physically and mentally healing from a car accident, she considered herself lucky:

> I was studying fluid dynamics, and I had a car accident . . . So, that was really hard because I couldn't really walk very well, and so it made me really sad because I just—I like to run, and so I wasn't able to, and I just didn't feel good about myself. So, it was hard to be in school and walk like that, and deal with my lawyer and deal with going to physical therapy, and for so long . . . now, I can walk. It wasn't that bad; I was super lucky. Super lucky, it could've been way worse.

Despite being faced with significant challenges, Valeria considered her situation fortunate: she was on her way to graduate the following year. Though Valeria's experience was unique, all participants had to surmount obstacles due to the COVID-19 pandemic—as we will see in the next section.

Pandemic as Both a Hurdle and an Opportunity. All the students in this study complained about the lack of personal interactions due to the COVID-19 pandemic. Additionally, some participants had trouble transitioning to the fully online format that was implemented near the start of the pandemic and continued for a few semesters. For these participants, the boundary between home and work contexts was blurred. This lack of structure showed to be a disadvantage for their study habits as it likely had a harmful effect on their well-being (Plakhotnik et al., 2021; U.S. Department of Education, n.d.). They had a hard time focusing and noted an increase in isolation. Erica stated: "It was kind of a little bit of both. I think mostly just studying. Again, whenever you're at home you kind of think of home as, like, your, almost, like, a safe place. You can just relax here, you don't have to be on top of everything in terms of being, like, go go go. You can kinda just take a step back and relax. But being at home constantly, for me personally, it messes with how I'm able to, I guess, categorize everything in my head." For some participants, a fully online environment led to poorer academic performance, at least temporarily. Erica recalled having to retake one or two classes due to the pandemic. Claudia described her struggle with the transition that occurred when the institution went to an entirely online format in 2020:

> I feel like I struggled so much, because there's, like, a lot of way—not always you can, like, actually study and there's not a lot of ways you can, like, learn. I'm a huge visual learner type of person. Not a lot of my classes were [interesting] notes and, like, droning on, and it was, like, really hard to pay attention, and I felt like it was, like, one of those times really, like, this is like a huge struggle. I was like "I don't know if I can, like, do well," if that makes sense.

However, participants also noted that the pandemic challenged them for the better. Claudia additionally described the COVID-19 pandemic as her biggest challenge yet, "and it was kind of like 'This is your biggest challenge. If you can do this, then you can definitely move forward, because it just gets easier than this.'" Like Valeria after her car accident, Claudia noted positive aspects about the obstacles in front her. In her case, her obstacle was viewed as temporary and an opportunity for growth. Such responses are how one can view resilience, through assets and not deficits (Gonzalez, 2020).

Engineering Provides a Path to a Good Career but Is Challenging

Two themes, "interest in STEM/engineering" and "STEM/engineering is a challenging environment," are discussed next. The first theme introduces motivations for the pursuit of engineering, while the second describes how social experiences influenced their pursuit.

Most participants went into engineering due to an interest in the field or a fascination with engineering, whether it was wanting to promote the latest engineering technology or being curious about "how things work and how things are built." Several participants also expressed the importance of their engineering degree leading to a stable career through financial security and/or gaining socioeconomic mobility. Valeria stressed the importance of the financial security that her degree would bring, in addition to her curiosity in engineering. When asked about family influence, Valeria highlighted that it was important to her to be financially independent enough to take care of her family and how engineering seemed to be the most direct route to achieve this: "I think the reason they had a big impact to me is because from a young age I just felt pressured to take care of my family, so I think, like, a lot of things I've done so far—even changing to a chemical engineering degree, I thought it was cool, yes, but the main reason I did it was because I knew the opportunities were just so much bigger, so much more than a chemistry degree which is what I was doing [prior to engineering]." Mari Jose also mentioned that the compensation of an engineering degree was worth the struggle: "That's why I'm pursuing it. [Laughter] Yes, even internships were super well paid. So yes, I've been making good money just doing the internships alone." Financial stability or security were common motivators for earning engineering degrees, whether for the financial stability of the participants or the opportunity to support their families—or both.

However, the students also discussed how challenging engineering was as a field, sometimes at great length. While most students noted that engineering is highly challenging academically, the theme "engineering is a challenging environment" actually refers to the social challenges highlighted by participants. "Intimidating, competitive, pretentious" and "lacks community, reciprocity, is unwelcoming" emerged as (sometimes overlapping) subthemes to describe this environment. Participants specifically described interactions within the classroom as "intimidating" or "competitive." Mari Jose mentioned how she liked to promote collaboration within civil engineering, particularly after noticing harmful competition in other engineering fields: "I've heard from, like, mechanical engineers, like, they'll just let you get it wrong because that will help them in the curve, so it's not that. [laughter] . . . Not my experience. I feel like no one puts energy into, like, trying to sabotage other people . . . I try to encourage, like, a collaborative environment because sometimes people can get a little like 'This is all about me and my grade.'" Although she did not experience sabotage within her field of engineering, it is interesting that Mari Jose nonetheless tried to promote community or more reciprocal interactions among her peers. Valeria mentioned that "nobody really wanted to study with me at the beginning," until she began to gain a reputation as a high-performing engineering student. Further details are described in the next section, which describes an overlapping theme that further highlights the social challenges faced by the participants.

Gendered and Racialized Experiences

"Gendered and racialized experiences" emerged as a theme largely present in the college context but not absent from the home/family context. Mari Jose described an encounter within class whereby she and other students experienced racial microaggressions from a professor during a discussion of a current event: "She was sending out some messages. As an ethnic minority, we all pick up on it, but, like, that, that was the most explicit." Lily shared her thoughts about gender and ethnic identities, particularly in the college context, and their importance within a student organization: "I probably, I didn't feel a sense of community, a Hispanic community, until I went to college, and I found [a Latine engineering organization]. And I think that's where I feel calm and I feel like I'm a part of something, and I'm not judged off of, I guess, just my ethnicity. [Interviewer: Yes. You're not the only one anymore.] I'm not the only one anymore. Exactly."

When asked to whom she attributed her persistence in engineering, Lily mentioned the Latine organization she had already referenced. She expressed finding solace in a group that knew what it meant to face issues "as a Latina," recognizing that such an identity comes with its own perspectives and challenges. Lily explained how student networks, such as college organizations, can provide a safe haven for women of color who may not feel a sense of belonging and identity in a

STEM field like engineering: particularly in contexts within their institution. This may be unsurprising, due to the discriminatory experiences experienced by some Latine students in engineering—such as the microaggression recounted by Mari Jose previously. Moreover, Lily mentioned how others have found it "a surprise when women actually do good." Valeria recalled a similar interaction from a male student: "I was like, 'That dude just didn't want to listen to me because of the way I look.' [laughter] He does not want to listen to me because he would be like—I would say something, and then he'd just be like, 'Mm-hm.' Then he'd talk to the other person, the other guy, and he wouldn't—he would ask him the same question that I just answered." These participants found that their presence was not taken seriously in engineering classes. Lily described the gendered environment she has experienced as a woman in mechanical engineering: "Engineering is progressing and allowing women in STEM, but there's—that's always going to be a problem is women having to fight for their opinion, having to fight for a seat at the table, having to prove themselves."

While only a couple of participants experienced this type of prejudice directly (or at least discussed it via interview), participants were aware of and/or concerned about the lack of female faculty, particularly women of color—not just in engineering, but also in all their institution's STEM departments. Some participants also noted how they gravitated toward Latine peers. Lily expressed how she yearned for and found a Latine community in which she fit. Claudia noted that two minority women engineering professors with whom she had taken classes were an "inspiration," because students did not often have leaders around them to whom they could relate.

Taken together, these findings suggest a desire among the participants for more women of color in engineering fields. Young Latinas want to see others like them—leaders to whom they can relate in engineering. Therefore, gender and ethnic identities are integrated in the way that one experiences engineering contexts, even at a diverse HSI. When expressing her satisfaction with continuing to ask questions, even when feeling unwelcomed by engineering professors and students, Mari Jose summed it up best: "Women in STEM, baby!"

DISCUSSION

The present study illustrates the importance of the college and home contexts (Gonzalez, 2020), particularly in the communities or relationships built while pursuing an engineering degree. Latina resilience is likely enhanced by these relationships, as Latina students must deal with difficult social contexts in their undergraduate education (Gonzalez et al., 2020), particularly within engineering (Ong et al., 2020).

There is substantial social and academic support among student networks, at least for those who have integrated themselves into such groups (Garriot et al., 2019). However, participants demonstrated that there was a lack of relationships with department leaders, such as faculty. This points to a disconnect between students and professors, particularly given that there were few Latina or female role models within the department, as noted by the participants. Some participants recommended an increase in faculty women of color (or any women), to see more relatable people in power. This is not surprising, given that participants were in male-dominated fields where they saw few like themselves and at times experienced hostile spaces. Engineering students who do not see gender or ethnic representation (such as women of color) may encounter harsh environments in engineering education that produce "social pain" (Ong et al., 2020), likely leaving them unsure of the space they are in. More female engineering faculty could play a significant role in Latina retention; in fact, engineering departments with more faculty who are women of color tend to award more degrees to undergraduate women of color (Main et al., 2020). There is also some evidence to suggest that Latine faculty have a positive effect on Latine graduation rates (Capers, 2019; Stout et al., 2018), though this may depend on the type of institution (Capers, 2019).

Family was also an important presence among Latinas' success strategies. In this case, several participants continued to live at home, and their home context, primarily their parents, greatly influenced their college persistence. Closeness to family is not unusual in Latine culture and is an integral part of life for many Latina students (Chlup et al., 2018; Trevino & DeFreitas, 2014). Parents encouraged their daughters to remain at home while attending university and/or the participants leaned toward living at home. Participants were likely taking advantage of "geography of opportunity," whereby students are more likely to apply to a college nearby, particularly if they can stay near family (Reyes et al., 2019) and there are several institutions from which to choose (Turley Lopez, 2009). Moreover, staying near the home is an example of how culture is integral to one's behaviors and values and thus influences their contexts, such as the college context (Gonzalez, 2020). It is also notable that at least one participant had family member(s) who were engineers, who therefore had an additional effect on participants' decisions to pursue engineering—a finding supported by Rodriguez, Pilcher, et al. (2019). Being a second-generation student also helped one study participant connect with professional networks early on in college.

However, other participants lacked these types of social capital, a finding that is not unusual among Latine and other students of color in college (Darling & Smith, 2007) and/or high school (Chlup et al., 2019). Given that our sample of Latina students was heterogenous and similar to that in Rodriguez et al.'s (2019) study, one may question whether some social and familial capital may be slightly higher among engineering students. Nonetheless, according to the Office of Institutional Research at this HSI, in 2021, there were 12,431 undergraduate Hispanic students, of which 8,280 (66.6 percent) possessed first-generation status. This proportion of first-generation students is similar to those in the present study (60 percent).

Regardless of status, Latina engineering students in this study demonstrated persistence by seeking support and overcoming various challenges, and thus strategies reflective of resilience. The COVID-19 pandemic introduced additional challenges that participants had to learn to overcome, but these were challenges that several viewed as an opportunity for growth. It could be argued that this positive attitude demonstrated self-efficacy, which "reflects confidence in the ability to exert control over one's own motivation, behavior, and social environment" (American Psychological Association, n.d.). Such a perspective may have been particularly helpful when participants were dealing with problematic social contexts. These contexts were characterized by competitiveness and sometimes prejudice due to gender and ethnic identities, similar to other findings (Garriot et al., 2019). It could be that such experiences were part of the motivation for these young women to seek communities, like those in student organizations. Furthermore, support from institutional stakeholders has the potential to help Latina students build resilience and motivation to overcome such situations, as we see in the section on recommendations and success strategies, later in the chapter.

Culture Is Relevant to the Latina Engineering Experience

As previously studied, Latine students in STEM bring culturally relevant assets to their college experience (Rendón et al., 2019; Rincón & Rodriguez, 2021). Not only were participants motivated by family encouragement and support, but they additionally sought out communities where they would find support and perhaps solace. This may be analogous to Latinas' ties to family, as support from peers and mentors could be viewed as an academic family (Castellanos & Gloria, 2007). It is not unusual for Latinas (Contreras Aguirre et al., 2020) and women of color more generally (Ong et al., 2011) to have strong roots or ties to family, as demonstrated by some of the participants in this study. Therefore, students may bring with them this familial capital (Yosso, 2005) or perspective/action toward kinship and community to their college context and experience, just as participants did through membership in close-knit student communities. The activation of familial capital by "extending cultural notions of family" to student peers has been previously reported (Fernández et al., 2021, p. 7).

Mentorship from departmental leaders is likely another avenue for creating familial or more communal spaces in engineering (Fernández et al., 2021). However, most participants in this study did not have regular interactions or many close encounters with faculty. This finding contrasts a study in which female students of color had more mentors than their male counterparts (Morales, 2008)—though it should be noted that this was not specific to engineering.

Resilience within Engineering

A communal, family-oriented perspective is likely integral to Latina resilience. Latinas bring their culture (values, behaviors, and knowledge; Gonzalez, 2020) to their college experiences and manifest it in various ways. In the case of the engineering students, one way was through membership and involvement in student organizations. These relationships or communities can also serve as places of solace, to help mediate the racial or gendered microaggressions encountered by engineering students.

The participants displayed resilience in various ways. For one, they framed difficult obstacles positively, such as seeing a challenge as an opportunity for improvement. This is relevant regarding the COVID-19 pandemic, as it reduced personal interactions among most students and challenged their academic performance. The participants in our study showed a positive attitude by, rather than remaining neutral, making it a goal to persevere through the pandemic's effects and perhaps be better for it—a direction in line with the definition of resilience as positive functioning and adaptability (Luthar & Cicchetti, 2000; Southwick et al., 2014). Part of this motivation came from maintaining their "eyes on the prize." Several participants expressed the importance of finishing their degree to have a stable and well-paid career. Similarly, Morales (2008) found that women of color from low socioeconomic backgrounds are more strongly motivated by their postgraduation goals than men.

More critical to the current study is how resilience was displayed in the different contexts and cultural tenets that define the Latine resilience model (Gonzalez, 2020). Not only did specific contexts (the college and home contexts) influence the Latina students in undergraduate engineering education, but the Latina students reciprocally influenced their contexts: specifically, their college context, as displayed by their collective attitudes and behaviors (such as forming a community within a student organization) and their resistance to gendered or ethnic microaggressions.

Recommendations and Success Strategies

Based on the data presented, some recommendations/success strategies are worth mentioning. First is to encourage Latinas to join and be active in a student organization, particularly one associated with Latine community building within STEM. This study and others (e.g., Banda & Flowers, 2017) found that Latinas seek out race-based and/or engineering organizations. In line with this, Espinosa (2011) found that women of color are more likely than White women to seek a community. Therefore, encouraging Latinas to join a student organization, specifically to be part of a community, should be encouraged, emphasized, and reintroduced (for example, beyond the first year of college). Moreover, Latinas in engineering can utilize minority/female STEM and engineering organizations as a success strategy, as such groups have great potential to strengthen their resilience during their studies (Rincón & Rodriguez, 2021).

Second, institutional stakeholders, such as faculty or administrators, can support Latina student resilience by reaching out to such students and letting them know that they are not alone. Students can be encouraged to expand their networks to include faculty/administrators. Stakeholders' advice on expanding such networks would be beneficial, as marginalized students may not perceive this as a choice. These recommendations are supported by recent work published by the

National Academies of Sciences, Engineering, and Medicine (2023) on advancing antiracism, diversity, equity, and inclusion among stakeholders and leaders in STEM organizations.

Limitations and Future Research

There may be unexplored psychosocial methods in terms of retention of female students of color in college. For example, counter-stereotype beliefs may be a success strategy and indicative of resilience, as they flip stereotypes or beliefs that could be debilitating to persistence into motivation. Nguyen and Riegle-Crumb (2021) have found that Latinas and Black girls in high school held greater counter-stereotypical perceptions—"perceptions that scientists are multi-faceted individuals with a variety of interests and talents who do not work in isolation" (2021, p. 2)—than male students of color. Moreover, these counter-stereotypical perceptions were associated with greater intentions to major in computer science and engineering, specifically for Black and Latine youth. This pattern translates to a form of resistant capital (Yosso, 2005) that female students of color can utilize. Further knowledge and research on how this type of resistant capital can be transformed into or utilized as a success strategy could benefit Latina persistence in STEM.

Authors' Note

We have no known conflict of interest to disclose. This research was funded by the National Science Foundation, CAREER Project 2045802.

NOTE

1. The authors elected to use "Latine" (or Latiné) rather than "Latinx" or "Latino" because it is (1) gender inclusive and (2) phonetically inclusive in Spanish and used among individuals with ties to Latin America (Salinas, 2020).

REFERENCES

American Psychological Association. (n.d.) "Self-efficacy" in APA dictionary of psychology. Retrieved February 4, 2022, from https://dictionary.apa.org/self-efficacy

Banda, R. M., & Flowers, A. M. (2017). Birds of a feather do not always flock together: A critical analysis of Latina engineers and their involvement in student organizations. *Journal of Hispanic Higher Education, 16*(4), 359–374. https://doi.org/10.1177/1538192716662966

Bhattacharya, K. (2017). *Fundamentals of qualitative research: A practical guide*. Routledge.

Bingham, A. J., & Witkowsky, P. (2021). Deductive and inductive approaches to qualitative data analysis. In C. Vanover, P. Mihas, & J. Saldana (Eds.), *Analyzing and interpreting qualitative research: After the interview* (pp. 133–148). SAGE Publications.

Bonanno, G. A., Galea, S., Bucciarelli, A., & Vlahov, D. (2007). What predicts psychological resilience after disaster? The role of demographics, resources, and life stress. *Journal of Consulting and Clinical Psychology, 75*(5), 671–682. https://doi.org/10.1037/0022-006x.75.5.671

Capers, K. J. (2019). Representation's effect on Latinx college graduation rates. *Social Science Quarterly, 100*(4), 1112–1128. https://doi.org/10.1111/ssqu.12639

Carlone, H. B., & Johnson, A. (2007). Understanding the science experiences of successful women of color: Science identity as an analytic lens. *Journal of Research in Science Teaching*, *44*(8), 1187–1218. https://doi.org/10.1002/tea.20237

Castellanos, J., & Gloria, A. M. (2007). Research considerations and theoretical application for best practices in higher education: Latina/os achieving success. *Journal of Hispanic Higher Education*, *6*(4), 378–396. https://doi.org/10.1177/1538192707305347

Cavazos, J., Johnson, M. B., & Sparrow, G. S. (2010). Overcoming personal and academic challenges: Perspectives from Latina/o college students. *Journal of Hispanic Higher Education*, *9*(4), 304–316. https://doi.org/10.1177/1538192710380744

Chlup, D. T., Gonzalez, E. M., Gonzalez, J. E., Aldape, H. F., Guerra, M., Lagunas, B., Yu, Q., Manzano, H., & Zorn, D. R. (2018). Nuestros Hijos van a la Universidad [Our sons and daughters are going to college]: Latina parents' perceptions and experiences related to building college readiness, college knowledge, and college access for their children—A qualitative analysis. *Journal of Hispanic Higher Education*, *17*(1), 20–40. https://doi.org/10.1177/1538192716652501

Chlup, D. T., Gonzalez, E. M., Gonzalez, J. E., Aldape, H., Guerra, M., Lagunas, B., Yu, Q., Manzano Sanchez, H., & Zorn, D. (2019). Latina/o high school students' perceptions and experiences obtaining information about going to college: A qualitative study for understanding. *Journal of Latinos and Education*, 1–13. https://doi.org/10.1080/15348431.2019.1568878

Contreras Aguirre, H. C., Gonzalez, E., & Banda, R. M. (2020). Latina college students' experiences in STEM at Hispanic-serving institutions: Framed within Latino critical race theory. *International Journal of Qualitative Studies in Education*, *33*(8), 810–823. https://doi.org/10.1080/09518398.2020.1751894

Coronella, T. T. (2021). First-generation Latina engineering students' aspirational counterstories. In E. M. Gonzalez, F. Fernandez, & M. Wilson (Eds.), *An asset-based approach to advancing Latina students in STEM: Increasing resilience, participation, and success* (pp. 115–130). Routledge.

Creswell, J. W. (2012). *Educational research: Planning, conducting and evaluating quantitative and qualitative research* (4th ed.). Pearson.

Darling, R. A., & Smith, M. S. (2007). First-generation college students: First-year challenges. In M. S. Hunter, B. McCalla-Wriggins, & E. R. White (Eds.), *Academic advising: New insights for teaching and learning in the first year* (pp. 203–211). National Resource Center for The First-Year Experience and Students in Transition.

Dedoose, version 9.0.46. (2021). Web application for managing, analyzing, and presenting qualitative and mixed method research data. SocioCultural Research Consultants, LLC. www.dedoose.com.

Delgado, R., & Stefancic, J. (2017). *Critical race theory: An introduction* (3rd ed.). NYU Press.

Espinosa, L. (2011). Pipelines and pathways: Women of color in undergraduate STEM majors and the college experiences that contribute to persistence. *Harvard Educational Review*, *81*(2), 209–241. https://doi.org/10.17763/haer.81.2.92315ww157656k3u

Excelencia in Education. (2015). *Finding Your Workforce: Latinos in Science, Technology, Engineering, and Math (STEM)*. Excelencia in Education. https://www.edexcelencia.org/media/348

Excelencia in Education. (2020). *25 Years of Hispanic-Serving Institutions: A Glance on Progress*. Excelencia in Education. https://www.edexcelencia.org/Excelencia-25-Yrs-HSIs-Glance-On-Progress

Fernández, É., Rincón, B. E., & Hinojosa, J. K. (2021). (Re)creating family and reinforcing pedagogies of the home: How familial capital manifests for students of color pursuing STEM majors. *Race Ethnicity and Education*, 1–17. https://doi.org/10.1080/13613324.2021.1997971

Garriott, P. O., Navarro, R. L., Flores, L. Y., Lee, H., Carrero Pinedo, A., Slivensky, D., Muñoz, M., Atilano, R., Lin, C., Gonzalez, R., Luna, L., & Lee, B. H. (2019). Surviving and thriving:

Voices of Latina/o engineering students at a Hispanic serving institution. *Journal of Counseling Psychology, 66*(4), 437–448. https://doi.org/10.1037/cou0000351

Gonzalez, E. (2020). Foreword: Understanding Latina/o resilience. *International Journal of Qualitative Studies in Education, 33*(8), 791–795. https://doi.org/10.1080/09518398.2020.1783016

Gonzalez, E., Contreras Aguirre, C. C., & Myers, J. (2020). Persistence of Latinas in STEM at an R1 higher education institution in Texas. *Journal of Hispanic Higher Education,* 1–14. https://doi.org/10.1177/1538192720918369

Gonzalez, E., Fernandez, F., & Wilson, M. (Eds). (2021). *An asset-based approach to advancing Latina students in STEM: Increasing resilience, participation, and success.* Routledge.

Gonzalez, E. M., & Forister, J. G. (2020). Qualitative research. In J. G. Forister & J. D. Blessing (Eds.), *Introduction to research and medical literature for health professionals* (5th ed., pp. 111–124). Jones & Bartlett Learning.

Greene, R. R., Galambos, C., & Lee, Y. (2004). Resilience theory: Theoretical and professional conceptualizations. *Journal of Human Behavior in the Social Environment, 8*(4), 75–91. https://doi.org/10.1300/J137v08n04_05

Hispanic Association of Colleges & Universities (2021). *2021 Hispanic higher education and HSIs Facts.* Retrieved January 18, 2022, from https://www.hacu.net/hacu/HSI_Fact_Sheet.asp

Ladson-Billings, G., & Tate, W. F. (1995). Toward a critical race theory of education. *Teachers College Record, 97*(1), 47–68. https://doi.org/10.1177/016146819509700104

Lincoln, Y. S. & Guba, E. G. (1985). *Naturalistic inquiry.* Sage Publications.

Lumpkin, L., Kolodner, M., & Anderson, N. (2021, April 18). *Flagship universities don't enroll proportionate numbers of Black and Latino students.* The Hechinger Report. https://hechingerreport.org/flagship-universities-fail-to-enroll-black-and-latino-high-school-graduates-from-their-state/

Luthar, S. S., & Cicchetti, D. (2000). The construct of resilience: Implications for interventions and social policies. *Development and Psychopathology, 12*(4), 857–885. https://doi.org/10.1017/s0954579400004156

Main, J. B., Tan, L., Cox, M. F., McGee, E. O., & Katz, A. (2020). The correlation between undergraduate student diversity and the representation of women of color faculty in engineering. *Journal of Engineering Education, 109*(4), 843–864. https://doi.org/10.1002/jee.20361

Martinez, A., & Christnacht, C. (2021, January 26). *Women making gains in STEM occupations but still underrepresented.* The United States Census Bureau. Retrieved May 13, 2021, from https://www.census.gov/library/stories/2021/01/women-making-gains-in-stem-occupations-but-still-underrepresented.html

Morales, E. E. (2008). Exceptional female students of color: Academic resilience and gender in higher education. *Innovative Higher Education, 33*(3), 197–213. https://doi.org/10.1007/s10755-008-9075-y

National Academies of Sciences, Engineering, and Medicine. (2016). The culture of undergraduate STEM education. In *Barriers and opportunities for 2-Year and 4-Year STEM degrees: Systemic change to support students' diverse pathways* (pp. 59–81). National Academies Press. https://doi.org/10.17226/21739

National Academies of Sciences, Engineering, and Medicine. (2023). *Advancing antiracism, diversity, equity, and inclusion in STEMM organizations: Beyond broadening participation.* The National Academies Press. https://doi.org/10.17226/26803

National Center for Science and Engineering Statistics. (2021). *Women, Minorities, and Persons with Disabilities in Science and Engineering: 2021* (Special Report NSF 21–321). National Science Foundation. https://ncses.nsf.gov/pubs/nsf21321

National Center for Science and Engineering Statistics (2023). *Diversity and STEM: Women, minorities, and persons with disabilities 2023* (Special Report NSF 23–315). National Science Foundation. https://ncses.nsf.gov/wmpd

National Science Board. (2022). *Higher education in science and engineering. Science and engineering indicators 2022.* (NSB-2022-3). National Science Foundation. https://ncses.nsf.gov/pubs/nsb20223/

Nguyen, U., & Riegle-Crumb, C. (2021). Who is a scientist? The relationship between counter-stereotypical beliefs about scientists and the STEM major intentions of Black and Latinx male and female students. *International Journal of STEM Education, 8*(28), 1–18. https://doi.org/10.1186/s40594-021-00288-x

Ong, M., Jaumot-Pascual, N., & Ko, L. T. (2020). Research literature on women of color in undergraduate engineering education: A systematic thematic synthesis. *Journal of Engineering Education, 109*(3), 581–615. https://doi.org/10.1002/jee.20345

Ong, M., Wright, C., Espinosa, L., & Orfield, G. (2011). Inside the double bind: A synthesis of empirical research on undergraduate and graduate women of color in science, technology, engineering, and mathematics. *Harvard Educational Review, 81*(2), 172–209. https://doi.org/10.17763/haer.81.2.t022245n7x4752v2

Patton, L. D. (2015). Disrupting postsecondary prose: Toward a critical race theory of higher education. *Urban Education, 51*(3), 315–342. https://doi.org/10.1177/0042085915602542

Plakhotnik, M. S., Volkova, N. V., Jiang, C., Yahiaoui, D., Pheiffer, G., McKay, K., Newman, S., & Reißig-Thust, S. (2021). The perceived impact of COVID-19 on student well-being and the mediating role of the University support: Evidence from France, Germany, Russia, and the UK. *Frontiers in Psychology, 12.* https://doi.org/10.3389/fpsyg.2021.642689

Rainey, K., Dancy, M., Mickelson, R., Stearns, E., & Moller, S. (2018). Race and gender differences in how sense of belonging influences decisions to major in STEM. *International Journal of STEM Education, 5*(10), 1–14. https://doi.org/10.1186/s40594-018-0115-6

Rendón, L.I., Nora, A., & Kanagala, V. (2014). *Ventajas/assets y conocimientos/knowledge: Leveraging Latin@ strengths to foster student success.* Center for Research and Policy in Education, The University of Texas at San Antonio.

Rendón, L.I., Nora, A., Bledsoe, R., & Kanagala, V. (2019). *Científicos Latinxs: The untold story of underserved student success in STEM fields of study.* Center for Research and Policy in Education, The University of Texas at San Antonio.

Reyes, M., Dache-Gerbino, A., Rios-Aguilar, C., Gonzalez-Canche, M., & Deil-Amen, R. (2019). The "geography of opportunity" in community colleges: The role of the local labor market in students' decisions to persist and succeed. *Community College Review, 47*(1), 31–52. https://doi.org/10.1177/0091552118818321

Rincón, B. E., Fernández, É., & Dueñas, M. C. (2020). Anchoring comunidad: How first- and continuing-generation Latinx students in STEM engage community cultural wealth. *International Journal of Qualitative Studies in Education, 33*(8), 840–854. https://doi.org/10.1080/09518398.2020.1735567

Rincón, B. E., & Rodriguez, S. (2021). Latinx students charting their own STEM pathways: How community cultural wealth informs their STEM identities. *Journal of Hispanic Higher Education, 20*(2), 149–163. https://doi.org/10.1177/1538192720968276

Rodriguez, S. Cunningham, K., & Jordan, A. (2019). STEM identity development for Latinas: The role of self- and outside recognition. *Journal of Hispanic Higher Education, 18*(3), 254–272. https://doi.org/10.1177/1538192717739958

Rodriguez, S., Pilcher, A., & Garcia-Tellez, N. (2019). The influence of *familismo* on Latina student STEM identity development. *Journal of Latinos and Education,* 1–13. https://doi.org/10.1080/15348431.2019.1588734

Saldaña, J. (2021). *The coding manual for qualitative researchers* (4th ed.). SAGE.

Salinas, C. (2020). The complexity of the "x" in Latinx: How Latinx/a/o students relate to, identify with, and understand the term *Latinx. Journal of Hispanic Higher Education, 19*(2), 149–168. https://doi.org/10.1177/1538192719900382

Schneider, B., Martinez, S., & Owens, A. (2006). Barriers to educational opportunities for His-
    panics in the United States. In F. Mitchell & M. Tienda (Eds.), *Hispanics and the future of
    America* (pp. 179–227). National Academies Press.

Southwick, S. M., Bonanno, G. A., Masten, A. S., Panter-Brick, C., & Yehuda, R. (2014). Resil-
    ience definitions, theory, and challenges: Interdisciplinary perspectives. *European Journal of
    Psychotraumatology, 5*(1), 25338. https://doi.org/10.3402/ejpt.v5.25338

Stout, R., Archie, C., Cross, D., & Carman, C. A. (2018). The relationship between faculty diver-
    sity and graduation rates in higher education. *Intercultural Education, 29*(3), 399–417.
    https://doi.org/10.1080/14675986.2018.1437997

Tajalli, H., & Ortiz, M. (2018). An examination of Hispanic college enrollment and graduation:
    Has the Texas closing the gaps plan been successful? *Journal of Latinos and Education, 17*(4),
    330–343. https://doi.org/10.1080/15348431.2017.1348301

Trenor, J. M., Yu, S. L., Waight, C. L., Zerda, K. S., & Sha, T.-L. (2008). The relations of ethnicity
    to female engineering students' educational experiences and college and career plans in an
    ethnically diverse learning environment. *Journal of Engineering Education, 97*(4), 449–465.
    https://doi.org/10.1002/j.2168-9830.2008.tb00992.x

Trevino, N. N., & DeFreitas, S. C. (2014). The relationship between intrinsic motivation and
    academic achievement for first generation Latino college students. *Social Psychology of Edu-
    cation, 17*(2), 293–306. https://doi.org/10.1007/s11218-013-9245-3

Turley, R. N. L. (2009). College proximity: Mapping access to opportunity. *Sociology of Educa-
    tion, 82*(2), 126–146. https://doi.org/10.1177/003804070908200202

U.S. Department of Education. (n.d.). *Supporting students during the COVID-19 pandemic: Maxi-
    mizing in-person learning and implementing effective practices for students in quarantine and isola-
    tion.* https://www.ed.gov/coronavirus/supporting-students-during-covid-19-pandemic

Verdín, D. (2021). "I learned how to divide at 25": A counter-narrative of how one Latina's
    agency and resilience led her toward an engineering pathway. In E. M. Gonzalez, F. Fernan-
    dez, & M. Wilson (Eds.), *An asset-based approach to advancing Latina students in STEM:
    Increasing resilience, participation, and success* (pp. 147–164). Routledge.

Yosso, T. J. (2005). Whose culture has capital? A critical race theory discussion of community
    cultural wealth. *Race Ethnicity and Education, 8*(1), 69–91. https://doi.org/10.1080/1361332052
    000341006

13 · *EMPODERAMIENTO A TRAVÉS DEL TESTIMONIO*

Learning about a College of Engineering's Mission to Become a Latinx-Serving College in a Predominantly White Institution

DIANA GARZA, ADRIANA FACUNDO, ULISES TRUJILLO GARCIA, AND ESTHER A. ENRIGHT

Colleges of engineering are accustomed to hearing the clarion call for US institutions of higher education to train not only more engineers but also a more representative engineering workforce. Often, the reason given for this call is that this diversity will enhance the United States' success, competing in an increasingly globalized economy (Chubin et al., 2005; Duderstadt, 2010; Varma, 2018). Just as important, if not as common a rationale, is that fostering a representative engineering workforce is essential to becoming a more just society (Duderstadt, 2010; Rodriguez et al., 2020; Varma, 2018). Yet, the representation of minoritized[1] and underserved populations in engineering education remains a persistent problem (Hagedorn et al., 2007; Hill, 2017; Rodriguez et al., 2019). In particular, the authors of this chapter are concerned about the limited representation of Latinx[2] people graduating in engineering and the apparent stagnation in Predominantly White Institutions (PWIs) around creating more inclusive and just conditions for Latinx students in engineering programs (Capers, 2019). Why does it seem like PWIs have relegated this work to Hispanic-Serving Institutions (HSIs)? Additionally, how might colleges of engineering within PWIs lead by example in creating a college-level identity as a Latinx-Serving College?

This qualitative case study investigated the articulation of a college of engineering's goal of becoming a Latinx-Serving *College* within a larger PWI, through the lenses of the reflections of administrators and faculty and lived experiences of Latinx alumni. The case study illustrates how colleges of engineering can learn to serve Latinx students from their Latinx students and alumni, by posing these research questions:

1. How do college-level administrators perceive their efforts toward leading by example in becoming a Latinx-Serving College of engineering in a PWI?
2. How do college of engineering faculty leaders perceive their own and college leaders' efforts toward serving Latinx students in a PWI?
3. How have Latinx alumni experienced navigating the college of engineering in a PWI?

LITERATURE REVIEW

Graduating more engineering students is necessary to keep up with evolving technological advances (Riegle-Crumb et al., 2019), yet there is limited research on colleges of engineering specifically at HSIs. Parity in retention rates is one response that would graduate more engineering students and support underserved groups such as Latinx students. Women and historically underrepresented students enroll in engineering programs at the same rate as white[3] men (who are currently overrepresented in engineering) yet fail to persist and achieve the same graduation rates (Beddoes, 2018; Riegle-Crumb et al., 2019). Therefore, a review of the literature tied to HSIs, servingness, and existing equity gaps for Latinx students on these campuses is what informed how we asked questions we asked of our Latinx alumni and engineering faculty and administrators.

The literature on HSIs shows the efforts these institutions are making to serve Latinx students both inside and outside the classroom (Contreras et al., 2008; Madsen Camacho & Lord, 2011; Rosenbaum et al., 2020). While federal grants allocated to institutions with this designation are plentiful, there is increased inquiry into whom those funds actually serve once they are received (Aguilar-Smith, 2021; Contreras et al., 2008; Cuellar, 2019). Furthermore, a lack of intentional efforts to increase institutional readiness for these populations includes issues of representation among peers, faculty, and staff—representation that has proven important for student retention and success (Hagedorn et al., 2007; Rincón, 2020). These efforts are critical: as discussed in an earlier chapter in this volume, Latinx students' engineering identity is important for them to persist (retention and success) and find a sense of place in engineering settings (Rodriguez & Espino, 2024 [ch. 2 in this book]). In addition, other scholars have highlighted not only the impacts of engineering identity on educational spaces but also on professional persistence (Godwin et al., 2016; Meyers et al., 2012; Pierrakos et al., 2009). The increasing number of HSIs and emerging HSIs (eHSIs) alone make it imperative that institutions begin to examine how their landscapes set Latinx students up for success in their respective fields before they arrive, while they are enrolled, and after they graduate.

Institutions can be defined as HSIs by enrolling at least 25 percent Latinx students (Cuellar, 2019; Garcia, 2019; Garcia, et al., 2019; Gonzalez et al., 2020) in contrast to Minority-Serving Institutions or Historically Black Colleges and Universities and Tribal Colleges and Universities that are focused on meeting the needs of the target populations they serve (Garcia et al., 2019). The distinction between "Latinx recruiting" versus "Latinx serving" becomes more pronounced when factors such as retention and graduation rates are compared to majority students and whether there is an intentional focus on Latinx culture and Latinx student belonging (Garcia, et al., 2019).

Latinx Student Servingness

There is an urgent need for colleges of engineering at PWIs to become Latinx-Serving Colleges. This begins by focusing efforts on increasing their readiness prior to receiving the designation at an institutional level (Garcia et al., 2019). Although some colleges believe that they are doing enough to support Latinx students, the reality is that they have limited Latinx representation and do not understand how students feel in these spaces.

While an institution can be federally designated as an HSI, it is not a foregone conclusion that Latinx students are being actively retained and served (Garcia et al., 2019; Vargas & Villa-Palomino, 2019). In an effort to dive deeper into this gap in student support, Garcia et al. (2019) conducted a systematic review of how HSIs conceptualized "servingness," which came down to four broad themes: outcomes (academic and nonacademic), experiences (student and nonstudent), internal organizational dimensions, and external influences. Noteworthy examples included sustainable efforts that found ways to institutionalize their servingness through scaling up and embedding more resources for existing federal grant programs; including existing Chicano studies courses into core curriculum requirements; student-centered pedagogy; and cultural validation and community found through representation among peers and faculty/staff (Hagedorn et al., 2007; Garcia et al., 2019; Rincón, 2020). However, this review also named barriers to Latinx students' full participation in their institutions, including a lack of representation in faculty, external state government pressures (depending on the state), and a lack of faculty preparedness for teaching nondominant populations (Garcia et al., 2019).

A recent critical qualitative study by Aguilar-Smith (2021) delved further into the divide between serving versus enrolling Latinx students by examining community colleges as well as public and private four-year institutions' use of these funds into categories targeted for addressing broad-based institutional needs, for signaling legitimacy, or for supporting all students. As a whole, they found institutions consistently spending designated federal HSI grants in race-evasive ways that helped them fulfill other budgetary constraints, equity gaps, or general student support needs on their campuses. Furthermore, there was a lack of mindfulness tied to their Title V grant applications, other than the fact that they met the bare minimum requirements for such funding.

Working to Become a Latinx-Serving College versus Waiting for HSI Designation

The authors of this chapter propose the idea of being a "Latinx-Serving College" as the ultimate goal for colleges of engineering, relevant at all stages of educational levels as institutions consider serving more Latinx students and working toward HSI status. The goal of HSI status should embrace meeting the needs of the students first and foremost, not merely measured through recruitment funnels and attainment of new sources of funding. The definition of a "Latinx-Serving College" begins with a focus on servingness. At a PWI, the definition includes a nuanced understanding that the students often do not share the same cultural background as those creating the university experience (Rodriguez & Espino, 2023). There is a need for Latinx students to see themselves represented, and colleges should have Latinx representation in faculty and administrative positions (Revelo & Mejia, 2024 [ch. 6, in this volume]). There should be a focus on being family oriented and involving Latinx families, as well as intentionally hiring staff dedicated to supporting the needs of these students every step of the way: for instance, having English and math specialists to ensure that the students are as prepared as their peers in these two essential subjects so that they can thrive in engineering fields. Curricula and instruction in engineering and computer science should also be designed to serve Latinx students in engineering colleges (Enright et al., 2020).

Conceptualizing a Systems Approach toward Engineering Servingness

Engineers are problem solvers, trained to rely on proven methods to tackle new challenges. Meeting the needs of Latinx students and their retention until obtaining a degree must become a focus in colleges of engineering. Approaching the work of servingness in a college of engineering can be considered through the more familiar lens of systems thinking, a commonly used engineering framework. Systems thinking is widely used in engineering and in higher-education spaces as a way to assess a problem through the interconnections of all parts of the system (Arnold & Wade, 2015; Association of Public & Land-Grant Universities & International Futures, 2021; Mosyjowski et al., 2019; Plate & Monroe, 2014). This idea can be extended when considering colleges that meet the needs of all students and must intentionally include the feedback loop in systems thinking—as the latter can help identify the pain points and address specific concerns. In particular, this paradigm can be used to address servingness.

Colleges of engineering should utilize a comprehensive systems thinking approach as the logical next step while they consider ways to improve the entire student ecosystem and becoming a Latinx-Serving College (Mosyjowski et al., 2019). The idea of serving students, if done well, will improve outcomes for all students—including Latinx students. ABET's (2022–2023) criteria on inclusive and equitable classrooms can guide this student-centered work and serve as an accountability measure for other colleges of engineering. This accountability measure combined with a gap in servingness efforts in colleges of engineering is what informed our research questions. In particular, there is a lack of research on the school or college level within PWIs. Research tends to focus on the institution level instead.

METHODS

This qualitative, single-case study was informed by the tradition in LatCrit research of "giving voice to silences, representing the other, reclaiming authority to narrate, and disentangling questions surrounding legitimate truth" (Delgado Bernal et al., 2012, p. 365). The study took place at a college of engineering at a large public university. The college of engineering leadership at this institution has publicly discussed their goal of becoming a Latinx-Serving College within a larger PWI. The case study examines their attempt to become a Latinx-Serving College from three stakeholder perspectives: college-level administrators (with a rank of associate dean and higher), mid-level faculty leaders (with a rank of associate professor), and Latinx alumni.

We chose *testimonios* to empower recent Latinx completers from the college of engineering to share their experiences as an act of critical reflection on and protest of their sociopolitical realities, as well as an opportunity to participate in affecting change in engineering for Latinx students (Blackmer Reyes & Curry Rodriguez, 2012). When gathering data from the Latinx alumni who participated in our study, we used *testimonios* as the methodology (Anzaldúa, 1990; Garza, 2022). We explain the use of *testimonios* as an approach to interviewing in our methods section as well as our choice to interview alumni rather than current students. To encourage a sense of parity between interviewer and participant, a current Latinx engineering student conducted the three Latinx alumni's *testimonio* interviews. These participants were recruited from a small pool of Latinx completers who were enrolled within the past six years, completed their coursework prior to the COVID-19 pandemic, and were accessible via social media or email. The interviewer created transcripts of each interview, which the research team later analyzed.

We also used a phenomenologically inspired approach in our interviews with the college-level administrators and mid-level faculty leaders. We structured the interview protocols to focus on the phenomenon of serving Latinx students in a college of engineering at a PWI, yet we sampled fewer participants in each group than we considered necessary for a true phenomenological study (six to twelve participants). Our sample satisfied our needs for the case study we were conducting, since our primary goal was to understand the similarities and differences across the three groups of participants' perspectives on the resources, structures, and activities related to serving Latinx students in the college of engineering. Phenomenologically inspired interviewing was deemed a more appropriate approach to data collection than the *testimonios* methodology we used to collect the Latinx alumni's data, as our goal here was not to elicit the truth telling of lived experiences to disrupt a socio-politically constructed silence. To encourage a sense of parity between interviewer and participant, a faculty member from a different college conducted the three faculty interviews. These faculty participants were recruited from different departments across the college, held the rank of associate professor, and occupied mid-level leadership positions, such as a director of a major laboratory, a program coordinator, and a department chair. All the faculty held majority racial and gender identities in engineering. Creating a similar sense of parity while interviewing college-level administrators was a challenge. We selected an interviewer from outside the college who is a professional staff member and has a background in qualitative research. The three administrators all identified as white; two identified as men and one as a woman. There was no need to select the administrators, since there were only three with the rank of associate dean or higher. The interviewers created transcripts of the administrator and faculty interviews for later analysis by the research team.

The research team conducted a thematic analysis of the transcript data (Saldaña, 2021). We coded the data from all three stakeholder groups for three central themes related to serving Latinx students in the college: needs, opportunities, and obstacles. First, "needs" as a theme are expressed as something missing that was required to advance the servingness of Latinx students in the college. Second, "opportunities" as a theme are extant resources, experiences, or practices that were viewed as helpful in advancing the college's work in serving Latinx students. Third, "obstacles" as a theme are existing or lacking supports that run counter to serving Latinx students in the college. The research team then conducted a second round of coding, examining the coded data for patterns as well as divergence in reported observations and experiences. In reporting our findings in the next section, we include multiple direct quotes from that *testimonio* data so that readers can hear the Latinx alumni's voices speaking their truths. While we use some direct quotes from the administrator and faculty data, we did not prioritize the use of their spoken language in quotes as we did with the Latinx alumni, which we felt was in line with our methodological choices.

FINDINGS

We identified three categories of data on Latinx servingness, on which we report here: college-level leadership, mid-level faculty leadership in the college, and recent alumni in the same college of engineering. We developed figure 13.1 to break down and illustrate our key findings.

College-Level Administrators' Perceptions of Their College's Latinx Servingness
Finding 1A[4]: Latinx students' needs, as seen by college-level administrators. Crucial to a Latinx engineering student's experience in the classroom is the experience of

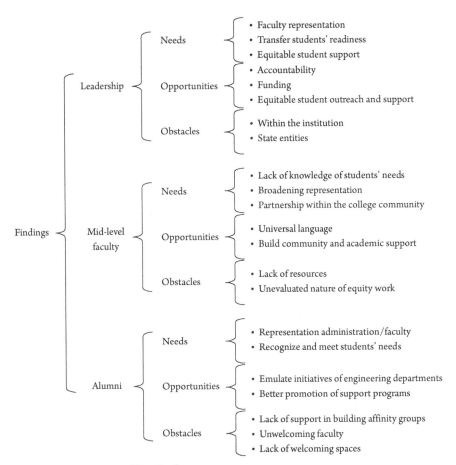

FIGURE 13.1. Summary of Key Findings

being seen and able to recognize themselves in others in the engineering domain (e.g., among faculty, staff, and peers; Hagedorn et al., 2007; Rincón, 2020; Rodriguez et al., 2018; Verdin & Godwin, 2018). Latinx students need to be validated in their lived experiences and for faculty to understand the community cultural wealth they bring to campus (Yosso, 2005). Part of this begins with having faculty who share the same Latinx identity. However, both administrators noted the lack of representation among their faculty, but neither pointed to active efforts to recruit or retain more faculty or staff of color in the college (Hagedorn et al., 2007). One administrator cited the importance of recognizing the pressure that family plays in a Latinx student's experience and the importance of making the curriculum relevant to their intended career goals, which included math self-efficacy: "[T]hat student has to just take it on faith that, you know, grinding through four semesters of calculus is going to get them something that's meaningful and I think we have a role to play and curriculum to show early and often why this matters" (Admin 2). Both administrators named the importance of a relevant curriculum that included flexi-bility for students' school and life management, as well as less rigidity tied to the more traditional input–output style of engineering education: "[I]t comes down to curriculum and it comes down to the leadership understanding that not everybody starts at the same place" (Admin 1). "You can't just teach them material, you have to be able to help them understand that they are good enough to do it and help them build confidence and self-advocacy" (Admin 2).

Both administrators also highlighted the need for the college to have stronger partnerships with community colleges in the area. However, they recognized that this required first finding out more about the current Latinx transfers that existed within their college. They also stated that this meant learning more about their state, especially considering that there were community colleges in the surround-ing areas that had already achieved HSI or eHSI status and were located in more rural areas. The administrators were committed to ensuring that they were ready to serve all transfer students within their college, not only through funding but through smooth course transfer processes.

Finding 2A: Opportunities for Latinx students, as seen by college-level administra-tors. We identified opportunities tied to external accountability, funding, and equitable student outreach and support. Both administrators pointed to creative ways of securing external funding for student support that went hand-in-hand with external accountability. However, they also acknowledged that they could do better at marketing funding opportunities for *all* students in their college. For example, they both named and were committed to disrupting the asymmetry of support provided by the different majors in the college. Sometimes this looked like providing funding through the college of engineering for students to partici-pate in or present at national conferences like the Society of Women Engineers or the Society of Hispanic Professional Engineers. Most recently, the college had secured funding for a New Student Success Center, to specifically address equity gaps within the college. One administrator also secured an NSF grant to allow scholarship funding to follow transfer students from their community college to their four-year institution.

Administrators also spoke of their strong awareness of steps being taken to insti-tutionalize the prioritization of inclusive teaching and practices that they felt were crucial to Latinx student success. This included projects that focused on addressing real-world problems, cultivating a community mindset so that students celebrated one another's successes, and increasing students' confidence in their math skills. "[W]hen we hire new faculty we always have a very succinct statement in the offer letter that they are expected to use evidence-based instructional practices in their classroom . . . they're (faculty) supposed to adopt a continuous improvement process, and so I think we've driven the culture toward inclusive teaching" (Admin 2). Yet, one administrator mentioned the impact of small but high-impact inclusive practices, like meeting with students one-on-one to discuss career options or taking the time to call on a student by name in the classroom.

Finding 3A: Obstacles for Latinx students, as seen by college-level administrators. Finally, we identified obstacles within the institution and from state entities that stood to challenge their college's Latinx servingness. One administrator pointed to a recent state board of education oral communication requirement. He also cited the dissolution of an introduction to engineering course that benefited Latinx and other students who already face compounding systemic barriers in higher education; courses like this not only helped demystify the assumptions about engineering majors but also served as a way to level the playing field for historically disadvantaged groups, particularly Latinx students. This course was crucial in helping all first-year engineering students understand what it meant to be an engineer and what courses would or would not smooth their path forward.

Both administrators noted that another institutional barrier was a lack of data on Latinx student success or enrollment as a whole. While gaining the federal designation of an HSI does not include counts across colleges, it is important for departments within universities to be aware of the existing equity gaps that may exist for Latinx students among their peers and the disparities that may exist from one department to another.

Mid-Level Faculty Leaders' Perceptions of Their College's Latinx Servingness

Finding 1F[5]: Latinx students' needs, as seen by mid-level faculty leaders. The coded data from the mid-level faculty leaders' interviews provided another set of perspectives on the needs, opportunities, and obstacles to becoming a Latinx-Serving College. During the secondary coding, we identified three subthemes related to the central theme of needs. First, the faculty participants expressed a deep sense of discomfort or possibly even shame about their lack of knowledge about the needs of their Latinx students. When faculty participants were prompted to share what they knew about the Latinx students and their needs, they responded,

> I know they [Latinx students in the college] exist. I have mentored and worked with a couple. . . . You know, to be honest, I don't think specifically about Latinx students. . . . I don't feel great about saying that. (Faculty Participant 1)

> Sitting here as department chair as you're asking me these questions, I don't even know how to answer them. I don't even know what the concerns of the Latinx student might be and how they might be different from a white student in my classroom. And that's kind of like pathetic to think about. Like, I don't even know what that perspective would look like. (Faculty Participant 2)

> God, I don't know if I have anything else to say there other than there isn't a lot of them [Latinx students in the college]. (Faculty Participant 3)

All three faculty participants indicated that their lack of knowledge about their Latinx students' needs was likely connected to their general lack of familiarity with those students. Additionally, the three participants reported knowing that there were not many Latinx students and inferred that it would likely be uncomfortable for the students.

Second, the faculty participants named broadening representation in colleges of engineering as a key need: specifically, more Latinx representation among the undergraduate and graduate student populations as well as in the faculty. Faculty Participant 2 likened the experience of Latinx students to women in his subfield of engineering, reporting that female students have shared with him that, in their entire experience as an engineering student in both their undergraduate and graduate degrees, they did not have a single female professor. The faculty member reflected that the experience of never sharing key social identities and related experiences with anyone on the faculty was "staggering" to consider and likely rather "isolating" to experience. None of the participants mentioned that there should be more Latinx representation on the college or university levels of leadership.

Third, the faculty participants highlighted connections and community within the college of engineering as a central need in order to serve Latinx students. The faculty members mentioned a Latinx professional engineering society and a mentoring program as having had an impact in the past, while wondering whether the mentoring program still existed. The faculty participants all connected the need for more Latinx representation among students and faculty as linked to the capacity of the college to build more opportunities for connections with and community for Latinx students. The faculty participants also referred to the one Latina staff member—either by name or racial and gender identities—in student services (on the leadership team) as a source of support for Latinx students.

Finding 2F: Opportunities for Latinx students, as seen by mid-level faculty leaders. We identified two subthemes related to the central theme of opportunities that currently help serve Latinx students in the college. The first subtheme was a generalized, "colorblind" emphasis on the opportunities that innovation creates within the college for all students and faculty. In Faculty Participant 1's responses, he referred to these opportunities as mostly universal and colorblind. When responding to the prompts about opportunities specifically designed for Latinx students, the faculty participants used general language about all students. All three faculty leaders mentioned that the university and college highly value innovation and welcome unorthodox research agendas. Faculty Participant 3 mentioned that there was a statement of inclusion in the college's mission statement, claiming that "In our college, everyone is welcomed and nurtured in the learning and the research that they choose to do."

The only opportunities identified by the faculty participants to better serve Latinx students specifically were external to the college of engineering. The Center for Teaching and Learning (CTL) was mentioned frequently by two of the faculty participants, who were involved in faculty learning communities on the university campus that were focused on inclusive teaching. All three faculty participants mentioned the push by the college leadership and the CTL in learning about and using evidence-based instructional practices. Faculty Participant 1 reported learning about decolonizing his syllabus through the CTL and suggested that it could be helpful in serving Latinx students in courses. The Society of Hispanic Professional Engineers was also mentioned as an opportunity to connect Latinx students and alumni.

In contrast to the universal language of the first subtheme, the second subtheme was the specificity of opportunities when discussing other affinity groups within the student population. For example, Faculty Participant 2 discussed how the college was "targeting students more from a first-generation perspective, so that's an opportunity that is available." Faculty Participant 3 provided examples of how the college leadership and faculty have intentionally built a community and supported specific populations of international students. Then, upon reflection, he raised the question as to why the college had not done that systematically and strategically with their Latinx students, since there were specific efforts to build community and academic supports centered around other identity-specific groups of students.

Finding 3F: Obstacles for Latinx students, as seen by mid-level faculty leaders. We identified two patterns related to the central theme of obstacles to serving Latinx students in the college: (a) a lack of resources and (b) the voluntary, unevaluated nature of equity work in the college. First, faculty participants viewed the lack of resources—specifically, funding—allocated to serving Latinx students as an obstacle to better serving them. They did not report, however, what they would do with additional resources. All three faculty participants stated multiple times that additional resources were necessary to better serve Latinx students.

Second, the faculty participants discussed the lack of leadership around integrating support for Latinx students into the promotion and tenure structure and policies. All three noted that it was up to faculty to decide whether to engage in the available programs targeting increased inclusion. They reflected on why most faculty do not volunteer for these opportunities and how the college leadership could change that lack of engagement:

> Conversely, what departments and programs are being successful for our students and why there might be differences amongst students from different backgrounds. (Faculty Participant 1)

> So just adding to faculty expectations without acknowledging or including that in the mechanisms that provide incentives for faculty to be involved in isn't going to be helpful . . . Faculty will resist (Faculty Participant 2).

> Broadening that awareness [of equity-based professional learning opportunities] and partially because one we're all super busy, and two, it sometimes it's not valued

as much by or perceived as being valued as much, probably both, by administration when it comes to tenure and promotion. People focus on what's going to be rewarded. Yes, it's kind of a natural tendency. (Faculty Participant 3)

The participants recognized that there are resources and opportunities to build capacity to be more Latinx serving within the college, yet they expressed frustration and concern that those efforts were all voluntary and not integrated into the college's formal evaluation processes.

Latinx Alumni *Testimonios* about Their College's Latinx Servingness

Finding 1S[6]: Needs identified by Latinx alumni in their testimonios. As with the interviews with the other participants, the interviews with the Latinx alumni also focused on the central themes of needs, opportunities, and obstacles: seen from their perspective as students and as young professionals or graduate students. We identified two subthemes connected to the first theme (needs): (a) Latinx representation within the college administration and faculty and (b) faculty to do the work to identify and serve Latinx students' needs.

All the Latinx alumni expressed an urgent need for more Latinx representation in the faculty and administrators. All the Latinx alumni mentioned one of the only Latinx faculty members in the college by name. In their *testimonios*, they shared that regular interactions (or the opportunity for those interactions) with Latinx faculty and administrators affected their experience in the college. For example, "When [Latinx faculty member] joined the college of engineering . . . even though I didn't take classes with him, he was there and anytime he would see me he would ask how I was doing. And how things are going. . . . [In contrast,] professors that I had two or three classes with, if they saw me in hallway, they didn't even say hi to me or even know my name" (Hector, Latinx Alum 1). In another example, Ana (Latinx Alum 2) expressed that, when she saw the Latinx faculty in the college of engineering she thought to herself, "[if] he can be a professor, you know like, why can't I?" Ana went on to pursue a doctoral degree in engineering at a different PWI, where she serves as a lab instructor. She shared that there, too, she has noticed how Latinx students seek her out for regular interactions and check-ins. All of the Latinx alumni interviewed also discussed the importance of having a Latinx staff member in student services in the college. Hector explained that, while they were in school, the Latinx staff member "was always there for support, and offering you know, even at least just to hear you out, and I don't know, it was just somebody that you could go and talk to."

In the second subtheme, the Latinx alumni shared an acute need for the faculty to do the work to identify and serve Latinx students' needs, rather than leaving it to Latinx students to figure out on their own. Ana stated that professors need to take the initiative to learn about and see what is going on in their students' lives when they see them struggling. All the Latinx alumni expressed their need to feel seen and cared for by the faculty, regardless of whether they matched the race of a professor.

For instance, Ana shared, "With [named an international faculty member] . . . , I remember one time the class was struggling and like we all had to go in and talk to him individually. . . . Small things like that show that the professor cared."

In another example, two Latinx alumni expressed a need for faculty to redesign the introductory engineering course to systematically teach what engineering is. They noted that their majoritized student counterparts were often exposed to engineers and the work of engineering within their affinity groups, yet the Latinx alumni were not, due to the lack of Latinx representation in those fields. As Hector stated, "They [majority students] know the ropes of how to get to college and what goes on."

The Latinx alumni did not reference any opportunities in their *testimonios*. For this reason, there are no additional findings reported for this theme.

Finding 2S: Obstacles identified by Latinx alumni in their testimonios. The Latinx alumni also identified three types of obstacles in the college to successfully serving Latinx students: (a) lack of college support in building Latinx affinity group community, (b) unwelcoming faculty, and (c) the lack of leadership from college administrators and faculty to create community and welcoming spaces for Latinx students. These obstacles negatively affected not only the Latinx alumni's feelings of belonging but also their access to other opportunities they viewed as accessible to their majority peers.

To begin, all the Latinx alumni reported that they had to independently seek out support and opportunities as well as build community. There were no initiatives created by administrators or faculty to build these affinity spaces, which they viewed as critical to their success. For instance:

> I would say the college in particular could put something together. I don't recall that being the case. (Hector)

> [T]hey would always like have gatherings or like coffee and things like that. I honestly do not really recall as many [engineering department of the student] events, and like, I actually don't recall really any other engineering events. (Ana)

> I guess if there's like a designated study space or group would be the things that might help. (Sara, Latinx Alum 3)

In addition, all the Latinx alumni expressed disappointment about the lack of initiative from professors to foster welcoming instructional and research spaces for Latinx students. Hector expressed hurt at having the same professor for several courses who never learned his name or made an effort to pursue connection. And Ana said, "I do think it's part of the professor's responsibility (right) to reach out." Building on this, Hector expressed discomfort in engaging with a professor to seek help, as he viewed the instructor as unapproachable. When he tried, he experienced the response from the professor as rude and unhelpful.

Finding 3S: Recommendations for culturally responsive servingness. The students offered some suggestions to alleviate these barriers. Ana shared that, in graduate school, they [the College community] have informal gatherings called "cafecitos," where students and faculty interact informally to get to know each other and build relationships. In addition, the college should bring in Latinx people from industry so that the students can learn more about them, see them as an example for their growth, and be able to include them in their group of connections. Hector expressed that the college leadership thus needs to be more intentional about serving Latinx students by celebrating Latinx cultures and creating more welcoming spaces and communities. Alumni also shared that the college leadership could ease these obstacles to serving Latinx students by preparing faculty and evaluating their efforts to serve *all* students as well as particular needs that arise for specific groups of students.

DISCUSSION

When approaching the idea of a Latinx-Serving College of engineering, the persistent questions that remain are as follows: Why is the leadership not responding to the suggestions from alumni and requests from faculty or staff? Or, when they do respond, why is it not well communicated to faculty as to what efforts are taking place? To what extent *are* activities taking place and what is the level of engagement by others? Clearly, there are opportunities for the college, the departments within the college, and the faculty to engage and support Latinx students. Other colleges of engineering should be asking themselves the same questions. Faculty need to take the initiative in assisting with these activities and encourage students to participate in them.

Moreover, the college, departments, and faculty need to establish a partnership with support programs around campus to champion Latinx students. There are self-directed students who take the initiative to find these programs, but what about those students who lack the time to seek opportunities or are taking a limited number of classes? Even academically talented students may miss out on opportunities if they are not more widely shared. In these instances, the students fall through the cracks, missing important opportunities to help them become the professor or staff member who looks like them—for example, through a federally funded program like the McNair Scholars. Campuses can provide a support system for students, starting with the federally funded yet locally operated educational opportunity programs such as the College Assistance Migrant Program or Upward Bound, which provide a support network of resources throughout the undergraduate level. While not all are STEM focused, they can provide critical mentoring and one-on-one support to historically marginalized student groups on campuses.

Latinx-Serving College of Engineering as the Mission

The goal to be a Latinx-Serving college of engineering must be explicit, shared with the entire organization. In our interviews with alumni, administrators, and faculty within a single academic college, it was evident that those who stood to gain the most from hearing about priorities around becoming a Latinx-Serving College had the least information about the leadership's efforts. The college should provide clear, transparent communication about initiatives aimed at increasing Latinx student readiness (Garcia et al., 2019). This goal within a college must be articulated at every turn, including sharing with the stakeholders who benefit from or create inclusive environments for Latinx students. Efforts can be made to expand funding and provide additional outreach and support for programs, but if these are not being communicated to the faculty, staff, or the students themselves, what good does this do? While accreditation guidelines can drive this process and emphasize the importance of this work (Handley & Marnewick, 2022), they do not intimately understand or mandate campus and faculty dynamics or interdepartmental communication efforts.

A Systems Approach to Latinx Engineering Servingness

Solutions can be found when thinking of the college as functioning like a system (Dugan et al., 2021). Participants were not steeped in the student success literature on supporting Latinx students, yet when asked, administrators, mid-level faculty, and alumni identified challenges and possible solutions. When approached strategically and addressed in a systematic way, efforts can lead to significant changes (Dugan et al., 2021). We argue, based on the literature and the efforts highlighted by administrators, that institutions tend to dance on the line between which methods can serve every student and which can serve Latinx students. However, the administrators outlined potential strategies that personally invited the lived experiences of students from historically marginalized communities into the engineering curriculum, which would ultimately uplift and honor every student. The administrators did not take the typical colorblind approach noted in the literature (Aguilar-Smith, 2021). Rather, they took the time to outline the barriers that they recognized could keep any engineering student from persisting beyond their first year. Moreover, they saw how changes in or removal of those barriers would uplift all Latinx engineering students.

Implications for College-Level Systems Change

Needs stated by both Latinx alumni and college of engineering administrators included the importance of connecting early engineering coursework to the engineering career as a whole. Alumni attributed this to having less exposure to the field as a whole, unlike their peers. Administrators attributed it to older engineering teaching methods that expected students to make those connections on their own, citing it as a teaching method that was being phased out through inclusive teaching requirements for new faculty. Yet, the low effort but high impact of intentional but simple inclusive practices, like asking students about themselves, providing a space to reflect, and taking time to learn every student's name, can go a long way in helping Latinx students feel like they matter and

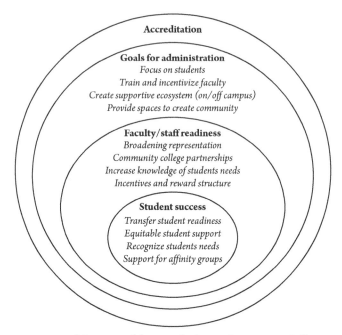

FIGURE 13.2. Wraparound Services for Latinx Engineering College Servingness. (Source: Garza, et al., 2023.)

belong (Gonzalez et al., 2020). This can enhance the confidence of Latinx students, as we saw with the alumni in our study who consistently named two Latinx staff members and two Latinx faculty members who made them feel comfortable— and who, in many instances, understood and validated what they were going through, due to their shared identities. All the alumni we interviewed were involved in high-impact practices (Finley & McNair, 2013) and shared examples of participating in student organizations, conducting research with faculty, and getting help from learning assistants (figure 13.2). However, with the limited Latinx representation in the college, not all Latinx students can meet or develop the same relationships with the few Latinx faculty and staff.

A culturally responsive curriculum in the classroom, along with incentives to support faculty involvement, are needed to support this critical work (Garcia, 2019). Capacity building (time and money) for minoritized faculty and staff at both PWIs and HSIs needs to be an expectation built into what it means to be a Latinx-Serving College or department. Expectations can be structured around inclusive teaching of *all* faculty and staff and make equitable student support and retention everyone's responsibility. Latinx faculty and staff member's lived experiences should be viewed as valuable and strong contributors to Latinx student support and extend the capacity and reach of an institution's ability to retain all Latinx students.

Here, students' needs are central, and the focus should be on meeting those needs equitably as transfers or members of an affinity group. A key component of the wraparound services is support from faculty and staff, with a focus on broadening participation—especially among historically minoritized individuals (Lane et al., 2024 [ch. 3, this volume]). There should be sufficient incentives and reward structures in place for faculty to be encouraged to do what can be labor-intensive work. Compensation structure and tenure processes should be aligned to prioritize this as a needed service for students (Perez-Felkner et al., 2020). The college leadership, deans, and associate or assistant deans should be putting the components in place for students to thrive. This includes creating spaces to gather and establish a connection with others (Garza, 2022).

Responding to accreditation can provide an additional incentive for institutions preparing for ABET visits. Administrators can create the infrastructure needed for visible efforts toward engineering servingness. Subtle or even inward-facing approaches may not create the meaningful ripple effect needed throughout the entire college. Meeting the needs of Latinx engineering students will help the entire student body and require wraparound support (Garcia et al., 2019; Gonzalez et al., 2020).

CONCLUSION

Colleges of engineering should embrace the idea of being a Latinx-serving college, rather than waiting for institutional initiatives toward becoming an HSI. Institutions working toward HSI status have time to reflect on and prioritize the needs of students and consider the imperatives associated with uplifting servingness. The moral imperative of proactively responding to Latinx student needs can be supplanted by the fiscal imperative, which will increase enrollments, assist with retention, and possibly even balance budgets at a time when enrollment declines are being seen across the United States. As Latinx-Serving Colleges, colleges of engineering could lead the campus conversation and position themselves at the forefront as leaders, as they prepare for the next wave of higher education enrollments with a surge of Latinx students. The strategies that must be put into place are easily within reach, requiring a focus on Latinx student voices while ensuring that the work must not fall on the Latinx students themselves. A systems thinking approach is a good starting point, as is taking the entire college ecosystem into account while securing diverse representation among students, faculty, and staff—with the latter two poised to respond to Latinx students' needs.

NOTES

1. We use Ogbu's (1978) typology of minority groups to define the term "minoritized" in this context as meaning involuntary minority groups who did not arrive in North America through voluntary migration but rather as a result of conquest, forced migration, or enslavement.

2. We define "Latinx" as a person of Latin American origin or descent. The three authors of this chapter who identify as Latinx opted to use this term as a gender-neutral or nonbinary alternative to Latino or Latina. The fourth author declined to make a decision as someone not holding that identity.
3. We have intentionally left "white" lowercase in order to de-center whiteness. We acknowledge that the de-colonization of research demands much more than the de-capitalization of letters, yet we see this action as a step forward in these efforts in our scholarship.
4. The "A" in this and the next two subheadings is in reference to the administrator interviews.
5. The "F" in this and the next subheading is in reference to the faculty interviews.
6. The "S" in this and the next subheading is in reference to the student (i.e., alumni) interviews. The authors of this chapter assigned pseudonyms to the Latinx alumni to protect their anonymity while preserving a sense of individuality. We did not do this for the faculty and leadership participants to further decenter whiteness. One of the Latinx-identified authors selected the pseudonyms to be culturally appropriate and gender specific.

REFERENCES

ABET Engineering Accreditation Commission. (2022–2023). *Criteria for accrediting engineering programs.* https://www.abet.org/wp-content/uploads/2022/01/2022-23-EAC-Criteria.pdf

Aguilar-Smith, S. (2021). Seeking to serve or $erve? Hispanic-serving institutions' race-evasive pursuit of racialized funding. *AERA Open.* http://dx.doi.org/10.1177/23328584211057097

Anzaldúa, G. E. (1990). *Making face, making soul/haciendo caras: Creative and critical perspectives of feminists of color.* Aunt Lute Books.

Arnold, R. D., & Wade, J. P. (2015). A definition of systems thinking: A systems approach. *Procedia Computer Science, 44,* 669–678. https://doi.org/10.1016/j.procs.2015.03.050

Association of Public & Land-Grant Universities & International Futures. (2021). *Design Thinking in the EEIDSP Conference Series.* Association of Public & Land-Grant Universities. Retrieved September 23, 2022, from https://www.aplu.org/wp-content/uploads/design-thinking-in-the-eeidsp.pdf

Beddoes, K. (2018). Selling policy short? Faculty perspectives on the role of policy in addressing women's underrepresentation in engineering education. *Studies in Higher Education, 43*(9), 1561–1572. https://doi.org/10.1080/03075079.2016.1266610

Blackmer Reyes, K., & Curry Rodríguez, J. (2012). *Testimonio:* Origins, terms, and resources. *Equity & Excellence in Education, 45*(3), 525–538. https://doi.org/10.1080/10665684.2012.698571

Capers, K. J. (2019). Representation's effect on Latinx college graduation rates. *Social Science Quarterly, 100*(4), 1112–1128. https://doi.org/10.1111/ssqu.12639

Chubin, D. E., May, G. S., & Babco, E. L. (2005). Diversifying the engineering workforce. *Journal of Engineering Education, 94*(1), 73–86. https://doi.org/10.1002/j.2168-9830.2005.tb00830.x

Contreras, F. E., Malcom, L. E., & Bensimon, E. M. (2008). Hispanic-serving institutions: Closeted identity and the production of equitable outcomes for Latino/a students. In M. Gasman, B. Baez, & C. S. V. Turner (Eds.), *Understanding minority-serving institutions* (pp. 71–90). State University of New York Press.

Cuellar, M. G. (2019). Creating Hispanic-serving institutions (HSIs) and emerging HSIs: Latina/o college choice at 4-year institutions. *American Journal of Education, 125*(2), 231–258. https://doi.org/10.1086/701250

Delgado Bernal, D., Burciaga, R., & Flores Carmona, J. (2012). Chicana/Latina testimonios: Mapping the methodological, pedagogical, and political. *Equity & Excellence in Education, 45*(3), 363–372. https://doi.org/10.1080/10665684.2012.698149

Duderstadt, J. J. (2010). Engineering for a changing road, a roadmap to the future of engineering practice, research, and education. In D. Grasso & M. B. Burkins (Eds.), *Holistic Engineering Education: Beyond Technology* (pp. 17–35). Springer. https://doi.org/10.1007/978-1-4419-1393-7_3

Dugan, K. E., Mosyjowski, E. A., Daly, S. R., & Lattuca, L. R. (2021). Systems thinking assessments in engineering: A systematic literature review. *Systems Research and Behavioral Science, 39*(4) 840–866. https://doi.org/10.1002/sres.2808

Enright, E. A., Justice, C., Loo, S. M., Taylor, E., Sample, C., & Shelton, D. C. (2020). Building capacity for systems thinking in higher education cybersecurity programs. *Journal of the Colloquium for Information System Security Education. 8*(1): 1–8. https://cisse.info/journal/index.php/cisse/article/view/122

Finley, A. & McNair, T. (2013). *Assessing underserved students' engagement in high-impact practices.* Association of American Colleges and Universities. http://hdl.handle.net/10919/87004

Garcia, G. A. (2019). *Becoming Hispanic-serving institutions.* John Hopkins University Press.

Garcia, G. A., Núñez, A.-M., & Sansone, V. A. (2019). Toward a multidimensional conceptual framework for understanding "servingness" in Hispanic-serving institutions: A synthesis of the research. *Review of Educational Research, 89*(5), 745–784. https://doi.org/10.3102/003465431986459

Garza, D. (2022). Rebirth through struggle: A qualitative study of Latinx students engineering identity. (Publication No. 29257618). [Doctoral dissertation, Northwest Nazarene University]. ProQuest Dissertations & Theses Open.

Godwin, A., Potvin, G., Hazari, Z., & Lock, R. (2016). Identity, critical agency, and engineering: An affective model for predicting engineering as a career choice. *Journal of Engineering Education, 105*(2), 312–340. https://doi.org/10.1002/jee.20118

Gonzalez, E., Ortega, G., Molina, M., & Lizalde, G. (2020). What does it mean to be a Hispanic-Serving Institution? Listening to the Latina/o/x voices of students. *International Journal of Qualitative Studies in Education, 33*(8), 796–809. https://doi.org/10.1080/09518398.2020.1751896

Hagedorn, L. S., Chi, W., Cepeda, R. M., & McLain, M. (2007). An investigation of critical mass: The role of Latino representation in the success of urban community college students. *Research Higher Education, 48*, 73–91. https://doi.org/10.1007/s11162-006-9024-5

Handley, H. A. & Marnewick, A. (2022). A diversity, equity, and inclusion model for engineering curriculums. *2022 IEEE International Conference on Recent Advances in Systems Science and Engineering (RASSE).* IEEE. https://doi.org/10.1109/RASSE54974.2022.9989693

Hill, A. J. (2017). State affirmative action bans and STEM degree completions. *Economics of Education Review, 57*, 31–40. https://doi.org/10.1016/j.econedurev.2017.01.003

Lane, T. B., Rincón, B., & Hernandez, R. (2024). "I don't think this is the place for you": The belonging cues that Latina/o/x students receive as they navigate within and across engineering environments. In L. Perez-Felkner, S. Rodriguez, & C. Fluker (Eds.), *Latin* Engineering Students: An Intentional Focus on a Growing Population.* Rutgers University Press.

Madsen Camacho, & Lord, S. M. (2011). Quebrando fronteras: Trends among Latino and Latina undergraduate engineers. *Journal of Hispanic Higher Education, 10*(2), 134–146. https://doi.org/10.1177/1538192711402354

Meyers, K. L., Ohland, M. W., Pawley, A. L., Silliman, S. E., & Smith, K. A. (2012). Factors relating to engineering identity. *Global Journal of Engineering Education, 14*(1), 119–131.

Mosyjowski, E. A., Daly, S. R., & Lattuca, L. (2019). *Defining and assessing systems thinking in diverse engineering populations.* Proceedings of the American Society for Engineering Education Conference. https://par.nsf.gov/biblio/10105632-defining-assessing-systems-thinking-diverse-engineering-populations

Ogbu, J. (1978). *Minority education and caste: The American system in cross-cultural perspective.* Academic Press.

Perez-Felkner, L., Shore, C., Dickens, T., & Yang, M. D. (2020). *Engineering resilience through research mentorship: Manufacturing pathways to careers* [Conference paper]. 2020 ASEE Virtual Annual Conference Content Access Proceedings, 34557. https://doi.org/10.18260/1-2 --34557

Pierrakos, O., Beam, T. K., Constantz, J., Johri, A., & Anderson, R. (2009, October). On the development of a professional identity: Engineering persisters vs engineering switchers. *2009 39th IEEE Frontiers in Education Conference* (pp. 1–6). IEEE.

Plate, R., & Monroe, M. (2014). A structure for assessing systems thinking. *The Creative Learning Exchange, 23*(1), 1–3.

Revelo, R. A., & Mejía, J. (2024). How Latinx students engage social and navigational capital to resist exclusionary engineering education. In L. Perez-Felkner, S. Rodriguez, & C. Fluker (Eds.), *Latin* Engineering Students: An Intentional Focus on a Growing Population.* Rutgers University Press.

Riegle-Crumb, C., King, B., & Irizarry, Y. (2019). Does STEM stand out? Examining racial/ethnic gaps in persistence across postsecondary fields. *Educational Researcher, 48*(3), 133–144. https://doi.org/10.3102/0013189X19831006

Rincón, B. (2020). Does Latinx representation matter for Latinx student retention in STEM? *Journal of Hispanic Higher Education, 19*(4), 437–451. https://doi.org/10.1177/1538192718820532

Rodriguez, S., Cunningham, K., & Jordan, A. (2019). STEM identity development for Latinas: The role of self- and outside recognition. *Journal of Hispanic Higher Education, 18*(3), 254–272. https://doi.org/10.1177/1538192717739958

Rodriguez, S. L., Doran, E. E., Friedensen, R. E., Martínez-Podolsky, E., & Hengesteg, P. S. (2020). Inclusion & marginalization: How perceptions of design thinking pedagogy influence computer, electrical, and software engineering identity. *International Journal of Education in Mathematics, Science, and Technology, 8*(4), 304–317. https://doi.org/10.46328/ijemst.v8i4.952

Rodriguez, S., Espino, M. L., Nichols, M., & Le, B. D. (2024). The importance of engineering identity for Latinx students. In L. Perez-Felkner, S. Rodriguez, & C. Fluker (Eds.), *Latin* Engineering Students: An Intentional Focus on a Growing Population.* Rutgers University Press.

Rodriguez, S. L., Lu, C., & Bartlett, M. (2018). Engineering identity development: A review of the higher education literature. *International Journal of Education in Mathematics, Science and Technology, 6*(3), 254–265.

Rosenbaum, R. P., Li, W., & Jeanetta, S. (2020). Hispanic-Serving Institution Host Communities and Latinx Community Development. *The Journal of Business Diversity, 20*(2), 50–60.

Saldaña, J. (2021). *The coding manual for qualitative researchers.* Sage.

Vargas, N., & Villa-Palomino, J. (2019). Racing to serve or race-in for money? Hispanic Serving Institutions and the colorblind allocation of racialized federal funding. *Sociology of Race and Ethnicity, 5*(3), 401–415. https://doi.org/10.1177/2332649218769409

Varma, R. (2018). U.S. science and engineering workforce: Underrepresentation of women and minorities. *American Behavioral Scientist, 62*(5), 692–697. https://doi.org/10.1177/000276 4218768847

Verdin, D., & Godwin, A. (2018). Exploring Latina first-generation college students' multiple identities, self-efficacy, and institutional integration to inform achievement in engineering. *Journal of Women and Minorities in Science and Engineering, 24*(3), 261–290. https://doi.org/10.1615/JWomenMinorScienEng.2018018667

Yosso, T. (2005). Whose culture has capital? A critical race theory discussion of community cultural wealth. *Race Ethnicity and Education, 8*(1), 69–91, https://doi.org/10.1080/1361332052000341006

14 · FUTURE DIRECTIONS AND IMPLICATIONS FOR LATIN* ENGINEERING

LARA PEREZ-FELKNER, CIERA FLUKER,
AND SARAH L. RODRIGUEZ

SUMMARY OF CHAPTER THEMES AND FINDINGS

This book advances our understanding of what an equitable and inclusive future for engineering students might entail. Empirically focused chapters offer opportunities to understand facets of Latin* students' experiences across an array of contexts, offering an evidence base around challenges and opportunities for the future of the field. Theoretically focused chapters explain, develop, and advance frameworks that, in turn, can be applied and tested in practice. Our too-often siloed interdisciplinary knowledge and professional communities can inhibit shared understandings about how to design, build, test, and scale this better future we hope to realize. Our collected chapters together offer blueprints toward this aim.

In engineering this model for Latin* students, we encourage readers to consider a shared finding across the chapters: that the solutions are not individual. Rather, they require investment in systems (see Garza et al., chapter 13). These systems may be bounded by classrooms, mentor networks, departments, colleges, postsecondary institutions and their policies, and various other configurations (see Lane et al., chapter 3; Revelo & Mejía, chapter 6; Beverly & Lattuca, chapter 9; Ozuna Allen et al., chapter 10; Torres & Peralez, chapter 11; and González & Pérez, chapter 12). Overall, the chapters demonstrate, with an array of distinct and aligned evidence points, that multiple stakeholders *can* and should be engaged to make engineering education more effective at recruiting and retaining Latin* students in the field. Beyond these common student success goals, given the nature of engineering and the motivations of those studying and working in the field, we hope to go further, toward making spaces that enhance opportunities for Latin* students to create, innovate, and thrive.

Together, the chapters also engage an intentionally nuanced and intersectional approach to Latin* engineering experiences and potential avenues for structural change. They each close with a focus on real and actionable implications, indeed, the educational possibilities—even in challenging political and legal climates—to facilitate equity-focused change in Latin* students' pathways through engineering education and the workforce.

Our call to action focuses on undoing the structural exclusion of Latin* students from engineering, both within and across institutional types and contexts. This discussion includes what readers could do in their own practice to enhance educational experiences for Latin* students in engineering. Implications also attend to faculty, departments, administration, and policy. Next, we highlight where scholars and practitioners might take this work and offer an overview of what this might mean for research and practice. We hope that the publication of our book might be an inflection point to further future research and innovation in this line of inquiry.

Theories Represented

The prior chapters established key theories and considerations that can be used to inform context-specific directions for policy and practice. Some chapters explaining and analyzing Latin* students' classroom, mentoring, and other training environments grounded their research in theories explaining the processes of stereotype threat, engineering identity development (on its own and in conjunction with other identities), and sense of belonging (Rodriguez et al., chapter 2; Lane et al., chapter 3; Doran et al., chapter 4; and Rodriguez-Simmonds et al., chapter 8). Critical frameworks—such as community cultural wealth, intersectionality, and critical race theory—were employed in multiple chapters, with a focus on uncovering structural patterns in how Latin* students work through constraints associated with socioeconomic disadvantages, resource disparities, and structural challenges faced in their pre- and post-college journeys through coursework and environments not designed with them in mind (e.g., Verdín, chapter 5; Revelo & Mejía, chapter 6; Perez-Felkner et al., chapter 7; and Torres & Peralez, chapter 11). Some authors extended theory—for instance, around socialization (e.g., Ozuna Allen et al., chapter 10) and resilience for first-generation college Latin* students (González & Pérez, chapter 12). The latter chapters and those with attention to institutional type (especially Hispanic-Serving Institutions [HSIs]) also attend to servingness and how institutions and their units can adapt in order to be more inclusive (Garza et al., chapter 13).

Methodology

The evidence base presented across and within these chapters is robust, whether primarily conceptual or empirical in aims. Conceptual chapters built on existing engineering literature and proposed new conceptualizations of engineering identity development (Rodriguez et al., chapter 2), Latinas' experiences of engineering instruction (Beverly & Lattuca, chapter 9), and critical analysis of disabled Latin* engineering students' experiences (Torres & Peralez, chapter 11). Empirical chapters included structural equation modeling (Verdín, chapter 5), semi-structured interviewing (e.g., Lane et al., chapter 3; Doran et al., chapter 4; Revelo & Mejía, chapter 6; Ozuna Allen et al., chapter 10; González & Pérez, chapter 12), large-scale data analysis, and fine-grain studies of engineering climates (e.g., Garza et al., chapter 13).

Some authors engaged in multiple methodologies within the same chapter (e.g., Verdín, chapter 5; Perez-Felkner at al., chapter 7), and one autoethnography presented a newer approach to capture deep insights, here with authors from distinct engineering disciplines and pathways into engineering education roles (Rodriguez-Simmonds et al., chapter 8). Grounding each chapter in theory and including authors' positionality statements (see Perez-Felkner et al., chapter 1; Secules et al., 2021) further anchors the work and renders visible the identities of our intentionally diverse slate of authors (see also chapter 1).

NEW FRONTIERS FOR RESEARCH AND PRACTICE

This book brings new frontiers for research and practice to enhance the experiences of Latin* students in engineering. We strove to be inclusive in our scoping of this book, but there are identities and nuances we were not able to cover as well as we would have liked. This is in part because the intersectionalities we sought are not yet in the published literature or are currently being developed and studied. We also acknowledge the multitude of practitioners working passionately to serve this community, and we cite the important contributions of scholars who were unable to participate in this book who are engaged in research on Latin* engineering students and who have given so much to the field. While we could not possibly speak to every issue touching the lives of our students, scholars across the disciplines can take from these chapters possible directions to guide their work and push further toward understanding. Likewise, the book's various chapters pose many implications for policy and practice for practitioners across our institutions and may serve as a point of departure for future efforts to improve outcomes for Latin* students in engineering.

Limitations and Future Research Directions for Scholars across Disciplines

Given that this text focuses on undergraduate Latin* engineering education, there is also much we could not observe or discuss with respect to these environments. For instance, we do not focus on graduate education, which differs significantly from the Latin* undergraduate student experience. Future work on Latin* graduate student experiences might extend existing scholarship focused on graduate students in engineering (Burt et al., 2019; Borrego et al., 2018).

Moreover, across the Latin* student population, there is considerable linguistic, cultural, and national origin variation, and international versus domestic student identities range as well among students (see also Rodriguez et al., chapter 2, and Doran et al., chapter 4). Some chapters were less able to engage in such nuance for reasons of data restrictions, availability, or other reasons specific to their studies. Still, it is important to stress that Latin* student experiences are not monolithic, and for this reason an edited volume is a powerful vehicle for connecting a series of perspectives together that capture different dimensions of students' experiences.

Similarly, there is insufficient attention to race and racialization for Latin* engineering students. Doran et al. (chapter 3) and Perez-Felkner et al. (chapter 7) establish the value of this lens. New work by emerging scholars like Fluker (2023) shows us how Afro-Latin* engineering students experience compounding challenges associated with isolation and limited resources (including time), while demonstrating resilience out of perceived necessity and carrying forward narratives from their families. In addition to the nuance of racialized intersections of their identity, more attention can be paid to Latin* engineering students' socioeconomic status, gender and sexual identities, abilities, and mentoring relationships as well as other dimensions of how they experience engineering climates.

Such limitations portend openings for future research attending to a range of identities and experiences such as Indigeneity in relation to Latin* engineering. Notably, Indigenous STEM research exists, but work on Latin* engineering indigeneity is still emergent. (For more on why this matters, see also Lopez, 2020; Campbell-Montalvo, 2021; Chow-Garcia et al., 2022; Castagno et al., 2023.) Additionally, researchers should work to disaggregate findings on Latin* engineering students by subethnic/racial and other social identities (e.g., Afro-Latin*, LGBTQ+, disabilities). The educational outcomes of Latin* engineering students may be captured more accurately and disparities brought to light with more self-reporting options and more nuanced data collection and analysis.

More robust and intersectional data collection and analyses are needed—both qualitative and quantitative—to facilitate mixed and single method analyses that can leverage critical frameworks to illustrate how to meaningfully make systemic change that would allow Latin* engineering students to pursue careers more readily and complete degrees in this field. Greater investment in such studies adds value and is essential for nuanced study of these topics (Doucet, 2021); it also allows the stories of Latin* and other underserved and underrepresented students in engineering to emerge and counter dominant narratives that support the persistence of the status quo. Future research may foster further understandings that may enhance innovation in research methodologies such as using mixed methods (see e.g., chapters 4 and 7), autoethnography (chapter 8), quantitative designs such as structural equation modeling (chapter 5), and emergent designs that align theory and method. Engineering education research might also incorporate critical conceptual frameworks and research approaches to conceptualizing race and ethnicity (Garcia-Louis, 2016; Holly Jr. & Masta, 2021).

Relatedly, these experiences are also conditioned and shaped by institutional types and their diverging missions and resources. There is also opportunity for more focused attention on institutions, especially given existing interest and funding aimed at institutional-level changes to broaden participation in engineering and related fields (see Camacho, foreword). This volume argues for the value of this literature with respect to engineering fields specifically, while attending to the racial and intersectional experiences of Latin* students studying in US institutions generally—and in Minority-Serving Institutions (MSIs) in particular. Several chapters discussed MSIs such as Hispanic-Serving Institutions (HSIs) (see Rodriguez et al., chapter 2; Doran et al., chapter 4; Verdín, chapter 5; and Ozuna Allen et al., chapter 10). Building on the work that has been done on engineering education within HSIs (Borrego et al, 2022) and STEM at HSIs more broadly (Hererra & Kovats Sánchez, 2022; Nunez, 2023) as well as within the servingness literature (Garcia et al, 2019), future research has already begun (Rodriguez & Berhane, in press) and can continue to explore how HSIs can create a culture of servingness for Latin* engineering students across the nation.

Implications for Policy and Practice

While chapters drew on a variety of theoretical frameworks, methodologies, and learning contexts, their implications shared commonalities. Our authors highlight the need to recognize and address the fact that Latin* students are embedded in communities, networks, intervention programs, disciplines, and systems that influence their engineering educational experiences.

Across the chapters, authors highlighted how Latin* students may encounter varying messages about their place within the engineering field, particularly from non-minoritized peers. Students contend with stereotype threats, including, but not limited to racial, ethnic, and gender identities. These threats are further complicated by factors like national origin, language, and the white-passing vs. consistently racially minoritized experiences.

The development of engineering identity and a sense of belonging among Latin* students play a pivotal role in their growth as engineers. This identity is shaped not only by their preexisting experiences but also by their encounters throughout their college journey. Specifically, chapter 2 offered a detailed outline of how specific activities (including for example enhanced advising) could align with the enhancement of engineering identity. Relatedly, as seen in chapters 3 and 4, which attended to the role of peers in engineering and related clubs, professional associations, and classrooms, by leveraging the relationships of Latin* peers, practitioners can propagate a sense of belongingness, which is crucial for immersing oneself fully in the engineering domain.

The chapters demonstrate that practitioners might look to asset-based approaches, drawing on Latin* engineering students' existing knowledges as well as forms of capital that they are bringing with them into engineering spaces. Chapters 5 and 9 offered implications for curricular and classroom spaces as sites of potential interventions to interrupt. Chapters 5 and 6 applied research on funds of knowledge to measure and demonstrate the resources available to students from their communities, including their own resilience and resources from their families (see also chapter 7). Latin* students bring valuable reservoirs of knowledge that can be harnessed to enrich their self-conception as engineers and guide their career paths. Connectedly, to combat exclusionary practices within engineering education, practitioners can assist Latin* students with leveraging forms of capital, such as social, navigational, and resistance capital. Witnessing a diverse array of faculty and peers, alongside cultivating mentor relationships proficient in engaging minoritized populations, may greatly benefit Latin* engineering students.

The book speaks to implications regarding Latina students, LGBTQ+ students, and Latin* students with disabilities. Indeed, chapter 8's collaborative autoethnography offers insights and recommendations grounded in the authors' experiences. Chapter 9 offers a framework for understanding how classroom environments are embedded in engineering culture, and especially offered opportunities for instructors to work proactively toward more inclusive climates. Chapter 11 highlights how Latin* students with disabilities in engineering have diverse experiences that are influenced by the sociopolitical climate and current events and their encounters encompass challenging transitions, issues of disclosure, and interpersonal struggles. To foster equity and inclusivity, educators and researchers must embrace a wide spectrum of identities within the engineering domain and work toward eliminating hostilities and barriers present within our institutions.

The experiences of Latin* students can vary based on the institutions and colleges they are situated within. Systems become a more explicit organizing focus as the book moves toward the end. The final chapters highlight the importance of creating structures that support the engineering experiences of Latin* students, with a focus on policies and practices that serve these students. Chapter 10 focuses on implications for dual credit, while chapter 12 suggests that more affirmation and validation of these identities would be useful for helping students. In chapter 13, the implications are more centrally organized around Latin* servingness, to advance institutions and individual engineering-related units within them to be organized such that Latin* students can be successful (Garcia et al., 2020). The concept of servingness is relevant for all engineering departments and colleges, irrespective of their designations and the demographic composition of their administration, faculty, and students. Addressing this requires enhanced representation of diverse faculty, targeted training, communal spaces, and curriculum designs that are centered around Latin* perspectives (Camacho, foreword, this book).

CONCLUDING REFLECTIONS

There is an urgent educational need to train US students in engineering skills to meet the demands of our increasingly technological workforce. Structurally excluding Latin* students from engineering precludes this goal, hindering their economic and educational possibilities. We encourage future researchers—including and beyond those represented as authors in this book—to further engage in these topics to inform policy and practice. Likewise, we are inspired by the work that so many practitioners already do to serve Latin* students in engineering and hope that this work can serve as a means to push this work even further toward equity. All in all, our book and its chapters offer suggestions for systematic efforts to address the continued underrepresentation of Latin* engineering students, such that this growing population is intentionally attended to, supported, and positioned for enhanced visibility and success in postsecondary engineering and its workforce.

REFERENCES

Beverly, S. P., & Lattuca, L. (2024). Studying Latinas' experiences in engineering classrooms: Toward a conceptual framework. In L. Perez-Felkner, S. Rodriguez, & C. Fluker (Eds.), *Latin* engineering students: An intentional focus on a growing population*. Rutgers University Press.

Borrego, M., Andrews, M. E., Kendall, M., & Rodriguez, S. L. (2022). Career plans of Latinx mechanical engineering undergraduates studying at Hispanic Serving Institutions. *Journal of Women and Minorities in Science and Engineering.* https://doi.org/10.1615/JWomenMinorScienEng.2022040409

Borrego, M., Knight, D. B., Gibbs Jr, K., & Crede, E. (2018). Pursuing graduate study: Factors underlying undergraduate engineering students' decisions. *Journal of Engineering Education, 107*(1), 140–163. https://doi.org/10.1002/jee.20185

Burt, B. A., McKen, A., Burkhart, J., Hormell, J., & Knight, A. (2019). Black men in engineering graduate education: Experiencing racial microaggressions within the advisor–advisee relationship. *Journal of Negro Education, 88*(4), 493–508. https://doi.org/10.7709/jnegroeducation.88.4.0493

Campbell-Montalvo, R. (2021). Linguistic re-formation in Florida heartland schools: School erasures of Indigenous Latino languages. *American Educational Research Journal, 58*(1), 32–67. https://doi.org/10.3102/0002831220924353

Castagno, A. E., Camplain, R., Ingram, J. C., & Blackhorse, D. (2023). "It hurts to do work like that": The nature and frequency of culturally based ethical barriers for Indigenous people in STEMM. *Science Education, 107*(4), 837-852. https://doi.org/10.1002/sce.21792

Chow-Garcia, N., Lee, N., Svihla, V., Sohn, C., Willie, S., Holsti, M., & Wandinger-Ness, A. (2022). Cultural identity central to native American persistence in science. *Cultural Studies of Science Education, 17*(2), 557-588. https://doi.org/10.1007/s11422-021-10071-7

Doran, E., Turochy, E. Perez, M., Poleacovschi, C., & Yuen, T. & (2024). The intersectional experience of Latina/o/x students in engineering with stereotype threats. In L. Perez-Felkner, S. Rodriguez, & C. Fluker (Eds.), *Latin* engineering students: An intentional focus on a growing population*. Rutgers University Press.

Doucet, F. (2021). *Identifying and testing strategies to improve the use of antiracist research evidence through critical race lenses*. William T. Grant Foundation. https://wtgrantfoundation.org/digest/identifying-and-testing-strategies-to-improve-the-use-of-antiracist-research-evidence-through-critical-race-lenses

Fluker, C. (2023). *Countering the Latin* monolith: The experiences of Afro-Latin* engineering undergraduate students* [Doctoral dissertation, Florida State University]. ProQuest Dissertations & Theses Global.

Garcia, G. A., Núñez, A.-M., & Sansone, V. A. (2019). Toward a multidimensional conceptual framework for understanding "servingness" in Hispanic-serving institutions: A synthesis of the research. *Review of Educational Research, 89*(5), 745–784. https://doi.org/10.3102/0034654319864591

García-Louis, C. (2016). Beyond multiculturalism: Acknowledging AfroLatina/o students. *Journal of Student Affairs, 25,* 21–27.

Garza, D., Facundo, A., Garcia, U. T., & Enright, E. (2024). Empoderamiento a través del testimonio: Learning about a college of engineering's mission to become a Latinx-serving college in a predominately White institution. In L. Perez-Felkner, S. Rodriguez, & C. Fluker (Eds.), *Latin* engineering students: An intentional focus on a growing population.* Rutgers University Press.

González, E., & Pérez, E. C. (2024). Latina resilience in engineering: Strategies of success in a Hispanic serving institution. In L. Perez-Felkner, S. Rodriguez, & C. Fluker (Eds.), *Latin* engineering students: An intentional focus on a growing population.* Rutgers University Press.

Herrera, F., & Kovats Sánchez, G. (2022). Curando la comunidad [Healing the community]: Community-centered STEM identity. *Journal of Hispanic Higher Education, 21*(2), 135–150. https://doi.org/10.1177/15381927211069543

Holly Jr., J., & Masta, S. (2021). Making whiteness visible: The promise of critical race theory in engineering education. *Journal of Engineering Education, 110*(4), 798–802. https://doi.org/10.1002/jee.20432

Lane, T. B., Rincón, B., & Hernandez, R. (2024). "I don't think this is the place for you": The belonging cues that Latina/o/x students receive as they navigate within and across engineering environments. In L. Perez-Felkner, S. Rodriguez, & C. Fluker (Eds.), *Latin* engineering students: An intentional focus on a growing population.* Rutgers University Press.

Lopez, J. D. (2020). Indigenous data collection: Addressing limitations in Native American samples. *Journal of College Student Development, 61*(6), 750–764. https://doi.org/10.1353/csd.2020.0073

Nuñez, A.-M. (2023). Examining organizational behavior of Hispanic-serving institution computer science departments: Toward servingness and equity in the field. *Journal of Women and Minorities in Science and Engineering, 29*(2), 75–96. https://doi.org/10.1615/JWomenMinorScienEng.2022038505

Ozuna Allen, T., Hall, C., & Matthews, C. (2024). The role of dual credits in socializing Latinx students for engineering fields. In L. Perez-Felkner, S. Rodriguez, & C. Fluker (Eds.), *Latin* engineering students: An intentional focus on a growing population.* Rutgers University Press.

Perez-Felkner, L., Fluker, C., & Templeton, D. S. (2024). A critical mixed methods analysis of Latin* students in diverse contexts. In L. Perez-Felkner, S. Rodriguez, & C. Fluker (Eds.), *Latin* engineering students: An intentional focus on a growing population.* Rutgers University Press.

Revelo, R. A., & Mejía, J. (2024). How Latinx students engage social and navigational capital to resist exclusionary engineering education. In L. Perez-Felkner, S. Rodriguez, & C. Fluker (Eds.), *Latin* engineering students: An intentional focus on a growing population.* Rutgers University Press.

Rodriguez, S., Espino, M. L., Nichols, M., & Le, B. D. (2024). The importance of engineering identity for Latinx students. In L. Perez-Felkner, S. Rodriguez, & C. Fluker (Eds.), *Latin* engineering students: An intentional focus on a growing population.* Rutgers University Press.

Rodriguez, S. L. & Berhane, B. (in press). Creating a culture of servingness for Latinx engineering students at community colleges. *New Directions for Community Colleges.*

Rodriguez-Simmonds, H. E., Vargas-Ordóñez, C., Kaufman-Ortiz, K. J., & Pollettini Marcos, L. (2024). Ser marica es pa' machos [Ser bicha é pra macho]: Agency, activism, and coping while engineering. In L. Perez-Felkner, S. Rodriguez, & C. Fluker (Eds.), *Latin* engineering students: An intentional focus on a growing population.* Rutgers University Press.

Secules, S., McCall, C., Mejia, J. A., Beebe, C., Masters, A. S., L. Sánchez-Peña, M., & Svyantek, M. (2021). Positionality practices and dimensions of impact on equity research: A collaborative inquiry and call to the community. *Journal of Engineering Education, 110*(1), 19-43. https://doi.org/10.1002/jee.20377

Tachine, A. R., Cabrera, N. L., & Yellow Bird, E. (2017). Home away from home: Native American students' sense of belonging during their first year in college. *Journal of Higher Education, 88*(5), 785–807. https://doi.org/10.1080/00221546.2016.1257322

Torres, L. E., & Peralez, K. (2024). What about disabled Latinos in engineering? In L. Perez-Felkner, S. Rodriguez, & C. Fluker (Eds.), *Latin* engineering students: An intentional focus on a growing population*. Rutgers University Press.

Verdín, D. (2024). Examining the funds of knowledge that support Latinx students engineering identity development and career certainty. In L. Perez-Felkner, S. Rodriguez, & C. Fluker (Eds.), *Latin* engineering students: An intentional focus on a growing population*. Rutgers University Press.

NOTES ON CONTRIBUTORS

SELYNA PÉREZ BEVERLY (she/her) is an assistant professor at Eastern Michigan University in the higher education and student affairs program. She was most recently a postdoctoral researcher at the University of Wisconsin, Madison, working on the National Science Foundation Eddie Bernice Johnson INCLUDES Aspire Alliance National Change Project, which focuses on promoting inclusive teaching and environments in STEM. She received her doctoral degree from the Center for the Study of Higher Education and Postsecondary Education at the University of Michigan, Ann Arbor. She has a bachelor of arts from UC Santa Barbara and a master of arts in student affairs administration from Michigan State University. Her research interests include teaching and learning in STEM with a focus on inclusive teaching and the effects on marginalized populations. In her work, she takes a critical approach to center the voices of those who are often marginalized in STEM contexts.

MICHELLE M. CAMACHO (she/her/ella) currently serves as dean of the College of Social & Behavioral Sciences, the largest college on the University of Utah campus. As a bilingual and bicultural Latina, Camacho is the first in her family to graduate from college. She brings over 30 years of deep personal commitment and advocacy to promoting greater equity within higher education. Camacho is trained as a cultural anthropologist with a Ph.D. from UC Irvine, in interdisciplinary social science. She was a professor for more than 20 years at the University of San Diego, teaching statistics, research methods, and courses in sociology. In 2019, the National Science Foundation invited her to serve as a program director in the Division of Undergraduate Education where she co-led the Hispanic Serving Institutions Program with a budget of $45 million. In 2020, she was promoted to acting deputy division director during which time she supervised national investments in support of innovation and transformative strategies to ensure broad participation in higher education. Camacho is a former fellow of the American Council on Education (ACE); Fulbright scholar; President's Faculty Fellow at UC San Diego; humanities research fellow at UC Irvine, and a research scholar of Sigma XI: the Society for Scientific Research.

ERIN DORAN (she/her/ella) is an associate professor of higher education and community college leadership at Iowa State University. Her research focuses on Latina/o/x students, the faculty who teach them, and Hispanic-serving institutions (HSIs). She has worked on a number of National Science Foundation-funded projects centered on broadening participation and revising curriculum for better teaching and learning experiences of students across STEM fields.

ESTHER A. ENRIGHT (she/they) is an assistant professor of educational leadership at the University of Maine. She is interested in the positioning of university-based professional preparation programs within the higher-education system and how a better understanding of that embeddedness could inform equity-based approaches to program recruitment, admissions, curriculum building, and student development. Additionally, she studies the role of youth voice and agency in the leadership of public systems within predominantly rural states. She has a background in K–12 teaching and higher-education student affairs.

MARIA L. ESPINO (she/her/ella) is a postdoctoral researcher at San Diego State University and a research analyst at the University of California, Los Angeles. Dra. Espino is a first-generation Queer Latina who is a proud daughter of immigrants. She earned her PhD in the Higher Education Administration Program at Iowa State University. She obtained her master's degree in educational policy and leadership at Marquette University and her bachelor's degree at the University of Wisconsin—Madison with a double major in community and nonprofit leadership and gender and women studies. She is a proud product of the Milwaukee Public school district in her hometown of Milwaukee, Wisconsin. As a qualitative researcher, she explores (in)equities in higher education, particularly focusing on the experiences of systematically excluded students. As a scholar, she believes that not only is it important to conduct research but it is also crucial to humanize, empower, and support the community. In STEM, she has supported various NSF-funded projects focusing on equity in STEM, in both students and faculty. To learn more about her and her work, visit www.marialespino.com.

ADRIANA FACUNDO (she/her) is the inaugural director of the Micron Student Success Center in the college of engineering at Boise State University and a doctoral student in the higher education leadership program at Colorado State University. She is originally from Michigan and attended Aquinas College in Grand Rapids, Michigan, and received a master of arts in higher education student affairs at Eastern Michigan University in Ypsilanti, Michigan. She is Mexican American and was a first-generation college student who grew up in a low-income household; her entire educational experience from kindergarten through her undergraduate career was spent navigating predominantly white spaces and environments. This combination of lived experiences has directly informed Adriana's commitment to institutionalizing equitable student support structures for BIPOC students at PWIs to foster their sense of belonging and mattering. Additional research interests are around Latinx student identity and voice development, BIPOC staff retention, and coalition building in predominantly white institutions. She has more than nine years of experience in student leadership development, orientation, transition, and retention work.

CIERA FLUKER (she/her/ella) graduated from the Higher Education Program at Florida State University in 2023. She served as a research graduate assistant at Florida

State University where her research explored policy and institutional strategies for improving underrepresented racial minority students' pathways through higher education and engineering. Through her mixed methods research, she centers Black and Latin* students' voices while underscoring their resiliency and success. In addition to education research, Ciera is a skilled training consultant with ten years of experience leading the design, development, implementation, and evaluation of adult education programs. She earned her bachelor's degree in business administration with a specialization in human resources from The Ohio State University and a master's degree in adult learning and development from Cleveland State University.

DIANA GARZA (she/her/ella) is the vice president of student affairs at Wenatchee Valley College, a Hispanic-serving institution serving a rural community. She is a practitioner scholar with two decades of experience supporting students in higher education. A first-generation college graduate, her educational interests include creating support networks for first-generation, rural, and historically minoritized students. She is passionate about helping students pursue higher education and creating environments to promote students' sense of belonging so they can thrive. She was named a 2022–2023 fellow of the fourth cohort of La Academia de Liderazgo of the Hispanic Association of Colleges and Universities. The fellowship is a one-year program to prepare the next generation of culturally diverse leaders for executive and senior-level positions in higher education.

ELSA M. GONZÁLEZ (she/her/ella) is an associate professor in the Department of Educational Administration and Human Resource Development at Texas A&M University (TAMU). She currently serves as program director at the National Sciences Foundation for the Improving Undergraduate STEM Education: Hispanic Serving Institutions (HSI) program, Louis Stokes Alliances for Minority Participation (LSAMP), and the Faculty Early Career Development Program (CAREER). Dr. Gonzalez is the regional editor of the *International Journal of Qualitative Studies in Education*, and a board member of the *Journal of Hispanic Higher Education*. She is the author of over one hundred publications. Her research interests include issues in higher education (HE) such as underrepresented students, Latinx students, access, resilience, retention, and graduation in STEM fields; HSI institutions; HE leadership and methodological issues in cross-language qualitative methodology. In 2020, Dr. Gonzalez was awarded a CAREER grant for her project "Broadening Participation in STEM: A Qualitative Analysis of Resilience Experiences and Strategies of Latina STEM Majors in HSIs" from the NSF. She is the author of the book *An Asset-Based Approach to Advancing Latina Students in STEM: Increasing Resilience, Participation, and Success*, which was awarded as the 2022 Book of the Year by the American Association of Hispanics in Higher Education (AAHHE). Recently, she was named by Diverse Issues in Higher Education as one of the twenty-five Most Outstanding Women in Academia 2023 in the United States.

CHRISTINE HALL (she/her/hers) is an assistant professor of professional practice in the Department of Educational Leadership and Higher Education at Texas Christian University. She completed her undergraduate studies at the University of California, San Diego, graduating with BA degrees in economics and history. Dr. Hall earned her MEd in student development administration from Seattle University and her PhD in educational leadership and policy studies from the University of Texas, Arlington. Her research interests include violence against women on college campuses as well as impacts of state and federal policies on colleges and universities.

RENÉ HERNANDEZ (he/him/él) is a Salvadoran first-generation graduate student in Virginia Tech's School of Education. He is pursuing his PhD in higher education with a cognate in engineering education. He has more than ten years of K–12 and higher-education experience, which he leverages toward his pursuits of helping others find success in education. He has an evolving research agenda focused on pathways, policy, and how it shapes undergraduate engineering education, with specific attention to first-generation college students, low-income, and immigrant populations. He is currently working on two NSF grants that focus on STEM and engineering students.

KEVIN JAY KAUFMAN-ORTIZ (he/him/él) is a graduate student pursuing a PhD in engineering education at Purdue University. He was born in Brooklyn, New York, and raised in Hormigueros, Puerto Rico. In May 2021, he received his BS in industrial engineering at the University of Puerto Rico, Mayagüez Campus (UPRM). During his undergraduate years, he obtained a teaching certification in math education at UPRM and became certified by the Department of Education of Puerto Rico. His current interests lie in belonging, acculturation, and migration of engineers from the US territories to the mainland. He has also previously explored broad topics like ecological classroom belonging, LGBTQ+ inclusion, knowledge transfer between physics and engineering concepts, and data analysis on protein expression. Kevin hopes to keep working with people from minoritized communities of non-visible identities and with efforts toward true inclusion in engineering.

TONISHA B. LANE (she/her) is an assistant professor of higher education in the School of Education at Virginia Tech. She received her PhD in higher, adult, and lifelong education from Michigan State University. She studies the experiences and outcomes of underrepresented groups in science, technology, engineering, and mathematics (STEM). Her research also focuses on the recruitment, retention, and well-being of Black students and professionals in higher education. Using a variety of methodological designs, and with support from the Spencer Foundation and the National Science Foundation, she has explored underserved, undergraduate researchers in remote research environments amid COVID-19; precollege, undergraduate, and graduate learners in STEM enrichment programs;

and the recruitment and retention of early career women of color faculty in STEM. Her published work can be found in *CBE-Life Sciences Education, Equity and Excellence in Education,* and *Urban Education.*

LISA R. LATTUCA (she/her) is the director of the Center for the Study of Higher and Postsecondary Education and a professor of higher education at the University of Michigan. She is also a member of the core faculty of the Engineering Education Research graduate program in Michigan's college of engineering and holds a courtesy appointment in its department of Integrative Systems and Design. Dr. Lattuca studies curriculum, teaching, and learning in higher-education institutions using a systems perspective to understand the many influences, both internal and external to institutions, that shape students' learning environments, experiences, and outcomes. Her engineering education research focuses on how engineering programs and the work of engineering faculty affect the educational outcomes of students. Her recent research projects include an examination of how engineering faculty and curricula present engineering practice to students and thus affect students' interest in the field, and a collaborative effort to develop the Teaching Engineering Equity Center at the University of Michigan. Her research has been funded by the US National Science Foundation, the Accreditation Board for Engineering and Technology, and the Helmsley Foundation.

BRIAN D. LE (he/him/his) is currently a PhD student in the Higher Education and Organizational Change (HEOC) program at UCLA. Brian holds a bachelor's degree in kinesiology and health from Iowa State University and a master's degree in student affairs in higher education from Marquette University. Prior to attending UCLA, Brian worked at Iowa State University for four years as a student's program coordinator for the Science Bound program, a precollege through college program focused on working with scholars from underrepresented backgrounds to pursue a degree in STEM. He has been a research affiliate on multiple NSF-funded projects surrounding equity in STEM. Brian's research interests are college access, retention, marginalized students, community colleges, first-generation, STEM education, STEM identity development, and engineering education.

COURTNEY MATTHEWS (she/her) is a coordinator for student activities in the Department of Student Activities at Tarrant County College. She earned her BA in mass communication/media studies at Northwestern State University. Dr. Matthews completed her MEd in Student Personnel in higher education at the University of Florida and her PhD in educational leadership and policy studies from the University of Texas at Arlington. Her research focuses on Black women aspiring to the college presidency and the successes and humanity of Black women leaders.

JANICE MEJÍA (she/her/ella) is an associate professor of instruction in the Department of Industrial Engineering and Management Sciences at Northwestern University. Her research interests focus on mixed methods in engineering education, curriculum assessment and development, and engineering identity. Dr. Mejía earned a BS in industrial engineering and management sciences, an MA in organizational leadership, an MA in higher education administration, and a PhD in mathematics and science education. Prior to joining Northwestern, she worked in for-profit and nonprofit sectors to optimize technologies, processes, and policies in organizations.

MORGAN NICHOLS (she/her/hers) is currently a PhD student in the Department of Special Education and Child Development at the University of North Carolina at Charlotte. Morgan holds a bachelor's degree in psychology and education from Rhodes College. She holds a master's degree in clinical psychology and a master's degree in counseling psychology from Capella University. Morgan is a board-certified behavior analyst, licensed behavior analyst in the state of Texas, licensed professional counselor in the state of Texas, and a certified clinical trauma professional. She has worked for inclusion programs in the K–12 educational setting for the past decade. She has also worked to help with transition programs in the residential settings and group home settings. Morgan has built inclusion and ABA programs in clinical settings as well as in-home settings over the past five years. Her research areas of interest include special education population, minority population, inclusion, transition, and behavior.

TARYN OZUNA ALLEN (she/her/hers) is an associate professor in the Department of Educational Leadership and Higher Education Leadership at Texas Christian University. Her research interests focus on the educational experiences of traditionally underrepresented students, particularly Latino students, as they access, transition, and enroll in higher education. Dr. Allen earned her BA in general family and consumer sciences and her MEd degree from Baylor University in Waco, Texas. She earned her doctorate in higher education administration from The University of Texas at Austin, with a specialization in Mexican American Studies. She is affiliated with Project Mentoring to Achieve Latino Educational Success (MALES) at The University of Texas at Austin and the Rutgers Center for Minority Serving Institutions.

KRYSTAL PERALEZ (she/her/hers) is a lecturer in the Department of Education Leadership and Policy studies at The University of Texas at San Antonio. Her professional and academic interests include education equity and access, college and career readiness, and workforce development for opportunity populations. Dr. Peralez has held positions in student and academic affairs, college and career readiness, and economic development over the last two decades and has spent the last several years building and expanding STEM-focused precollege and work-based learning programs and initiatives in San Antonio and across the country. As a first-generation college student, Dr. Peralez earned a BA in English from Trinity University, an MEd in counseling and guidance from Texas State University, and her doctorate in education administration at The University of Texas at Austin.

EMMA CLAUDIA PÈREZ (she/her/ella) is a postdoctoral research fellow and project manager in the Department of Educational Administration and Human Resource Development at Texas A&M University. In this role, she coordinates the design, execution, and management of the NSF-funded project "Broadening Participation in STEM: A Qualitative Analysis of Resilience Experiences and Strategies of Latina STEM Majors in Hispanic Serving Institutions." She received her PhD in behavioral neuroscience and cofounded the Society for Advancement of Chicanos/Hispanics and Native Americans in Science (SACNAS) chapter at the University of Houston during her PhD. Her personal experience and expertise as a Latina in sciences support the project as she assists in advancing knowledge and understanding of the Latine student population. Her interests include community cultural wealth, STEM culture, and the Black and Brown experience in STEM.

MICHAEL A. PEREZ (he/him) is an associate professor at Auburn University in the Department of Civil and Environmental Engineering. He received his PhD and MS from Auburn University and his BS from Florida State University. His primary research is in the management of construction and post-construction stormwater runoff. In addition, he has interests in engineering education and the role it plays in Latin students.

LARA PEREZ-FELKNER (she/her/ella) is a professor of higher education and sociology at Florida State University (FSU) and senior research associate with FSU's Center for Postsecondary Success. Her research focuses on the mechanisms shaping social disparities in postsecondary access and persistence. This work investigates racial–ethnic, gender, and socioeconomic disparities in postsecondary pathways to educational attainment and scientific careers, especially in computing and engineering fields. Dr. Perez-Felkner coedited and authored a *New Directions in Institutional Research* volume on undergraduate women in STEM and a special issue in *International Journal for Gender, Science, and Technology*. She is actively engaged in and has held various professional leadership roles in societies such as the American Educational Research Association and the Association for the Study of Higher Education. She has been a key campus player in institutional change efforts, including APLU ASPIRE IChange; the President's Council for Equity, Diversity, and Inclusion, past cochair of FSU's Latinx Faculty and Staff Collective, and a WT Grant Foundation Institutional Challenge Mid-Career Fellow. She is now the PI of an NSF ADVANCE grant team at FSU. Dr. Perez-Felkner has also been a Student Experience Research Network (SERN) Mid-Career Fellow and Institute in Critical Quantitative, Computational, & Mixed Methodologies (ICQCM) NSF Quantitative Critical Methodologies Scholar. She has won a series of awards at FSU for excellence in teaching, advising, and mentoring.

CRISTINA POLEACOVSCHI (she/her) is currently a professor at Iowa State University. Her work focuses on addressing issues of microaggressions in engineering education. She uses mixed methods approaches to identify how microaggressions affect marginalized students and proposes curriculum interventions to address them.

LEONARDO POLLETTINI MARCOS (he/him) is currently a PhD student in the engineering education program at Purdue University. He received a bachelor's degree in materials engineering and a master's degree in materials science and engineering from the Federal University of São Carlos (UFSCar), Brazil. Leonardo also served as one of the national directors for the student branch of the Brazilian Society of Engineering Education from 2019 to 2020. His research interests lie in global perspectives on the accreditation of engineering programs, curriculum development, and assessment.

RENATA A. REVELO (she/her) is a clinical associate professor at the University of Illinois, Chicago, in the department of Electrical and Computer Engineering. She was a first-generation college student and migrated from Ecuador to the United States as a teenager with her parents and sister. She is the first in her family to obtain a PhD. Her research focuses on shifting the culture of engineering via the study of engineering identity, which centers students of color and examines systemic change.

BLANCA E. RINCÓN (she/her/ella) is an associate professor of higher education at the University of Nevada, Las Vegas. She received her PhD in education policy studies from the University of Illinois at Urbana-Champaign. She identifies as a proud Mexicana/Latina, a daughter of immigrants, and is the first in her family to attend college. Her research agenda aims to advance educational equity for historically underserved and marginalized college students in STEM, and specifically Latina/o/x college students. With support from the National Science Foundation, her research investigates three interrelated lines of inquiry: (1) the origins, sustainability, and impact of STEM enrichment programs, (2) the institutional environments that shape college student experiences and outcomes, and (3) the culturally derived assets that college Students of Color mobilize to navigate and persist in STEM. Her research has been published in the top journals in her field, including the *Journal of Hispanic Higher Education*, the *Journal of College Student Development*, and *Teachers College Record.*

SARAH L. RODRIGUEZ (she/her/hers) is an associate professor of engineering education and an affiliate faculty member with the Higher Education Program at Virginia Tech. Her engineering education research agenda centers on engineering and computing identity development of historically marginalized populations at higher-education institutions. In her research, she concentrates on identifying and asking urgent questions about systemic inequities such as racism, sexism, and classism that marginalized communities experience as they transition to and through their engineering and computing higher education experiences. Currently, Dr. Rodriguez is involved with several large-scale interdisciplinary research projects focused on institutional environments and STEM identity development sponsored by the National Science Foundation (NSF) and the Kapor Center. In recent years, she was selected as an Early Career Awardee and Faculty Fellow with the American Association of Hispanics in Higher Education (AAHHE) and a NASPA Emerging Faculty Leader. She also received the Barbara Townsend Early Career Scholar Award by the Council for the Study of Community Colleges (CSCC) and gave the distinguished ASHE-CAHEP Barbara Townsend Lecture. To learn more about her current projects, visit http://sarahlrodriguez.com/.

HÉCTOR E. RODRÍGUEZ-SIMMONDS (he/him/él) was raised in South Florida and born in Mexico. Half Colombian and half Mexican: proud Mexilombian. Currently, he is a visiting assistant professor in engineering at Boston College. Before receiving his PhD in engineering education, he earned his master's degree in electrical and computer engineering. Héctor's research primarily investigates how students negotiate their visible and less visible identities as they form their professional engineering identity, specifically at the intersection of their racial/ethnic, sexual orientation, gender, and engineering identities. Héctor's research projects include collaborative autoethnographic inquiries investigating how engineering students intertwine and leverage the power of their varied identities in engineering, neurodivergence, and disability in engineering, and examining the structural factors that impact student experiences in computer engineering courses. Héctor has taught first-year engineering courses for several years and is invested in showing learners he cares about them and their future success. He creates a space where learners can feel safe to experiment, iterate, and try different problem-solving approaches while encouraging learners to be critical of their professional practice so they create effective, holistic solutions that work for a broader range of individuals.

DA'SHAY TEMPLETON (she/her/ella) is a proud Afro-Latina who was raised in Los Angeles by her mother with her nana, uncles, and cousins. Da'Shay is a tenure-track assistant professor at California Lutheran University. She received her doctorate from Florida State University in higher education and quantitative methodology. She is a critical race theorist and a critical quantitative methodologist, meaning that she subscribes to the key tenets of critical race theory and aims to produce socially just research informed by critical theories like critical race theory, TribCRT, and DisCrit. Her scholarship focuses on the intersection of theory and methodology with a specialization in experimental and quasi-experimental research designs. Her research considers how material and ideological forces of white supremacy impact public school students, specifically Black American, disabled, and gender expansive students. Her current research focuses on how school discipline negatively impacts vulnerable American school children as well as how physical fitness in schools can improve health and academic outcomes. Her work has been funded by the National Academies of Science, Engineering, and Medicine, the American Education Research Association, and the International Philanthropic Education Organization.

LISETTE E. TORRES (she/her/ella) is a trained scientist and disabled scholar-activist who is a senior researcher at TERC, a nonprofit made up of teams of math and science education and research experts. She is also the director of operations and communication for the new national NSF AISL equity resource center called the Reimagining Equity and Values in Informal STEM Education (REVISE) Center. Torres has a PhD in education with a certificate in social justice from the School of Education at Iowa State University and an MS in zoology with a certificate in ecology from Miami University. Her academic research focuses on addressing racialized gender justice and disability in science and higher education. She is an active member of Science for the People, a cofounder of Sines of Disability: Dismantling ableism in mathematics and beyond, and a cofounder and former executive board member of the National Coalition for Latinxs with Disabilities (CNLD). Torres is also an advisory board member of Science Friday's Breakthrough Dialogues Program and the Invisible Disability Project (IDP). Lastly, she has been identified as a Kavli Foundation Sponsored Network Leader for Inclusive Science Communication.

ULISES TRUJILLO GARCIA (he/him/él) is pursuing a PhD in engineering education systems and design at Arizona State University. He has held a number of leadership positions during his undergraduate career, which earned him a variety of accolades. These experiences helped him identify his passion which is rooted in supporting Latina/o/x students with migrant farm working backgrounds in higher education, especially in engineering spaces. Currently, Ulises is working on a project titled "Empowering Children of Migratory/Seasonal Farmworkers with Gamification and Culturally-Responsive Engineering Design Instruction." He is a fellow for the National Science Foundation Graduate Research Fellowship Program, a former fellow for the Station1 Frontiers Fellowship, the Micron Academy for Inclusive Leadership, and HACU ¡Adelante! Leadership Institute. In the summer of 2021, Ulises started a scholarship for migrant students in eastern Oregon to pursue higher education, raising over $12,000 in scholarship dollars for this population. As a first-generation, low-income Latino from a farm-working family, he is passionate about helping his community and devotes his time to diverse educational and leadership causes.

ELIZABETH TUROCHY (she/her/hers) is pursuing a PhD in civil and environmental engineering at Auburn University. While completing her undergraduate degree at Auburn, she participated in data collection, analysis, and writing on an NSF project focused on the influence of stereotypes and stereotype threats on Latina/o/x students in engineering. This research aligns with her passion for seeking a career that helps improve the lives of others, especially in underserved communities and populations. Elizabeth also volunteers with THE FIFTY FUND, a nonprofit organization that primarily focuses on assisting others in making their communities safer, cleaner, and more beautiful. Currently, her research at the National Center for Asphalt Technology focuses on developing asphalt pavement

designs and maintenance techniques that are environmentally friendly, economically conscious, and socially aware.

CRISTIÁN VARGAS-ORDÓÑEZ (he/él) is a Colombian PhD student in engineering
education at Purdue University focusing on epistemic justice, diversity, and multicultural engineering education. As a first-generation professional, gay, yoga teacher,
dancer aficionado, and English as a second language (ESL) student, his research has
been focused on social justice in engineering education, compassion resulting from
interdisciplinary education, and international engineering graduate students' well-
being reflect this awareness. Vargas-Ordóñez has expertise in educational programs
serving first-year engineering students, museums, and underrepresented precollege
students. His educational background includes an MA in education from the University of Los Andes (Colombia), an MSc in science, technology, and society from
the National University of Quilmes (Argentina), and a BSc in chemical engineering
from the University of America (Colombia). He has been recognized through various awards and grants, including the Estus H. and Vashti L. Magoon Award for Excellence in Teaching, the CILMAR Seed Grant program, and the Bryon Fellowship.

DINA VERDÍN (she/her/ella) is an assistant professor of engineering at the Ira A.
Fulton Schools of Engineering at Arizona State University. She graduated from San
José State University with a BS in industrial systems engineering and from Purdue
University with an MS in industrial engineering and a PhD in engineering education. Her research program is dedicated to promoting equity and inclusion in engineering by confronting the pervasive barriers facing minoritized students. Her
work seeks to address the challenges Latinx, first-generation college students, and
women in engineering face by focusing on two research strands: access and persistence. Dr. Verdín seeks to create culturally responsive opportunities that broaden
access to engineering while simultaneously dismantling the systemic obstacles that
hinder minoritized students' persistence. With this work, Dr. Verdín is committed
to transforming engineering education and ensuring that all minoritized students
can thrive. She has won several awards, including the 2022–2023 Outstanding
Research Publication Award by the American Educational Research Association
(AERA) Division I, the 2022 ASEE ERM Apprentice Faculty Grant, and the 2018
ASEE/IEEE Frontiers in Education Conference Best Diversity Paper Award.

TIMOTHY YUEN (he/him) is the associate dean for undergraduate studies in the
College of Sciences at the University of Texas at San Antonio. His research investigates best practices in broadening diversity and improving student success in
computer science and engineering through culturally responsive and transformative teaching practices. Dr. Yuen holds faculty appointments as a professor in
interdisciplinary learning and teaching, computer science, and electrical and computer engineering at UTSA. He is currently a PI for a grant from the National Science Foundation that aims to increase Latinx participation in computer science
through in-service teacher professional development.

INDEX